SOCIETY FOR EXPERIMENTAL BIOLOGY

seminar series: 32

DEVELOPMENTAL MUTANTS IN HIGHER PLANTS

Developmental mutants in higher plants

Edited by

HOWARD THOMAS

Welsh Plant Breeding Station, Aberystwyth

and

DONALD GRIERSON

University of Nottingham School of Agriculture,
Sutton Bonnington

CAMBRIDGE UNIVERSITY PRESS

Cambridge

London New York New Rochelle

Melbourne Sydney

Published by the Press Syndicate of the University of Cambridge
The Pitt Building, Trumpington Street, Cambridge CB2 1RP
32 East 57th Street, New York, NY 10022, USA
10 Stamford Road, Oakleigh, Melbourne 3166, Australia

First published 1987

Printed in Great Britain at the University Press, Cambridge

Library of Congress cataloging in publication data

Developmental mutants in higher plants.
(Society for Experimental Biology seminar series; 32)
Includes index.
1. Plant mutation – Congresses. 2. Plants – Development – Congresses. I. Thomas. Howard, 1948– II. Grierson, Donald. III. Series: Seminar series (Society for Experimental Biology (Great Britain))
QK981.D47 1987 581.1′592 86–31055

British Library cataloguing in publication data

Developmental mutants in higher plants.
(Society for Experimental Biology seminar series; 32)
1. Crops – Genetics 2. Plant mutation breeding
I. Thomas, Howard II. Grierson, D.
III. Series
631.5′3 SB123

ISBN 0 521 32844 6

CONTENTS

CONTRIBUTORS

J. BAILEY-SERRES, Department of Botany, The King's Buildings, Mayfield Road, Edinburgh EH9 3JH.

B. BATHGATE, Department of Physiology & Environmental Science, University of Nottingham School of Agriculture, Sutton Bonington, Loughborough, LE12 5RD.

R. CARPENTER, John Innes Institute, Colney Lane, Norwich NR4 7UH.

E.S. COEN, John Innes Institute, Colney Lane, Norwich NR5 7UH.

D.J. COVE, Department of Genetics, University of Leeds, Leeds LS2 9JT.

V.S. FORD, Department of Genetics, University of California, Davis 95616. U.S.A.

T. GERATS, Department of Genetics, Free University of Amsterdam, Amsterdam. Netherlands.

L.D. GOTTLIEB, Department of Genetics, University of California, Davis 95616, USA.

D. GRIERSON, Department of Physiology & Environmental Science, University of Nottingham School of Agriculture, Sutton Bonington, Loughborough, LE12 5RD.

S.P.C. GROOT, Department of Plant Physiology, Agricultural University, Nl-6703 BD Wageningen, The Netherlands.

C.L. HEDLEY, John Innes Institute, Colney Lane, Norwich NR4 7UH.

R. HORGAN, Department of Botany & Microbiology, The University College of Wales, Aberystwyth SY23 3DA.

P.G. ISAAC, Department of Botany, University of Edinburgh, Edinburgh EH9 3JH.

C.M. KARSSEN, Department of Plant Physiology, Agricultural University, Nl-6703 BD Wageningen, The Netherlands.

J.E. KNAPP, Department of Physiology & Environmental Science, University of Nottingham School of Agriculture, Sutton Bonington, Loughborough, LE12 5RD.

C.D. KNIGHT, Department of Genetics, University of Leeds, Leeds LS2 9JT.

M. KOORNNEEF, Department of Genetics, Agricultural University, Nl-6703 BD Wageningen, The Netherlands.

M. KREIS, Biochemistry Department, Rothamsted Experimental Station, Harpenden, Herts. AL5 2JQ.

C.J. LEAVER, Department of Botany, University of Edinburgh, Edinburgh, EH9 3JH.

C. MARTIN, John Innes Institute, Colney Lane, Norwich NR3 6UH.

M.E. PURTON, Department of Physiology & Environmental Science, University of Nottingham School of Agriculture, Sutton Bonington, Loughborough LE12 5RD.

J.A. ROBERTS, Department of Physiology & Environmental Science, University of Nottingham, Faculty of Agricultural Science, Sutton Bonington, Loughborough LE12 5RD.

P.R. SHEWRY, Biochemistry Department, Rothamsted Experimental Station, Harpenden, Herts. AL5 23Q.

I.D. SMALL, Department of Botany, University of Edinburgh, Edinburgh EH9 3JH.

J.L. STODDART, Welsh Plant Breeding Station, Aberystwyth, Dyfed SY23 3EB.

I.B. TAYLOR, Department of Physiology & Environmental Science, University of Nottingham, School of Agriculture, Sutton Bonington, Loughborough LE12 5RD.

H. THOMAS, Plants & Soils Division, Welsh Plant Breeding Station, Plas Gogerddan, Aberystwyth, Dyfed SY23 3EB.

T.L. WANG, John Innes Institute, Colney Lane, Norwich NR4 7UH.

M.S. WILLIAMSON, Biochemistry Department, Rothamsted Experimental Station, Harpenden, Herts. AL5 23Q.

PREFACE

The study of plant development poses fundamental scientific questions with important implications foa griculture and horticulture. Al Although the general framework of developmental processes has been established by anatomical and physiological studies it is clear that we require other means of investigating these problems if we are to expose the underlying regulatory mechanisms. There is a growing conviction among plant scientists that the genetic approach, coupled with the techniques of biochemistry and molecular biology, offers unparalleled opportunities for unravelling the molecular basis of plant development. With this in mind the 32nd Symposium in the Society for Experimental Biology's Seminar Series took place in March 1986 at the University of Nottingham. The meeting was a joint venture between the Plant Development and Plant Molecular Biology Groups under the title "Developmental Mutants in Higher Plants". This volume comprises twelve papers given at the meeting plus (appropriately) a thirteenth which had been scheduled but which fate, in the shape of a fractured leg, prevented from being delivered.

Research on mutants in micro-organisms and animals has provided a tremendous boost to our understanding of the regulation of metabolism and development and it is reasonable to expect that the study of plant mutants should be equally rewarding. Indeed, it is worth recalling that the rules of inheritance were established by studies of the garden pea and transposable elements were detected first in maize. These days experimental approaches to the developmental biology of plants range over the fields of morphology, physiology, biochemistry and molecular biology. A major aim of this volume is to give a taste of this diversity and to illustrate the value of the study of mutants. Thus the thirteen papers in the book describe mutations that affect one or more stages of the life cycle, beginning with flower formation and including seed development and fruit ripening, germination, root and shoot development and responses to the environment, hormone synthesis, tissue culture, and leaf morphogenesis and senescence. The organisers also wished to present a different perspective on some aspects of the subject: thus, transposable elements are discussed, but in Antirrhinum rather than the widely publicized maize; organelle genetics is represented not by the chloroplast, which would require a whole book to itself, but by the mitochondrion in the context of male sterility. It was our hope not only that some general biological principles would be discernable, but also that some of the unique features of plant developmental processes and their genetic regulation would be apparent, in order to emphasise that plants are not just slow green animals.

One object of meetings of this sort is to provide an opportunity for scientists in different disciplines to engage in productive debate to the benefit of all sides. In practice this requires careful orchestration by informed chairmen. We were fortunate in being able to call on the services of Harry Smith, John Bryant, Helen Ougham, and, as scene-setter and summer-up, Harold Woolhouse. We were also lucky to assemble a group of contributors requiring only relatively mild intimidation before they agreed to submit their manuscripts (more or less) promptly. Speakers' expenses were met by generous contributions from the Society for Experimental Biology and we are grateful to John Noble-Nesbitt and John Bryant for their help with the financial administration. Leslie Gottlieb attended the meeting while in receipt of support from the Underwood Fund of the Agricultural and Food Research Council. Thanks are due to Peter Silver and his colleagues at Cambridge University Press for their help throughout the editorial and production process. We are grateful for Margaret Mack's help in preparing the Index.

Donald Grierson
Howard Thomas

GENETIC AND DEVELOPMENTAL STUDIES OF THE ABSENCE OF RAY FLORETS IN *LAYIA DISCOIDEA*

L.D. Gottlieb and V.S. Ford
Department of Genetics, University of California,
Davis 95616.

INTRODUCTION

Genetic analyses in plants have identified two categories of morphological evolution (Gottlieb, 1984). Discrete differences in structure, shape, architectural arrangement or orientation, and presence versus absence are often governed by one or a few genes. Multiple or polygenic systems most often govern continuous changes in dimensions, weight, or number, with many of the genes probably operating indirectly via general effects on physiological processes at the tissue, organ, or whole plant levels.

If one or two gene substitutions are responsible for differences between Phenotype A and Phenotype B, then, in principle, a developmental analysis should identify only one or two primary changes, regardless of the number or extent of consequent or secondary changes brought about by the interaction of unchanged genetic information with genetically modified substrates, structures, or processes. If the phenotypic differences are controlled by numerous genes, which individually have relatively small effects, numerous primary changes are expected, and it will be extremely difficult to distinguish them from consequent changes.

Given the present slim knowledge of plant development, we believe that the appropriate emphasis for developmental genetics should be discrete character changes because of the greater likelihood of identifying the primary ontogenetic steps leading to divergence of the mature structures. Developmental analysis must be integrated with genetic analysis because only data from progeny segregations can reveal whether a morphological character difference results from few or many primary (i.e., genetic) changes and reveal which changes are functionally related.

For example, hooded barley, a single dominant mutation causes the development on the lemma of two extra florets separated by a rachis-like structure bearing two lateral appendages (Stebbins and Yagil, 1966). Prior to the genetic analysis, this complex structure, which typically shows an inverted polarity, had been thought to reflect numerous genetic changes. The developmental study by Stebbins and Yagil revealed that the entire structure differentiated from a proliferation of relatively small epidermal and subepidermal cells on the lemma surface which were oriented in several planes rather than parallel to the lemma axis as in normal development. The mutant gene appeared to interfere with the timing and orientation of cell divisions, and the consequent formation of the primordium-like tissue mass apparently called into play already available genetic information that led to the differentiation of the unusual hood structure. The mutant gene did not in any sense specify the additional florets or appendages.

A second example involves the determination of fruit shape in red pepper (Capsicum annuum). Sinnott and Kaiser (1934) found that varieties with elongate fruits and those with nearly spherical ones had similar ovary shapes until fertilization. Then, in the elongate type, the allometric relationship between length and diameter changed sharply whereas the spherical fruit maintained the original allometry throughout its enlargement. The F_2 progeny between the two types segregated plants with two classes of immature fruits: 31 plants showed the allometric change and 9 plants maintained the constant rate, suggesting a single locus controlled the difference (Kaiser, 1935). In addition to their growth allometry the shape of mature fruits depended on growth duration (fruits with an allometric constant greater than unity became more elongate if they grew for longer durations) and environmental factors (e.g., nutrition, shading, water availability, planting density, etc.). The result was that the mature F_2 fruits were continuously variable. In this example, the developmental analysis revealed that the primary determinant of shape differences had to do with a change of allometric slope at a specific stage during fruit enlargement. The genetic analysis validated the developmental finding because it demonstrated that a single gene substitution was the major operative factor accounting for the allometric change.

Such integrated analyses of genetics and development have rarely been undertaken for morphological character differences that distinguish plant species. This is unfortunate because they are essential to understand how new characters evolve and to decide the relative roles of discrete and continuous morphological changes in plant evolution. Since annual plants can be grown in large numbers and under uniform conditions suitable for genetic studies, we have begun a genetic and developmental analysis on two diploid species of Layia, a genus of spring-flowering annual plants in the Sunflower family (Compositae), native to California. This paper describes our initial genetic and morphological results, and discusses several developmental models to account for a striking morphological difference between the species.

THE TWO LAYIA SPECIES

Layia discoidea is endemic to a small remote region of serpentine soils in the inner South Coast Ranges of California and its close relative, *L. glandulosa*, which we believe is its progenitor (Gottlieb *et al.*, 1985), is a widespread species of sandy soils at lower elevations in the far western U.S.A. The two species differ in several discrete morphological characters of their reproductive heads. The head of *L. glandulosa* has eight showy white or yellow ray florets on the periphery of its central disk of tubular florets. The head is surrounded by the same number of involucral bracts or phyllaries, each of which completely encloses the ovary of the adjacent ray floret. An additional ring of individual receptacle bracts separates the ray florets from the disk florets. The disk florets have a pappus of long, glistening white bristles. In contrast, the head of *L. discoidea* has no ray florets, its involucral bracts have slightly curved rather than clasping bases, it lacks a second ring of bracts, it has fewer disk florets which produce smaller achenes, and its pappus consists of short, tan-coloured stubby scales. Disregarding the differences in pappus and in number of disk florets, the mature head of *L. discoidea* closely resembles a head of *L. glandulosa* from which the ray florets and accompanying phyllaries have been removed.

Under uniform conditions, the species also differ in a number of quantitative traits of vegetative organs. Plants of *L. discoidea* exhibit smaller and fewer rosette leaves than those of *L. glandulosa*, and are shorter in stature with fewer stem nodes and shorter internodes. Although the differences between the species involve many aspects of their phenotypes, they are easily crossed and the F_2 and backcross progenies are fully fertile (Clausen *et al.*, 1947; Gottlieb *et al.*, 1985; unpub.).

Electrophoretic analysis of enzymes encoded by 21 genes revealed that populations of the two species from the same geographical region have a very high genetic identity, $\bar{I} = 0.90$, on a scale of 0 to 1.0, with the same allele generally exhibiting the highest frequency at each locus (Gottlieb *et al.*, 1985).

The high genetic identity of the species, their complete interfertility, and the limited geographical distribution and specialized habitat of L. discoidea all suggest that it was probably a relatively recent derivative of L. glandulosa (Gottlieb et al., 1985). Since their extensive morphological differentiation is displayed on what appears to be a relatively similar genomic background, they provide an attractive model to study how large morphological differences evolve.

We are studying the most apparent difference between them, the presence versus absence of ray florets and their accompanying clasping phyllaries. Since L. discoidea is the only species of Layia and of its subtribe (Madiinae, a group of 10 genera) that lacks these structures, the absence is certainly a derived condition. Thus, we have to be concerned only with how they were lost, and not with the possibility that their presence in L. glandulosa represents a gain.

Our goal is to answer the following questions: (1) What is the genetic basis for the loss of ray florets and their enclosing bracts; i.e., how many genetic factors govern the change to this alternate character state and what are their separate inputs? (2) How has the course of development of the typical Layia head composed of ray and disk florets, as exemplified in L. glandulosa, been modified? What are the primary changes and what are the consequent or secondary changes at the morphological level? (3) What are the genetic and developmental interactions between presence versus absence and number of rays? (4) And an important question, mostly for the future, what biochemical changes might cause these morphological changes?

THE LAYIA GROWTH PATTERN

The two Layias generally follow a similar pattern of growth and differentiation. Following germination, a seedling forms a rosette of leaves and, after about 40-60 days, depending on photoperiod, light intensity, and temperature, the apical meristem switches first to the production of bracts (these resemble leaves but lack axillary meristems), and then differentiates into a reproductive head. The primary branches form in the axils of the highest leaves. The apical meristem on each primary branch follows a similar pattern, producing leaves, bracts, and then a reproductive head. Eventually, the internodes of the stem elongate, thereby raising the head above the rosette and, at the same time, the internodes of the upper primary branches also elongate. Elongation of the stem and of each primary branch continues until all disk florets on its apical head have opened.

Following maturation of the head and dispersal of achenes, the branch or stem beneath the head dies back to the next lower node that bears a branch. The dead branch parts are not shed so larger older plants are composed of an assemblage of living and dead organs. A large plant growing in favourable greenhouse conditions may display as many as seven orders of branches, but mature individuals in nature generally have only a few branches. The terminal head on the stem is usually the largest one on the plant and has the highest number of florets. Since the terminal heads of all plants are developmentally comparable, most counts and measurements were made on them.

HEAD DEVELOPMENT IN LAYIA

The early ontogeny of the heads of the two species has been examined by scanning electron microscopy. We describe here a summary of our observations beginning with the head of *L. glandulosa*. The vegetative apex has a shallow dome and initiates leaf primordia and axillary buds in a sequential spiral. The switch to a reproductive mode begins with the initiation of bracts instead of leaves. This is associated with an increase in the height and diameter of the apical dome, and the accelerated growth and development of the highest axillary buds on the stem. As the apex, now termed a receptacle, continues to enlarge, phyllary primordia are initiated, followed by the initiation of receptacle bracts and disk florets (Figures 1-3). These appear to originate from "common primordia" which then bifurcate transversely near their middle (Figure 4) so that the lower portion gives rise to a receptacle bract and the upper portion to a disk floret. The production of the common receptacle bract/disk floret primordia continues until they fill the periphery of the receptacle above the phyllaries. Then, higher on the receptacle, the disk florets continue to be initiated but without accompanying bracts.

The ray floret primordia, each in the axil of an involucral bract, become apparent by the time distinct disk floret/receptacle bract primordia are evident. Their precise time of initiation remains to be determined, but it is evident that their early growth is greatly delayed with respect to their corresponding bracts, in contrast to the simultaneous growth of the receptacle bract/disk floret pairs. The eight ray floret primordia appear more or less in synchrony whereas the involucral bracts, common primordia, and later disk floret primordia develop sequentially. (Light microscope studies are in progress to confirm the SEM observation that the

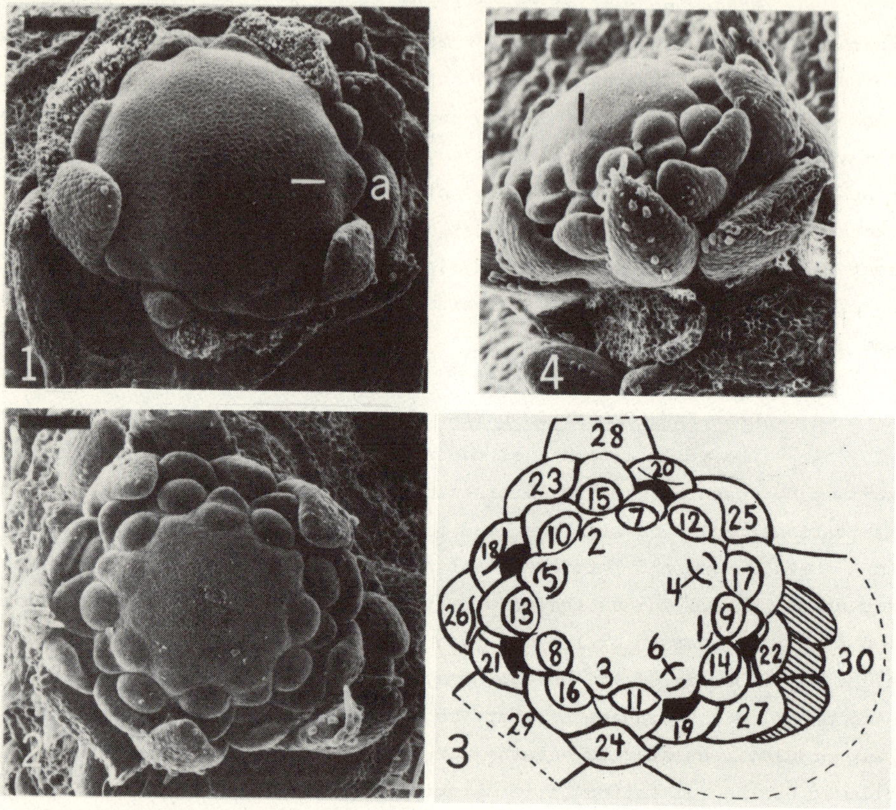

Figures 1-4. Two stages in development of head of L. <u>glandulosa</u>. Bar = 100μ. (1) Pointer indicates a phyllary with no visible growth of ray floret primordium in its axil. This phyllary is the 13th primordium initiated above the highest axillary bud, labelled (a). No more than ten of the intervening primordia are stem bracts and, consequently, the designated structure can be identified as one of the phyllaries. (2) Top view of head showing initiation of primordia of receptacle bracts and disk florets. Note that the ray primordia (five are visible, marked in black in accompanying diagram in (3)) are smaller than the outer disk primordia. Thus, compare ray at p18 with disk floret at p17. The visible primordia are denoted p1, p2, ..., with p1 being the most recently initiated. Note the apparent bifurcation of the common primordium at p5. (3) Diagram of (2). (4) Side view of same head illustrated in (2). Primordium 5 is indicated by pointer.

receptacle bracts and associated disk florets arise from common primordia rather than as closely adjacent primordia, to determine when the ray floret primordia are initiated, and to investigate the apparent synchrony of their growth.)

The mature ray and disk florets are different in many respects. The ray floret ovary and subsequent achene are glabrous, pappus is lacking, anthers are not produced, and the corolla tube is short (3-4 mm) and surmounted by a large (13-16 mm length) strap-shaped three-lobed ligule. The disk floret has a long corolla tube (about 7 mm) which is topped by five short radially symmetrical lobes, the ovary and achene are densely pubescent, the achene carries a pappus of numerous bristles, and the style is surrounded by a tube made up from the five connate anther sacs.

The vegetative apex of L. discoidea closely resembles that of L. glandulosa. Stem bracts are initiated similarly, but the outgrowth of the highest axillary bud is delayed. The initiation of stem bracts is immediately followed by the formation of pairs of involucral bract/disk floret primordia which appear to originate simultaneously (Figures 5-7). When the periphery of the receptacle is fully occupied by them, subsequent primordia give rise only to disk florets. Thus, during head development in L. discoidea, no primordia having the temporal relationship of phyllary and ray primordia are seen. The course of differentiation proceeds directly from stem bract to common primordia, with no intervening structures and no receptacle area bypassed or left open, and without interruption of the spiral sequence of appearance of primordia.

GENETIC STUDIES

The genetic basis of the loss of ray florets and enclosing phyllaries was originally examined by Clausen et al. (1947), who also determined in the same study, that L. discoidea was actually a Layia. This unusual situation occurred because when L. discoidea was discovered in 1940, its single row of involucral bracts, and lack of ray florets and receptacle bracts suggested it belonged to the tribe Helenieae, but it could not be assigned to any known genus. Its phylogenetic relationships became apparent following cytological studies and the results of experimental hybridizations. The species was found to have eight chromosome pairs similar to many species of Layia, and hybrids between it and L. glandulosa proved to be vigorous and fertile.

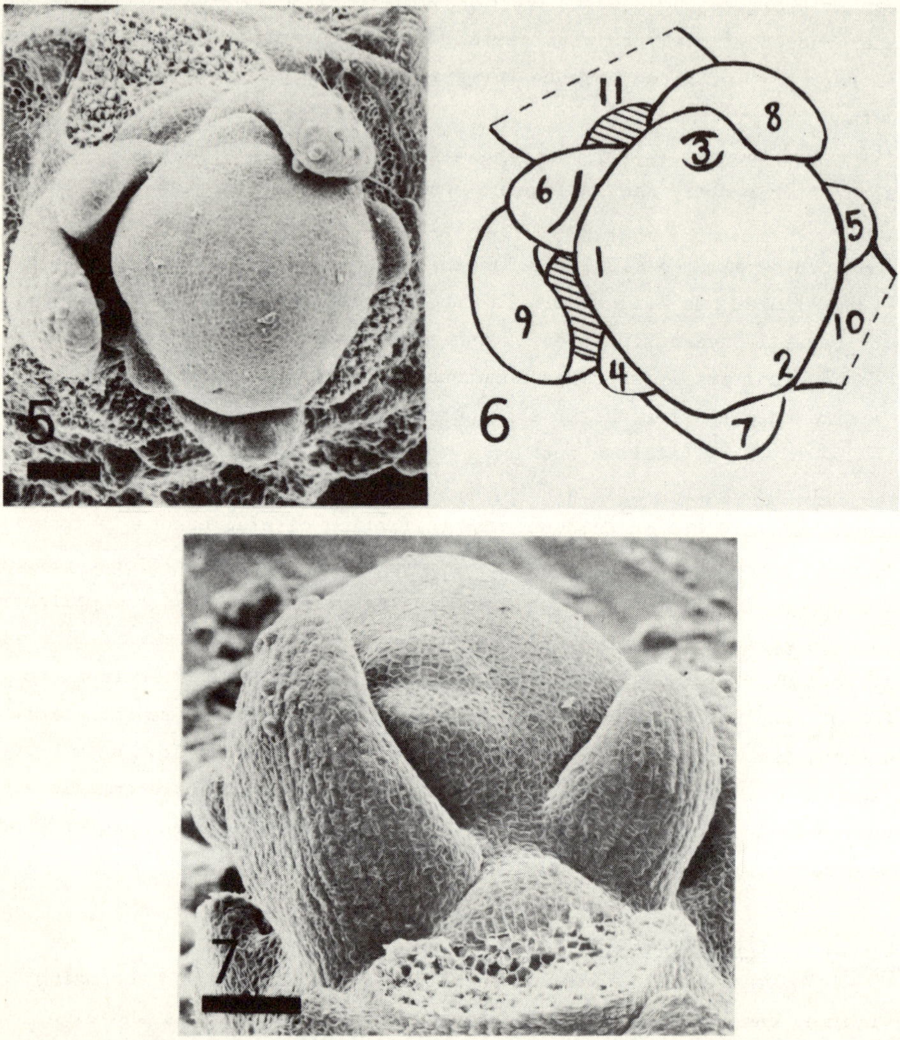

Figure 5-7. An early stage in the development of the same head of L. discoidea. Bar = 50μ. (5) shows the top view, (6) a diagram, and (7) the side view. Five stem bracts intervene between the top axillary bud (p9) and the first pair of receptacle bract/disk floret primordia at p3. This same pair is illustrated in side view in (7). Note the apparent simultaneous outgrowth of floret and bract.

Clausen *et al.* (1947) also analyzed a large F_2 progeny and on the basis of an approximate 13:3 ratio suggested that the loss of ray florets and phyllaries was governed by two loci. The Clausen group concluded that the unusual morphology did not justify generic status and treated *L. discoidea* informally as a subspecies of *L. glandulosa*. Later it was designated a species (Keck, 1958).

Our initial genetic studies involved primarily the synthesis of backcrosses between interspecific F_1 hybrids and each of the parents. In these crosses, we used both subspecies of *L. glandulosa*: subsp. *glandulosa*, with white rays (the one previously used by Clausen), and subsp. *lutea*, with golden yellow rays, which has a limited geographical distribution in the inner South Coast Ranges near that of *L. discoidea* but on sandy soils at lower elevations (Gottlieb *et al.*, 1985). Since the progenies between *L. discoidea* and both subspecies of *L. glandulosa* gave the same results, we describe here only the results of the crosses with subsp. *glandulosa*.

In the F_1, all plants had at least several ray florets (Figure 8). The associated bracts were variable, but generally clasped less completely around the florets than in *L. glandulosa*. The number varied from 2 to 7, with a very strong mode at 5 rays.

F_1 plants with intermediate ray numbers (usually 5) were backcrossed to both *L. discoidea* and *L. glandulosa*. The BC_1 to *L. discoidea* segregated with a ratio of three plants without rays to one with rays (Table 1), consistent with a two-locus model. On this model, ray florets are present only when the genotype at each locus is heterozygous, like that of the F_1. Thus $\underline{A}^G\underline{A}^D\ \underline{B}^G\underline{B}^D$ has rays, but $\underline{A}^G\underline{A}^D\ \underline{B}^D\underline{B}^D$, $\underline{A}^D\underline{A}^D\ \underline{B}^G\underline{B}^D$, and $\underline{A}^D\underline{A}^D\ \underline{B}^D\underline{B}^D$ do

Table 1. Number of plants segregating in backcross 1 and backcross 2 progenies of *L. discoidea* x *L. glandulosa* backcrossed to *L. discoidea* for presence versus absence of ray florets on terminal head of central stem.

	Number of plants			
Cross	≥1 Ray	0 Ray	Total	X^2
BC_1 D x (D x G)	21	46	67	1.44
BC_2 D x [D x (D x G)]	23	59	82	0.40
BC_2 D x [D x (D x G)]	19	60	79	0.04

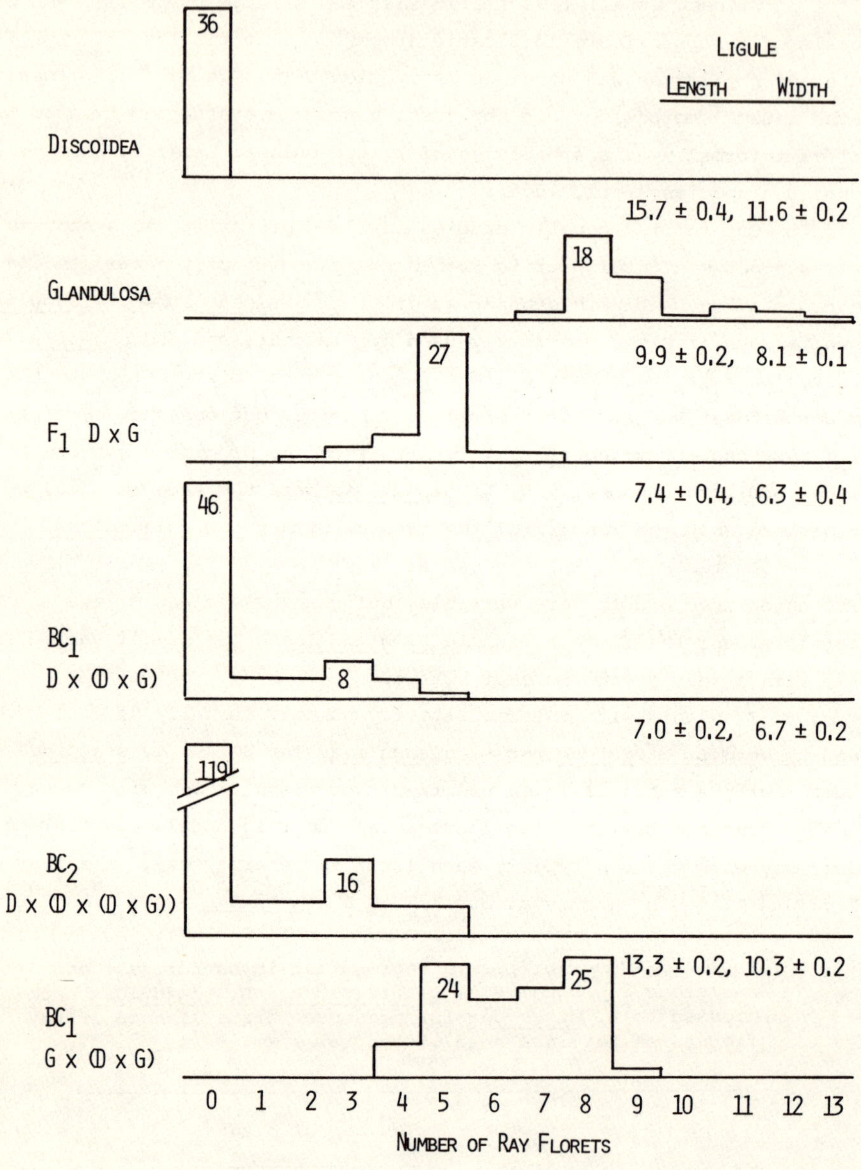

Figure 8. Number of ray florets and mean ligule length and width of L. glandulosa and L. discoidea, and F_1 progeny between them, BC_1 and BC_2 to L. discoidea, and BC_1 to L. glandulosa. The number of plants in each modal class is identified on the histograms. The length and width data are mm ±s.e.

not have them. Among the plants with rays, the modal number was 3 (Figure 8).

The ratio of 1:3 ray floret presence to absence in the backcross progeny is not consistent with the 13:3 presence to absence model proposed by Clausen for the F_2. On the basis of the photograph of representative heads from the F_2 published by Clausen et al. (1947), we believe that the inconsistency results from incorrect classification of many F_2 individuals. We have observed that heads on F_1 and backcross plants occasionally display "chimeras"; these appear to be disk florets and could be confused with small ray florets because one or more of their outside corolla lobes are enlarged and these are white rather than yellow. The published photograph shows a number of heads with one or two chimeric structures but not normal rays. We think it likely that such plants were classified improperly as having ray florets.

BC_1 individuals with at least three rays were used to construct the BC_2 to L. discoidea. Two independent BC_2 progenies were grown out, and both of them also segregated with a ratio of three plants without rays to one with rays (Table 1). As in BC_1, the modal number of rays was at 3 (Figure 8). These backcross ratios indicate that the two loci assort independently.

The backcross results also have several important implications for genetic control of ray number. (1) The shift in ray number and modal class between the F_1 and the BC_1 or BC_2 suggests that L. discoidea possesses an additional genetic factor(s) that modifies ray floret number without affecting presence/absence. This follows because the F_1 and the 1/4 of the backcross progenies with rays had identical doubly heterozygous genotypes at the two major loci, but the number of rays and their distribution into classes changed. (2) The similarity of ray number distribution in BC_1 and BC_2 suggests that all the L. glandulosa alleles relevant to ray floret presence and number were successfully brought along in the plants used to make the crosses and, thus, that the number of genes is few. (3) It is worth noting that 8, 5, and 3 are Fibonacci numbers, and it is probably not coincidental that the modal classes centre on them. Whether this itself reflects geometric considerations or genetic factors remains to be determined.

The ray achenes of both backcross generations were slightly to moderately pubescent (but not so densely as disk achenes on the same heads). The pubescence did not always clothe the entire achene, but frequently was distributed only 1/4 to 1/2 around its circumference and, when this was the case, the pubescence was on the inner or adaxial surface. Although all ray achenes were not examined, about 30% of those studied also had one or more pappus scales similar in shape to those on the disk achenes, but generally fewer in number.

Figure 8 also presents the progeny means for ligule length and width. The ligule length on F_1 plants averaged 9.9 mm, more similar to that of L. glandulosa than predicted by the arithmetic mean. In BC_1 and BC_2 the mean ligule length for the plants with ray florets was 7.4 and 7.0, respectively. A significant positive correlation was found between ray number and ligule length for BC_1 but not BC_2. The reduction in ligule length between the F_1 and the backcrosses suggests, as with ray floret number, that there are additional loci within L. discoidea that modify ligule length (even though the plants of this species do not have ligules).

The BC_1 to L. glandulosa showed two modal classes of 5 and 8 ray florets (Figure 8). The lower number presumably reflects the influence of alleles from L. discoidea, but additional crossing must be carried out to determine whether loci controlling presence/absence or only number are involved.

DEVELOPMENTAL MODELS

The genetic results support Clausen's original finding that the presence versus absence of ray florets and their enclosing bracts are controlled by two genes. We do not yet know what phenotype will be specified when each gene is changed independently of the other (i.e., A^GA^G B^DB^D and A^DA^D B^GB^G), but this information will be obtained from progenies now being grown. We also identified an additional gene or genes in L. discoidea that modifies ray number but does not appear to affect their presence/absence.

The analysis of head development in L. glandulosa by scanning electron microscopy showed that the manner of initiation and development of the two floret types and associated bracts is very different. Ray florets and their bracts are initiated independently from adjacent primordia and grow out at different times. The outer disk florets and their bracts appear to be initiated from common primordia that bifurcate transversely, but the subsequent development of the upper and lower portions proceeds simultaneously. Disk florets higher up on the apical dome are initiated without subtending bracts.

Other studies of head development in radiate composites (e.g. Lawalree, 1948; Phillipson, 1946, 1948) have noted that florets are initiated acropetally and that young rays grow more slowly than young disk florets. The initiation of ray floret primordia is markedly delayed relative to the development of disk floret primordia in Cosmos bipinnatus (Molder and Owens, 1973).

The origin of disk florets and receptacle bracts from common primordia has been reported in Helianthus (Marc and Palmer, 1978). In another plant family (Saururaceae), Tucker (1981) described an inflorescence in which the first flowers and bracts arise from separate primordia with later flowers and bracts originating from common primordia.

The developmental analysis in L. discoidea was able to rule out several potential explanations for its lack of ray florets and bracts. Their absence does not result from suppression of primordia because in the early stages of head development, all primordia are initiated in a continuous sequence and none is aborted. Their absence is also not a consequence of L. discoidea having a smaller vegetative apex which develops into a smaller receptacle with less "space" for rays and clasping bracts, or with a smaller volume that limits the production of some "morphogen". This is because the vegetative apices of the two species cannot be distinguished in size and because receptacles produced on older plants of L. glandulosa are frequently smaller than those of L. discoidea but still develop ray florets.

A useful model for the loss of ray florets is suggested by the observation that the peripheral disk florets of L. discoidea grow out simultaneously with their associated bracts. If these florets and bracts prove to arise from common primordia, then the separate pairs of phyllary/ray floret primordia characteristic of L. glandulosa have been "omitted" in the development of L. discoidea. In this case, their origin from common primordia identifies the involucral bracts of L. discoidea as homologous to the receptacle bracts of L. glandulosa.

Alternatively, the peripheral disk florets and bracts of L. discoidea could arise from independent primordia as in L. glandulosa, but the time of floret outgrowth might be greatly accelerated so that it is simultaneous with that of its associated bract. Light microscope examination of sectioned material will reveal the morphological sequence of events. In either case, the growth of the peripheral florets relative to their bracts begins sooner in L. discoidea than in L. glandulosa.

We propose that this timing change causes the peripheral florets to develop as disk florets rather than ray florets, perhaps because now all florets become subject to the same biochemical and biophysical inputs. By contrast, the delay in the growth of peripheral floret primordia in L. glandulosa permits them to receive the appropriate signals to become rays.

The apparent synchrony of growth among the early ray primordia may also be functionally important.

The model suggests that the genetic changes governing the loss of rays have to do primarily with processes that affect the timing of localized spurts of growth. Changes in the relative time of initiation of primordia have been clearly documented in the Saururaceae (Tucker, 1981) and Leguminosae (Tucker, 1984a). Simple changes in gene regulation could effect the required change in timing, for example, by modifying rates of cell divisions. The model is interesting because it implies that *L. discoidea* has not lost the genetic information to "make" rays, but only that the information is not utilized because conditions are not appropriate. In this regard, we know that *L. discoidea* has genes that modify the colour, length, and number of rays (Clausen *et al*., 1947; this paper). Finally, the model also suggests an explanation for the fact that ray florets of plants in the hybrid progenies often have features characteristic of disk florets such as varying amounts of ovary pubescence and pappus scales. If these characters were variously regulated by the timing of floret growth, then intermediate times might result in a mixture of ray and disk traits. The prediction is that ray florets with pubescence and pappus develop sooner or more rapidly than the normal glabrous, epappose ray florets of *L. glandulosa*. Since we are backcrossing *L. glandulosa* alleles governing the development of rays into *L. discoidea*, it will be possible to study their effects on timing and other ontogenetic attributes during head development.

CONCLUSIONS

In order to understand how changes in morphology evolve, we believe it is necessary to carry out combined genetic and developmental studies. For any given character change, we want to know both the number and type of operative genetic changes and how they are translated during ontogeny into different shapes and structures. We have chosen to study morphological differences that characterize species. Unlike induced or spontaneous mutations in cultivated plants, we expect the different character expressions to be well integrated into their respective developmental systems and, most likely, to have ecological significance.

But we don't have many of the advantages of working with cultivated plants such as true-breeding lines, self-fertility to simplify breeding programs, genetic maps, the ability to manipulate chromosomes, for example, to construct aneuploids to test the effect of single gene doses, nor an accumulated record of previous studies. These advantages mean that an understanding of plant development will probably progress more rapidly in cultivated species, and the first connections between developmental processes and molecular biology will be achieved with them. However, we believe that much of what is learned with cultivated plants is also true for wild species (we suspect that developmental processes are reasonably similar in all angiosperms, Gottlieb, 1984), and that we can apply it to character differences in natural populations to account for the evolution of plant diversity.

Many excellent studies have been made in wild plants but, in general, they are either genetic (reviewed in Gottlieb, 1984; Hilu, 1983) or developmental (Guerrant, 1982; Kaplan, 1980, 1984; Kirchoff, 1983; Lord, 1981, 1984; Tucker, 1984a, 1984b; Tucker *et al.*, 1985). With the notable exception of Bachmann's (1983) studies of achene pubescence and pappus number, very few morphological differences in wild plants have been approached from both genetic and developmental standpoints.

We selected the presence versus absence of ray florets in *L. glandulosa* and *L. discoidea*, respectively, because the direction of evolution was certain, the genetic basis of the change had previously been shown to be simple, and, since the *Layias* are diploid and fully interfertile, an appropriate breeding program could identify and isolate the relevant genes and backcross them to alternate backgrounds in order to observe their effect(s) in the presence or absence of potential modifiers. In addition, the character difference is very significant in the taxonomy of the Compositae at the tribal level (Cronquist, 1955, 1977; Jeffrey, 1977) and often is used to distinguish species, for example, in *Bahia*, *Rudbeckia*, *Encelia*, and *Chrysopsis*, from the California flora.

What we now know reflects results from both the genetic and developmental analyses. By demonstrating that changes in only two genes result in the absence of ray florets and their bracts, we inferred that their absence must result from only a few changes during ontogeny (which means, in principle, that the changes could not be complex). We inferred from the segregation of the backcross to *L. discoidea* that additional factors affect the number of rays and their length. The finding of pubescence and pappus on the ray florets of backcross plants and the ray/disk floret chimeras suggested that the two floret types were probably similar in many aspects of their development (indeed, their initial ontogeny is similar). The discovery that they were initiated at different times relative to their associated bracts suggested that the absence of ray florets did not mean that a structure was "lost", only that a course of development was redirected.

Working back and forth between the genetic and developmental analyses stimulated many ideas that could not have emerged from either single analysis. In addition, an idea from one analysis could be evaluated and rejected from the perspective of the other analysis.

REFERENCES

Bachmann, K. (1983) Evolutionary genetics and the genetic control of morphogenesis in flowering plants. Evolutionary Biology 16, 157-208.

Clausen, J., Keck, D.D. & Hiesey, W.M. (1947) Heredity of geographically and ecologically isolated races. Amer. Naturalist 81, 114-133.

Cronquist, A. (1955) Phylogeny and taxonomy of the Compositae. Amer. Midl. Naturalist 53, 478-511.

Cronquist, A. (1977) The Compositae revisited. Brittonia 29, 137-153.

Gottlieb, L.D. (1984) Genetics and morphological evolution in plants. Amer. Naturalist 123, 681-709.

Gottlieb, L.D., Warwick, S.I. & Ford, V.S. (1985) Morphological and electrophoretic divergence between *Layia discoidea* and *L. glandulosa*. Syst. Botany 10, 484-495.

Guerrant, E.O. (1982) Neotenic evolution of *Delphinium nudicaule* (Ranunculaceae), a hummingbird-pollinated larkspur. Evolution 36, 699-712.

Hilu, K.W. (1983) The role of single-gene mutations in the evolution of flowering plants. Evolutionary Biology 16, 97-128.

Jeffrey, C. (1977) Corolla forms in Compositae, some evolutionary and taxonomic speculations. *In*: The biology and chemistry of the Compositae, Vol. 1, ed. V.H. Heywood, J.B. Harborne and B.L. Turner. London: Academic Press.

Kaiser, S. (1935) The factors controlling shape and size in *Capsicum* fruits: a genetic and developmental analysis. Bull. Torrey Bot. Club 62, 433-454.

Kaplan, D.R. (1980) Heteroblastic leaf development in *Acacia*: morphological and morphogenetic implications. La Cellule 73, 135-203.

Kaplan, D.R. (1984) Alternative modes of organogenesis in higher plants. *In*: Contemporary problems in plant anatomy, ed. R.A. White and W.C. Dickison. New York: Academic Press.

Keck, D. (1958) Taxonomic notes on the California flora. Aliso 4, 101-114.

Kirchoff, B.K. (1983) Allometric growth of the flowers of five genera of the Marantaceae and in *Canna* (Cannaceae). Bot. Gaz. 144, 110-118.

Lawalree, A. (1948) Histogénèse florale et végétative chez quelques composées. La Cellule 52, 214-294.

Lord, E.M. (1981) Cleistogamy: a tool for the study of floral morphogenesis. Bot. Rev. 47, 421-449.

Lord, E.M. (1984) Cleistogamy: a comparative study of intraspecific floral variation. *In*: Contemporary problems in plant anatomy, ed. R.A. White and W.C. Dickison. New York: Academic Press.

Marc, J. & Palmer, J.H. (1978) A sequence of stages in flower development in the sunflower. Proc. 8th Inter'l. Sunflower Conference. pp. 130-137.

Molder, M. & Owens, J.N. (1973) Ontogeny and histochemistry of the intermediate and reproductive apices of Cosmos bipinnatus var. Sensation in response to gibberellin A_3 and photoperiod. Can. J. Bot. 51, 535-551.

Phillipson, W.R. (1946) Studies in the development of the inflorescence. I. The capitulum of Bellis perennis L. Ann. Bot. 10, 257-270.

Phillipson, W.R. (1948) Studies in the development of the inflorescence. IV. The capitula of Hieracium boreale Fries and Dahlia gracilis Ortg. Ann. Bot. 12, 65-75.

Sinnott, E.W. & Kaiser, S. (1934) Two types of genetic control over the development of shape. Bull. Torrey Bot. Club 61, 1-7.

Stebbins, G.L. & Yagil, E. (1966) The morphogenetic effects of the hooded gene in barley. Genetics 54, 31-40.

Tucker, S.C. (1981) Inflorescence and floral development in Houttuynia cordata (Saururaceae). Amer. J. Bot. 68, 1017-1032.

Tucker, S.C. (1984a) Unidirectional organ initiation in leguminous flowers. Amer. J. Bot. 71, 1139-1148.

Tucker, S.C. (1984b) Origin of symmetry in flowers. In: Contemporary problems in plant anatomy, ed. R.A. White and W.C. Dickison. New York: Academic Press.

Tucker, S.C., Stein, O.L. & Derstine, K.S. (1985) Floral development in Caesalpinia (Leguminosae). Amer. J. Bot. 72, 1424-1434.

THE CONTROL OF FLORAL PIGMENTATION IN ANTIRRHINUM MAJUS

Cathie Martin (1), R. Carpenter (1), E.S. Coen (1),
Tom Gerats (2)
(1) John Innes Institute, Colney Lane, Norwich NR4 7UH
(2) Department of Genetics, Free University of Amsterdam, Amsterdam

INTRODUCTION

At first glance, the control of floral pigmentation may not seem to be a key topic for a symposium on development in plants. Although biochemists and even physiologists have long recognised metabolic change as part of development, other biologists have tended to use the term "development" in a more restricted sense, to refer specifically to morphogenesis.

Anthocyanin production is not involved in the formation of structure. However, examination of species such as *Antirrhinum* shows that there is a wealth of variation in pattern and shade of flower colour amongst commercial lines and natural variants. Such variation reflects differences in anthocyanin production. Anthocyanin production represents a simple model system for understanding morphogenetic regulation because the biosynthetic pathway is regulated with respect to developmental time and developmental pattern. An examination of this type of simple system will facilitate our understanding of morphogenesis, which must depend on the co-ordinated interaction of biochemical pathways that are regulated in this way. An integration of disciplines such as cell biology, physiology, biochemistry, molecular biology and genetics is a necessity in investigating plant development. One reason why the anthocyanin biosynthetic pathway can be used as a model system is that it is well characterised biochemically and genetically, so that such an integrated approach is possible.

The aim of this chapter is to emphasise how genetic mutants can be used to understand the regulation of anthocyanin biosynthesis. As anthocyanin biosynthesis is a non-essential function of plant development (at least in terms of individuals under laboratory conditions) it may be possible, in the long term, to derive a near complete set of mutants in the functioning and regulation of this pathway.

Antirrhinum is a particularly suitable plant to study because of the presence of biological "mutagenising agents", transposable elements, in two of the structural genes of the anthocyanin biosynthetic pathway. These elements have assisted research in two very important ways; firstly by facilitating the cloning of both genes (Weinand *et al.*, 1982; Sommer *et al.*, 1985; Martin *et al.*, 1985) and secondly because they have given rise to a very large number of allelic mutants (Fincham and Harrison, 1967; Carpenter *et al.*, 1986). Such mutants can vastly expand our understanding of how genes work. In maize, a large number of genes involved in anthocyanin biosynthesis have unstable alleles due to transposable element insertions (McClintock, 1948; Emerson, 1921). The investigation of these genes complements the research on *Antirrhinum*. For example, the flower is a relatively sophisticated structure and the regulation of its pigmentation is complex. Similar complexities are not generally obvious in the pigmentation of single organs of maize. There are both similarities and differences in the regulation of pigmentation in maize and *Antirrhinum*. The genetic systems of maize have been reviewed recently (Coe, 1985) but where comparison is useful, reference will be made to mutations known in other species such as maize and *Petunia*.

THE REGULATION OF ANTHOCYANIN BIOSYNTHESIS IN THE FLOWER

Types of control

Modifications in cellular levels of anthocyanins may theoretically be achieved by varying either synthesis or degradation, or both. However, it has been reported that anthocyanins turn over very little (Gerats, 1985), and that pattern and degree of pigmentation is largely a function of changes in biosynthesis. This discussion will therefore be restricted to the control of anthocyanin biosynthesis. Two levels of control may operate:-

1. Coarse control. This term is intended to include factors that determine the biosynthesis and degradation of the biosynthetic enzymes. Such factors set limits for rates, patterns and times of flux along the pathway. For anthocyanin biosynthesis in *Antirrhinum* flowers, coarse control may be of three types -

a a) Developmental - including tissue specificity, induction during organogenesis and hormonal regulation.

b) Environmental - including induction by light, low temperature and drought.

c) Genetic - including *cis*-acting functions of structural genes and *trans*-acting regulator genes.

2. Fine control. This term includes metabolic control systems that determine the *actual* rates of flux along the pathway. Very little is known of fine

control of anthocyanin biosynthesis. However, it is likely that some environmental stimuli, such as drought, may take effect, by modifying fine control.

The control unit

In order to assess the effects of such potential regulatory influences on anthocyanin biosynthesis, the whole biochemical pathway should be analysed. The price to be paid for such a holistic approach is that the pathway is very complex (Figure 1), even in *Antirrhinum* where nature has pruned the biochemical steps to the minimum. If the regulatory unit to be considered is restricted to the specialised metabolic branch involved in flavanoid biosynthesis, the activity of at least seven enzymes is involved.

The sequence of reactions in anthocyanin biosynthesis was elucidated primarily by using mutants. Analysis of intermediates that accumulate in particular mutants, and precursor feeding experiments to by-pass genetic blocks, established the sequence of reactions shown in Figure 1. This work was the result of a great number of experiments in maize and other species (Coe, 1957; Harborne, 1967; Hahlbrock and Grisebach, 1975; Styles and Ceska, 1977; McCormick, 1978). The sequential ordering of the enzymic steps and genetic blocks in *Antirrhinum* was the work of Harrison and Stickland (1974, 1978, 1980) and Stickland and Harrison (1974).

THE ANTHOCYANIN BIOSYNTHETIC PATHWAY

Synthesis and subsequent modification of the basic flavanoid ring structure

1. The first step involves condensation of p-coumaroyl CoA with 3 molecules of malonyl CoA to form chalconaringenin. The enzyme that catalyses this reaction (Table 1) is chalcone synthase and the genetic block at this stage in *Antirrhinum* is known as *nivea* (Spribille and Forkmann, 1982). *Nivea* null mutations are white-flowered as they produce no pigment at all.
2. The second step is an isomerisation catalysed by the enzyme chalcone isomerase to form naringenin. This step can also occur spontaneously, *in vitro*, and there may be spontaneous isomerisation *in vivo* as well. Forkmann and Dangelmayr (1980) found no chalcone isomerase activity in *Dianthus* lines recessive for the gene *i*. These flowers make a small amount of anthocyanin probably as a result of spontaneous isomerisation *in vivo*. No mutations lacking this enzyme have been found in *Antirrhinum*.

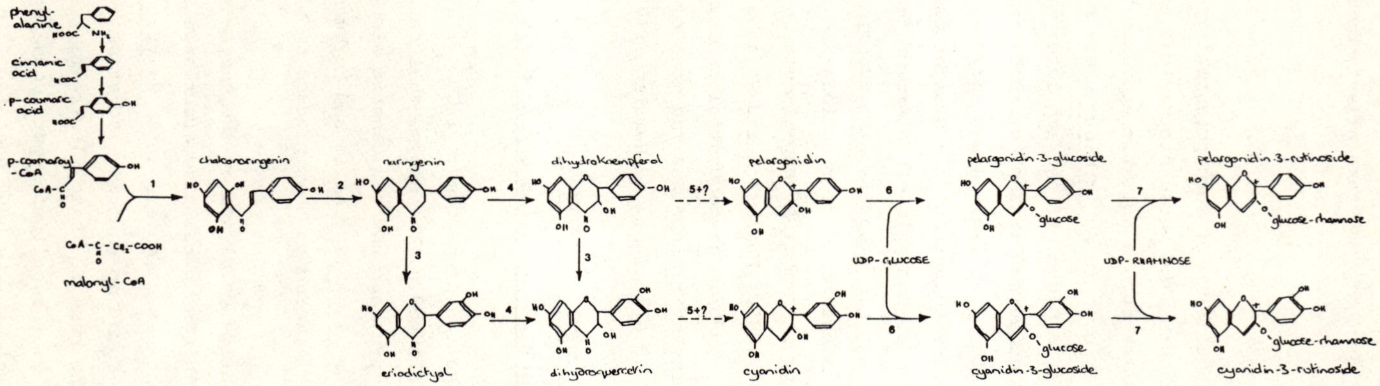

Figure 1. Sequence of biochemical reactions involved in the formation of anthocyanins. The enzymic steps (1-7) are labelled as they are referred to in the text. The early stages involved in synthesis of malonyl-CoA and p-coumaroyl-CoA are considered in more detail in reviews such as that by Hahlbrock and Grisebach (1975).

1. CHALCONE SYNTHASE - Nivea.

p-coumaroyl CoA + 3 malonyl CoA ⟶ Chalconaringenin.

2. CHALCONE ISOMERASE.

Chalconaringenin ⟶ Naringenin.

3. FLAVONOID-3'- HYDROXYLASE. - Eosina.

Naringenin ⟶ Eriodictyol.

Dihydrokampferol ⟶ Dihydroquercetin.

4. FLAVONOID-3-HYDROXYLASE. - Incolorata.

Naringenin ⟶ Dihydrokaempferol.

Eriodictyol ⟶ Dihydroquercetin.

5. ? - Pallida

Dihydrokaempferol - - ⟶ pelargonidin.

Dihydroquercetin - - ⟶ cyanidin.

6. UDP-GLUCOSE FLAVONOID GLUCOSYL TRANSFERASE.

pelargonidin ⟶ pelargonidin-3-glucoside.

cyanidin ⟶ cyanidin-3-glucoside.

7. UDP-RHAMNOSE:ANTHOCYANIDIN-3-GLUCOSIDE, 6" RHAMNOSYL TRANSFERASE.

Pelargonidin-3-glucoside + UDP-rhamnose ⟶ pelargonidin-3-rutinoside

Cyanidin-3-glucoside + UDP-rhamnose ⟶ cyanidin-3-rutinoside

Table 1. Details of enzymic steps involved in anthocyanin biosynthesis in Petunia the po mutation lacks chalcone isomerase in the anthers and the pollen is white (van Weely et al., 1983). The corolla does contain chalcone isomerase activity, however, and Po may well be a regulatory gene rather than a structural one.

The next two steps involve hydroxylations of the flavanoid rings, and are catalysed by 3-hydroxylase and a 3'-hydroxylase.

3. If the 3'-hydroxylase is inactive, anthocyanin is made but only with a single hydroxyl group on the third ring. This anthocyanin is called pelargonidin and gives pink flowers. This mutation is known as eosinea (Forkmann and Stotz, 1981; Stotz and Forkmann, 1981). The production of pelargonidin is in contrast to normal wild type Antirrhinum where most of the anthocyanin made carries two hydroxyl groups on the third ring (cyanidin) to give a magenta colour.
4. If the 3-hydroxylase is inactive no anthocyanins are made and the flowers are ivory. This mutation is known as incolorata (Forkmann and Stotz, 1981; Stotz and Forkmann, 1981).

Both the 3-hydroxylase and the 3'-hydroxylase may hydroxylate a number of substrates and the exact order of their activity in the pathway may vary.

Production of colour

The second stage in the anthocyanin biosynthetic pathway involves the formation of coloured anthocyanidins. Colour is formed when the flavanoid ring structure acquires charge. This may also be the stage at which the compounds are transferred across the tonoplast into the vacuole. The biosynthetic steps at this stage are not well characterised. In maize, at least 2 genes are thought to be involved, a_1 and a_2 (Coe, 1955; Kirby and Styles, 1970; Reddy and Reddy, 1971). The products of these genes have not yet been characterised enzymatically although results from barley suggest that 2,3trans-3,4cis leucocyanidin may be an intermediate between the two steps (Kristiansen, 1984). The same work in barley suggests that there may be as many as 16 genes which act at this stage, although it is unlikely that they are all structural genes: some may be genes involved in membrane transport of anthocyanidins or regulatory genes.

In Antirrhinum we have one block at this stage, pallida (Stickland and Harrison, 1974; Harrison and Stickland, 1974), which is thought to encode a biosynthetic enzyme, although its exact function is as yet unknown.

Stabilisation of the anthocyanidins

Anthocyanidins are unstable compounds which are stabilised in plants by formation of glycosides. In Antirrhinum this involves two steps:

6. In the first step a glucose residue from UDPG is added to the 3-hydroxyl group by a glycosyl transferase (UFGT)(Figure 1.6).
7. In the second step rhamnose is added to the glucose from UDP-rhamnose, by a rhamnosyl transferase to form the anthocyanidin-3-rutinoside (Figure 1.7).

No genetic blocks are known in these steps in Antirrhinum. The bz-1 (bronze-1) locus of maize is a lesion of the glycosyl-transferase (Larson and Coe, 1968; Dooner and Nelson, 1977; Larson and Coe, 1977; Dooner, 1980). In bronze mutants a complex murky pigment is formed in the aleurone as a result of condensation of the unstable unglycosylated anthocyanidins (Kirby and Styles, 1970). In many other species there may be further modifications of the anthocyanins, such as methylation and acylation. The methylating enzymes have been studied in Petunia (Wiering and de Vlaming, 1977; Jonssen et al., 1982, 1983, 1984) and in some other species (Poulton, 1981; de Luca and Ibraham, 1982). Some other modifying enzymes have also been reported (Kamsteeg et al., 1980b). These steps are absent in Antirrhinum.

In summary, there are four characterised genetic lesions of the anthocyanin biosynthetic genes in Antirrhinum; nivea, incolorata, eosinea and pallida (Table 2). Two of these genes, nivea (chalcone synthase) and pallida, have been cloned and multiple allelic series also exist at these loci. For these two genes a very detailed study of the regulation of

GENETIC LESIONS IN ANTHOCYANIN BIOSYNTHESIS IN Antirrhinum majus.

GENES OF THE BIOSYNTHETIC PATHWAY.

GENETIC BLOCK.	ENZYME.	PHENOTYPE.
nivea	Chalcone Synthase.	White.
eosina	Flavonoid-3'- Hydroxylase.	Pink.
incolorata	Flavonoid-3- Hydroxylase.	Ivory.
pallida	?	Ivory.

expression is possible.

For the other two genes, incolorata (3-hydroxylase) and eosinea (3'-hydroxylase), no clones are yet available, and there are no allelic series. However, enzymic assays have recently been published (Forkmann and Stotz, 1981; Stotz and Forkmann, 1981) and so more detailed study may be possible in the future. Similar progress may also be possible with the other enzymes for which assays are available; chalcone isomerase (Moustafa and Wong, 1967), UDP-flavonoid glycosly transferase (Larson and Coe, 1968; Dooner, 1980) and UDP-flavonoid rhamnosyl transferase (Kamsteeg et al., 1980a).

TYPES OF REGULATION OF ANTHOCYANIN BIOSYNTHESIS

1. Developmental regulation

a. Tissue specificity

There appears to be tissue specificity in anthocyanin biosynthesis. In most species anthocyanins are formed mainly in the epidermis, although in some species there may also be restricted pigment formation in the parenchyma (Wheldale, 1916). In Antirrhinum flowers the pigment appears to be wholly epidermal. No biochemical studies have yet shown whether individual enzyme activity is restricted to the epidermis of Antirrhinum. Results from Petunia suggest that enzymes such as chalcone isomerase that act early in the pathway may be active in both epidermal and parenchymal tissue (Jonssen et al., 1984). Enzymes acting later in the pathway such as UFGT and methylating enzymes are active only in the epidermis. Recent data from Hahlbrock's laboratory (Hahlbrock et al., 1985) using immunofluorescence suggests that even very early enzymes such as phenylanine ammonia lyase (involved in formation of paracoumaryl CoA) may normally be most active in the epidermis, although not completely restricted to it.

No mutations that alter tissue specificity have yet been found in Antirrhinum. In maize, loci which affect pigmentation of specific tissues have been identified (for review see Coe, 1985). For example, the B (Booster) locus directs pigmentation of all vegetative parts of the plant (Styles et al., 1973). In the recessive form anthocyanin production is restricted to the tassel husks and the seedling. The pigmentation of reproductive tissues including anther walls, silks, pericarp and aleurone is independent of the B locus and is regulated by the R locus. The R locus has several alleles which affect pigmentation of specific tissues. The r-r allele blocks pigmentation only in the aleurone. The R-g allele blocks pigmentation in the silks, tassel glume bars and the leaftips. Recently the R locus has been cloned (Dellaporta et al., 1985). There appear to be at least three duplicated genes at the complex locus, each affecting pigmentation of different tissue types. Thus, in maize the regulation of tissue specificity appears to be under the control of particular regulatory genes. It is possible that similar regulatory genes induce the expression of the structural genes in the epidermis of Antirrhinum.

b. Expression during organogenesis

There is a large increase in anthocyanin biosynthesis during flower development. This is preceeded by a significant increase in the amounts of steady-state transcripts of at least two structural genes nivea (chalcone synthase) and pallida. Transcript levels are low in very young buds and remain low whilst the cells of the bud are undergoing rapid cell division. Later, during the period of rapid cell expansion nivea (chalcone synthase) and pallida transcripts increase rapidly and co-ordinately. Anthocyanin levels rise a little later (Figure 2).

The regulation of the induction of gene expression during flower development has not yet been fully examined. The existence of a multiple allelic series of the pallida locus may facilitate research in this area.

As a result of transposable element insertion in the promoter of the pallida gene in some Antirrhinum lines, a number of allelic mutations have been derived (Fincham and Harrison, 1967). Of particular interest with respect to developmental regulation are mutations which have lost the transposable element with a concomitant small alteration of the promoter structure (Coen et al., 1986). Some of these may be described as self-coloured, showing a uniform pigmentation at a reduced intensity. Others show distinct patterns of pigment production within the flower. Two types of patterning may prove interesting in connection with developmental regulation. One pattern allele called pallida varicolor restricts pigment to the face of the front lobes, the edge of the back lobes and the base of the corolla tube (Figure 3C). The other type of pattern mutant restricts pigmentation to varying extents to the corolla tube. Three of these alleles are known (Figure 3D,E,F). When the pallida gene transcript from these alleles is compared to the transcript from $Pallida^{+}$ wild types on Northern blots, no differences are observed (Coen et al., 1986). This strongly suggests that the changes in the pattern of pigment production observed in the phenotypes are the result of changes in control of gene expression rather than changes in the gene product.

The pallida varicolor line shows reasonably high levels of pallida gene expression (about 20% full-red levels). The expression of this allele has been compared to expression of a self-coloured allele (pallida carnea) with similar levels of transcript and to a full-red commercial line over the course of flower development (Figure 4). The self-coloured pale pallida carnea and the patterned pallida varicolor show similar levels of transcript and a very similar time course of expression. It therefore seems unlikely that change in timing of pallida expression is the means of

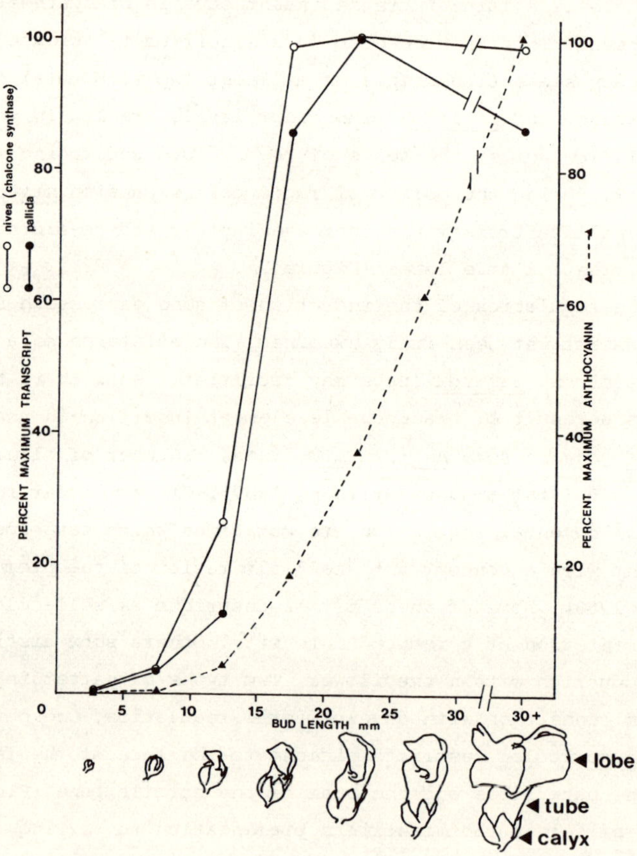

Figure 2. Time course showing the steady-state levels of nivea (chalcone synthase) and pallida transcripts during Antirrhinum bud development. The subsequent production of anthocyanin is also shown (from Coen et al., 1986).

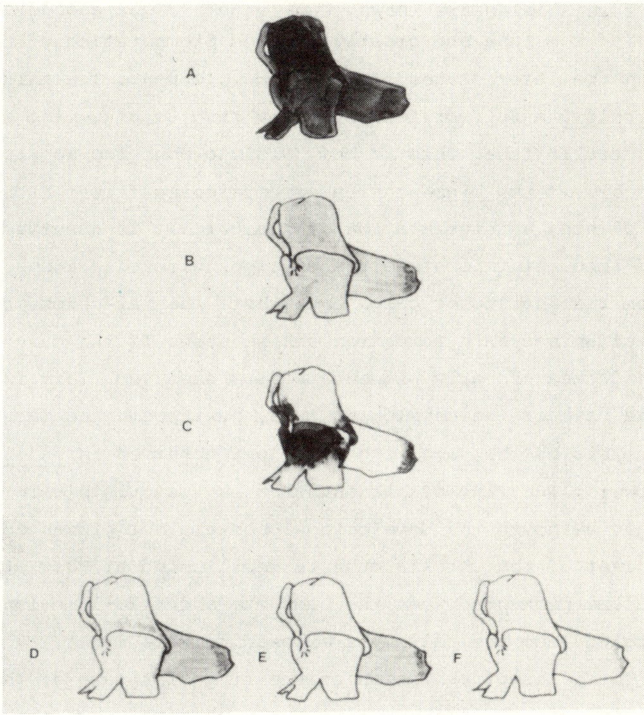

Figure 3. Diagrammatic representation of pallida alleles

A. Full Red Pal^+/Pal^+ type

B. pallida carnea (JI 37) a self-coloured pale allele

C. pallida varicolor (JI 41) a patterned allele with pigment concentrated on the face and at the base of the tube

D. JI 33; a pallida allele with pigment concentrated in the tube

E. JI 32; a pallida allele with less pigment than JI 33, again concentrated in the tube

F. pallida tubocolorata (JI 15) with pigment restricted to a ring at the base of the tube deriving the varicolor pattern. Perhaps more hopeful in this respect are the pallida alleles which show differential pigmentation of the corolla tube. Three such alleles are known. Two (stocks JI 32 and 33) show pigmentation of the tube but greatly reduced pigmentation of the lobes especially in the later stages of flower development. The third allele, pallida tubocolorata is ivory except for a ring of pigmented tissue at the base of the corolla tube. This is only visible when the sepals are pulled back (Figure 5). If the pigmentation of developing flower buds is examined in in full-red plants, no pigmentation of the corolla is observed until the bud is about 6mm long. At this stage the developing corolla lobes are just emerging from the enveloping calyx and, where they are exposed to light, the first tinge of anthocyanin formation can be seen. If the calyx is peeled back, a second area of early pigmentation is observed. This is the ring of tissue at the base of the developing tube, just above the insertion point of the anthers. This is the same ring of colour observed in pallida tubocolorata (pal^{tub}) flowers. The ring of pigmentation is just visible in pal^{tub} flowers, at this stage, although the developing lobes are not pigmented. In full-red plants, the rest of the corolla tube is unpigmented at this stage, and only starts to become pigmented when the buds are about 10-15mm long and the tube tissue, emerging from the calyx, is exposed to more light.

Thus, the pigmentation of the ring of tissue at the base of the corolla tube is rather different from the pigmentation of the rest of the corolla tube. It occurs earlier, and may well occur independently of the light stimulus that influences pigmentation of the rest of the flower. If

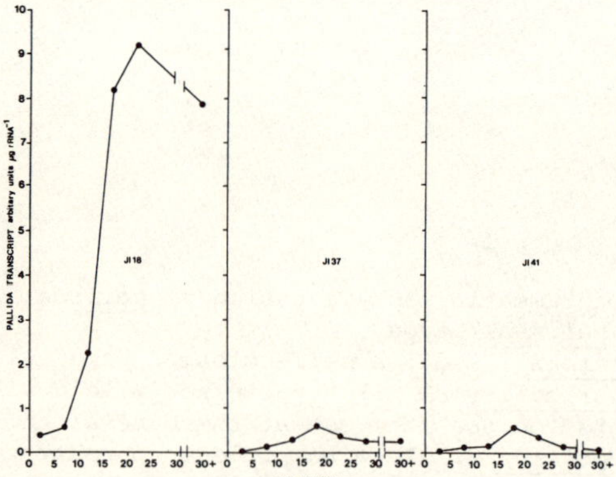

Figure 4. Changes in steady-state levels of pallida transcript in Full Red (JI 18), pallida carnea (JI 37) and pallida varicolor (JI 41) during bud development

pigmentation of the tube occurs very late in flower development, (as when the *delila* block is overcome for example), the tissue of the tube, shielded by the calyx, does not become pigmented (see section on *delila*). So late tube pigmentation appears to be light dependent. The *pal*^{tub} allele only shows pigmentation of the ring of tissue at the base of the tube. This allele carries a small alteration in the *pallida* gene promoter (Coen *et al.*, 1986). It may be that the promoter region contains sequences important for the timing of anthocyanin biosynthesis in the flowers, or for light-dependent anthocyanin biosynthesis in the flowers (i.e. a light enhancer). The identification of light enhancer sequences at the *nivea* (chalcone synthase) locus is discussed more fully later. These mutants may therefore help us identify sequences of the *pallida* promoter involved in regulation of timing of gene expression.

2. Environmental regulation

The environment exerts a very great influence on the biosynthesis of anthocyanins. Unfortunately at the physiological level this is not well understood. The two best documented environmental stimuli are temperature and light.

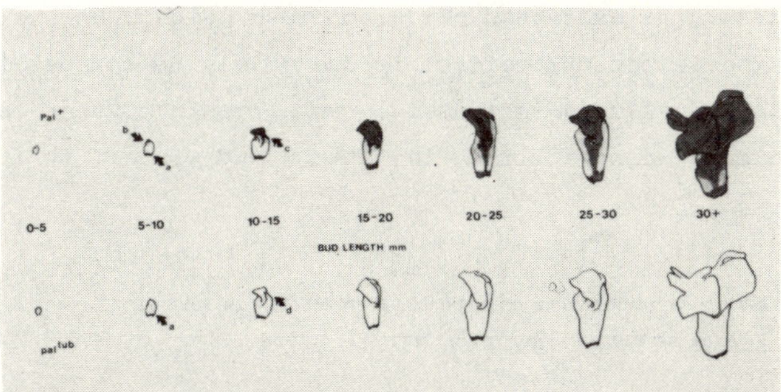

Figure 5. Comparison of pigment development in Full Red (*Pal*$^+$) flowers and *pal*^{tub} flowers, during bud development. The calyx is not shown for clarity.

a. ring of pigment just apparent at base of tube below calyx
b. pigment forming on lobes as they start to project through the surrounding calyx
c. pigment beginning to form at top of tube as it emerges through the calyx
d. no pigment formed in the lobes or at the top of the tube in the *pal*^{tub} allele

a. Temperature

Anthocyanin levels in plants appear to increase as temperature drops. It is not clear whether the main effect of temperature is:-

i. Metabolic. Sucrose levels may well rise at lower temperatures which could stimulate anthocyanin biosynthesis by fine control.

ii. On turn-over. The rate of anthocyanin breakdown may increase as temperature rises whilst biosynthesis is unaffected.

iii. On biosynthesis. The rate of anthocyanin biosynthesis may increase at lower temperatures due to an increase in production of the biosynthetic enzymes.

In Antirrhinum there is a strong effect of temperature on the anthocyanin levels in the flowers. Comparison of plants grown at 15°C and 25°C shows that the plants grown at 15°C contain 50% more pigment in their flowers (Table 3). The relative increase in anthocyanin content is somewhat greater in the corolla tube than in the corolla lobes. However, temperature also has a very marked effect on flower size. When plants grown at 15°C are compared to those grown at 25°C, the flowers at 15°C show about 50% greater fresh weight. The temperature effect on anthocyanin production may, therefore, be a function of changes in organ size and anthocyanin biosynthetic capacity. Part of the difference in anthocyanin levels at 15°C and 25°C is lost if anthocyanin is expressed per gram fresh weight. However, it is unlikely that the temperature effect is due solely to changes in the physiology of the flower cells as increases in anthocyanin content in the cold can be observed in many species during the autumn and winter, in fully expanded

EFFECT OF TEMPERATURE ON ANTHOCYANIN FORMATION IN FULL RED Pallida REVERTANT LINE (STOCK JI7).

		TEMPERATURE. 15°C	25°C	RATIO 15°C:25°C
Anthocyanin content	LOBES	0.985	0.668	1.5
umol $flower^{-1}$.	TUBES	0.167	0.058	2.9
Fresh weight (g)	LOBES	0.257	0.147	1.7
	TUBES	0.129	0.080	1.6
Anthocyanin content	LOBES	3.864	4.544	0.9
umol g fresh $weight^{-1}$	TUBES	1.295	0.707	1.8

as well as in young leaves.

Obviously, this effect requires considerable research to be understood. No temperature-insensitive mutants have been found in Antirrhinum. It is likely that the effect of temperature on anthocyanin levels is complex, operating through a number of influences.

b. Light

The importance of light for synthesis of anthocyanins has been recognised for a long time (Wheldale, 1916) although the mechanism of regulation has yet to be elucidated. Three types of light, red, blue and ultra-violet, have been identified as stimulating anthocyanin production. This indicates that three separate photoreceptors may be involved in the regulation, including phytochrome (associated with the red light response and possibly to some extent with the blue light response). The relative importance of each type of light stimulus varies from species to species. In Antirrhinum it has been suggested that phytochrome is the most important stimulus (Kaulen et al., 1986). The study of light induction of anthocyanin biosynthesis has been almost exclusively confined to leaves. It is clear that anthocyanin biosynthesis in the flower is largely, but not totally, dependent on light. Hahlbrock et al. (1976) and Schröder et al. (1979) have demonstrated that the enzymes involved in flavonoid glycoside biosynthesis (including chalcone synthase) are induced by ultra-violet light in parsley tissue culture cells. In this case, they suggest that the rate-determining step is phenylalanine ammonia lyase, an enzyme active very early in the pathway (Figure 1).

Most current research is focusing on the light-regulated expression of chalcone synthase. Kaulen et al. (1986) have shown that chalcone synthase is induced in tobacco leaves by 24h sunlight. In order to investigate light regulation further, they have fused 1.2kb of the 5'-promoter of chalcone synthase from Antirrhinum to the bacterial neomycin phosphotransferase gene which confers resistance to kanamycin. By transforming tobacco with this construct they have demonstrated uv-induced expression of neomycin phosphotransferase. This evidence suggests that the sequences involved in uv-light regulation of chalcone-synthase expression lie within the 1.2kb promoter fragment used for transformation. Sommer and Saedler (1986) have suggested, from sequence analysis of the chalcone synthase promoter, that a 28bp motif between bases -662 and -690 of the promoter region may be involved in light regulation. This 28bp sequence shows 70% homology with the light regulatory sequence of the small subunit of ribulose bisphosphate carboxylase from pea.

In the future, mutants may offer a means to increase our understanding of light regulation. In *Antirrhinum*, the transposable element, Tam3 is inserted in the chalcone synthase gene, in an unstable *nivea* line, (JI 98) just upstream of the TATA box. This transposable element causes a large number of structural alterations especially when it excises. Amongst these mutants are some which appear to lack parts of the upstream promoter sequences but which still retain some gene expression. It will be very interesting to determine whether these alleles retain or lose light inducibility. These mutants offer a means of examining light inducibility of the gene *in situ* without the position effects and other artefacts that result from transformation experiments.

In maize there exists one particularly interesting mutant, *pl*. In the dominant form, *Pl*, anthocyanin can be formed in the plant independent of light. In the recessive homozygote, anthocyanin is formed only in response to light. The action of this gene might be as a coupler (or uncoupler) of anthocyanin biosynthesis to developmental light responses. The regulatory effect of *Pl* is not well understood in maize, nor whether it affects a single rate-limiting step of anthocyanin biosynthesis, or all the steps of the pathway (Gerats *et al.*, 1984).

3. Genetic regulation

The term 'genetic regulation' is meant to refer to the hierarchy of endogenous regulatory factors, encoded by DNA, that determine the extent of activity of the biosynthetic enzymes. Obviously this group will include genes also considered in the other two categories of control.

For analytical purposes the genetic control of anthocyanin biosynthesis can be divided into two types. The first type includes *cis*-acting control systems involving regulatory sequences normally assigned to the promoters of structural genes. In the simplest models such regulatory sequences are considered likely to be binding sites for regulatory proteins. The second type of genetic control involves *trans*-acting regulatory genes which modify the expression of the structural genes.

a. The role of *cis*-acting DNA sequences in modifying anthocyanin biosynthesis

The analysis of the role of *cis*-acting control in *Antirrhinum* has only been possible for the two structural genes of the pathway that have been cloned, *nivea* (chalcone synthase) and *pallida*.

Unstable mutations that are the result of insertion of the transposable element, Tam3, exist for both loci. In both cases the site of Tam3 insertion is just upstream of the TATA box (Coen *et al.*, 1986; Sommer *et al.*, 1985). In both cases, as a result of transposable element activity, a number of phenotypic mutants have been derived (Fincham and Harrison, 1967; Carpenter *et al.*, 1986). As these appear to be due to alterations in promoter structure they help to elucidate the cis-acting sequences involved in the regulation of gene expression (Sommer *et al.*, 1985; Coen *et al.*, 1986; Sommer and Saedler, 1986).

In the unstable *Antirrhinum* line (JI 98) Tam3 is inserted at the *nivea* locus about 30 bp upstream of the TATA box and 65 bp from the start of transcription (Figure 6). Tam3 is 3.5 kb long. Tam3 is therefore located very close to the putative gene regulatory signals and it physically separates the CAAT box and other upstream regulatory signals from the more proximal signals. Despite the insertion, gene expression is only partially blocked. This can be seen most clearly when somatic excision of Tam3 is inhibited genetically using a *Stabiliser* gene (Harrison and Fincham, 1968)

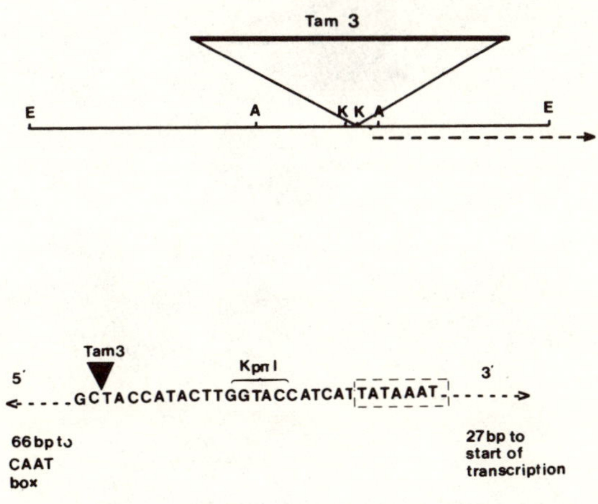

Figure 6. The position of transposable element (Tam3) insertion in the chalcone synthase locus of *Antirrhinum majus*.

a. The position of Tam3 in relation to the coding sequence (shown by dashed line). Restriction enzyme sites are shown E = EcoRI, K = KpnI, A = AvaII
b. Fine detail of Tam3 position. The KpnI restriction enzyme site is shown and the TATA box is ringed. From Sommer *et al.* (1985) and Sommer and Saedler (1986).

(Figure 7). Under these conditions the transposable element sits tightly at the locus. The insertion is seen to cause a reduction in the amount of pigment produced. Precursor feeding experiments have shown that this is due to a reduction in active chalcone synthase (Carpenter, unpublished results).

A second feature of the insertion is most interesting. The reduction in pigmentation is not even. Pigmentation is almost completely lost in the corolla tube, except at the base. There remains a strong flush of anthocyanin in the lobes and at the base of the corolla tube.

In conclusion, some apparently normal gene expression is maintained after separation of 5' regulatory sequences including the CAAT

Figure 7. Diagrammatic representation of the phenotype of the nivea allele (JI 98) carrying the Tam3 insertion
A = Full Red (Niv^{+})
B = nivea recurrens (JI 98) grown at about 20°C. Many Full Red sites of colour are seen on a background of flushed pigmentation in the lobes and at the base of the tube. The sites are the result of somatic transposable element excision
D The phenotype caused by Tam3 insertion can also be seen when excision is reduced genetically with the Stabiliser gene (Harrison and Fincham, 1968). Again the flush of pigment on the lobes and at the base of the tube is visible

box from the TATA box and the start of transcription. Thus, it is possible that the 5' signals can continue to exert some influence across an extra 3.5 kb inserted in the promoter. Alternatively the upstream regulatory signals may not be involved in producing this residual gene activity, or the transposable element may provide alternative promoter signals for expression. In this context, the work of Kaulen et al. (1986) is most interesting. Using chimeric constructs of the Antirrhinum chalcone synthase promoter and neomycin phosphotransferase (neo) transformed into tobacco, they found no neo gene expression with a promoter consisting of the 39 bp adjacent to the start of transcription, and only 20% maximal activity with the proximal 357 bp of the promoter. This evidence suggests that sequences distal of -357 are important in regulating chalcone synthase gene expression. From sequence analysis Sommer and Saedler (1986) have identified a 28 bp sequence between bases -662 to -690 that shows 70% homology to the light enhancer sequence of the small subunit of ribulose biphosphate carboxylase from pea. This may be one of the important upstream regulatory sequences.

The second interesting feature of the Tam3 insertion at the nivea (chalcone synthase) locus is that it imposes a pattern of gene expression that is not evident in unmutated flowers. The patterned phenotype could be a function of gene expression regulated by sequences within the transposable element. A number of stable allelic mutants have been derived from this unstable line. These mutants show the same pattern of pigment production concentrated on the lobes and at the base of the tubes (Figure 8). Pigment production is reduced relative to full red. At the molecular level these mutants show complete excision of Tam3 and concomitant loss of small regions (60-80 b.p.) of promoter sequence, probably including the TATA box but not including the start of transcription (Figure 9). As the patterned phenotype persists after Tam3 excision, the generation of pattern would appear to be an inherent property of sequence alterations within the promoter. The amount of pigment produced is reduced as well as patterned in its distribution, and the patterned phenotype might be the result of the normal distribution of chalcone synthase production in the different cells of the flower. Under this theory, when the chalcone synthase gene is unmutated, transcription of it does not limit anthocyanin production and pigmentation is uniform. When chalcone synthase gene expression is reduced, pigment production becomes limited by the amount of chalcone synthase produced and pigment distribution reflects relative chalcone synthase gene expression across the flower.

This type of argument cannot explain all the patterned alleles at nivea. The existence of other types of patterned allele suggests that cis-acting sequences may also carry information concerning the pattern of cellular gene expression although the formation of new DNA sequences, rather than simple deletions, may be important in generating these alleles.

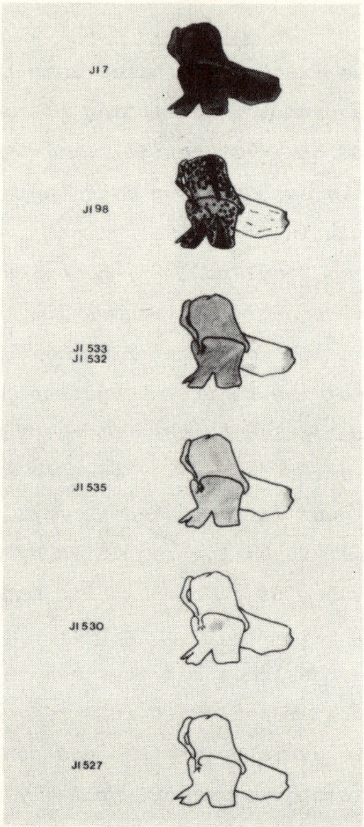

Figure 8. Diagrammatic representation of alleles of nivea (chalcone synthase) resulting from deletions induced by Tam3
JI 7 Full Red (Niv$^+$)
JI 98 nivea^{recurrens}
JI 533, 530. Stable alleles with a dark flush of pigment on the lobes and at the base of the tube. A small amoung of pigment may be formed in the upper parts of the tubes in lower temperatures
JI 535. A stable allele with a pale flush on the lobes and at the base of the tube
JI 530. A stable allele with a hint of pigment on the face of the flower and at the base of the corolla tube. No pigmentation is detected towards the outer edges of the lobes
JI 527. A white allele with no detectable pigment formed

A third stable allele derived by Tam3 excision from nivea shows a very pale flush of pigment again concentrated in the lobes and at the base of the tube (Figure 8 - Jl 535). At the molecular level this allele shows a 140 b.p. deletion. It is likely that this deletion includes both the TATA box and the putative CAAT box. A fourth mutation shows an even bigger deletion in the promoter of about 670 b.p. The phenotype of this mutant shows a

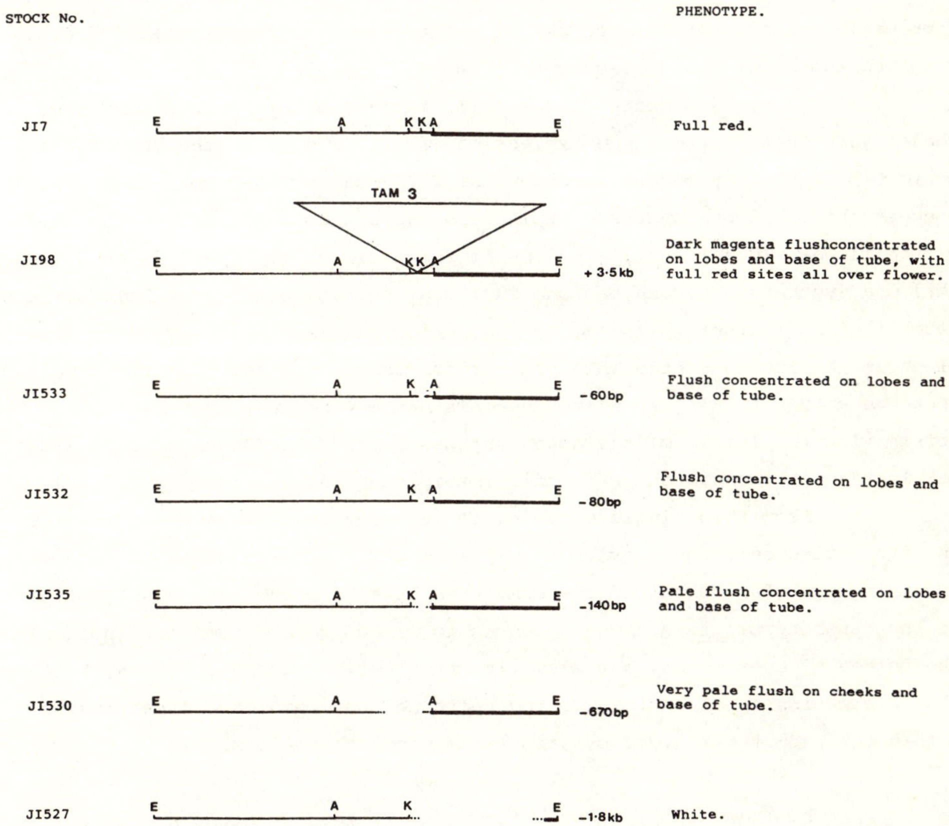

Figure 9. Maps of structural alterations involved in the formation of the nivea alleles listed in Figure 8. The approximate size of the alterations is given. Restriction sites are given; E = EcoRI, K = KpnI, A = AvaII. The bold line represents the transcribed region of the gene

trace of pigment just detectable on the face of the corolla lobes and at the base of the tube (Figure 8 - Jl 530). The promoter in this case must have lost all of the putative promoter signals including the TATA box, the CAAT box and part of the putative light enhancer. Interestingly, the pigment in this allele is strongest at the base of the corolla tube where pigment production may be somewhat light independent as normally the sepals shield the developing tube from incident light.

Finally, confirming that these mutations do not affect the chalcone synthase coding sequence but its regulation, we have derived four completely white mutations and all of these show loss of the start of transcription and other coding sequences (Figure 9 - Jl 527).

Thus, in summary, the allelic mutants of *nivea* (chalcone synthase) show that limited gene expression may occur even in the absence of conserved regulatory signals such as the TATA box and CAAT box. They also suggest that chalcone synthase expression may not be equal across the cells of the flower. They identify regions of the promoter important in enhancing chalcone synthase gene expression, including 60 b.p. around the TATA box and other regions further upstream. By sequencing alleles with very small deletions in the region of the TATA box, it has been suggested that the repeated sequence motif TACCAT just 5' to the TATA box may be very important in determining the levels of chalcone synthase expression (Bonas *et al.*, 1986). Our results are consistent with this interpretation.

From this single example, it is clear that transposon-directed *in vivo* mutagenesis can provide a very sensitive and artefact-free method of examining promoter function. These mutants have broadened our appreciation of the capacity of *cis*-acting sequences to direct gene expression, not only the degree of expression, but also its distribution within an organ. Part of the information involved in creating patterned gene expression may reside within the promoter regions of structural genes themselves.

b. The role of *trans*-acting factors in regulating anthocyanin biosynthesis

It is very hard to establish whether a particular mutant is a mutation of a regulatory gene or an interesting allele of a previously undiscovered structural gene. As we find out more about structural genes and their mutants such distinctions become more blurred.

An example that illustrates this well is the close phenotypic similarity between the *pallida varicolor* mutation (a *cis* mutation of a structural gene) and the trans-acting dominant Eluta mutation which has long been considered a regulatory mutant (Figure 10).

Table 4 lists a number of reported lesions that might affect regulatory genes of anthocyanin biosynthesis in Antirrhinum. This list is not definitive. Stubbe (1966) has catalogued a number of other mutants that affect flower colours. The mutations listed have been selected because they are clearly defined phenotypically and limited in their effects to alterations in anthocyanin production especially in flowers. The aim is to consider only mutants which affect anthocyanin biosynthesis directly.

Diluta and Vitrix. Amongst the regulatory genes listed are two which are reported to affect the intensity of flower pigmentation, Diluta and Vitrix. Baur (1910) described two intensities of anthocyanin pigmentation in commercial lines. The paler is generally referred to as magenta, and Baur called the darker Schwartzrot (black-red). A number of commercial varieties show similar dark red flowers (Figure 11). When pigment is extracted from the darkly pigmented commercial types and compared to anthocyanin levels from magenta lines, the intensely pigmented lines can contain between 3- and

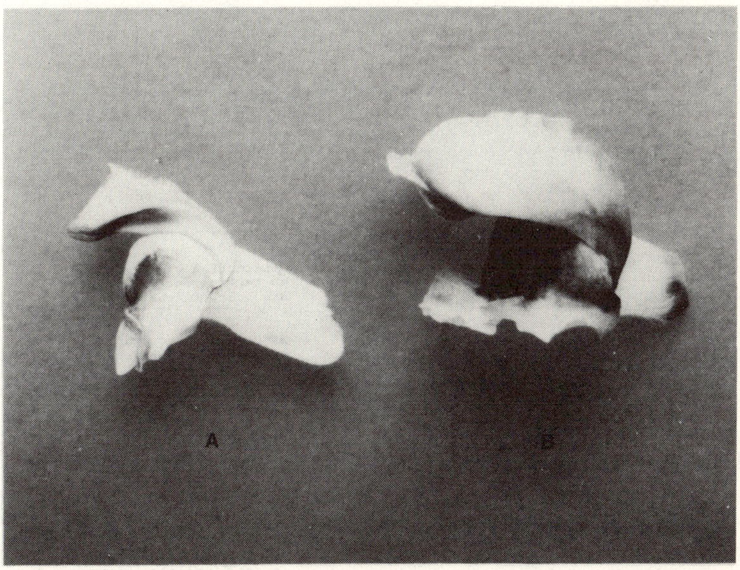

Figure 10. Comparison of phenotypes of A) a regulatory mutant, Eluta. This flower is also recessive for delila and so the pigmentation at the base of the tube is not evident. B) pallida varicolor, a change in the structural gene, pallida.

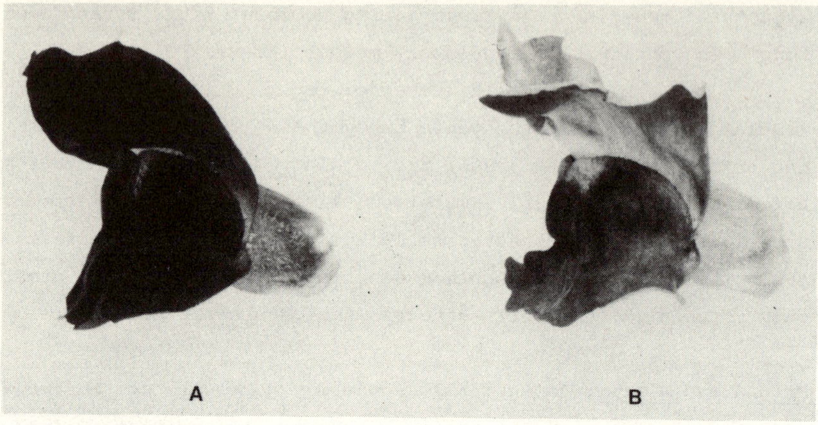

Figure 11. Flowers showing different anthocyanin intensities, which might be the result of the action of intensifier genes. A) A very dark commercial line. b) A normal Full-Red flower (JI 7).

TABLE 4

MUTATIONS IN ANTHOCYANIN BIOSYNTHESIS IN ANTIRRHINUM MAJUS.

REGULATORY MUTANTS.

MUTATION.		PHENOTYPE.
Eluta	semi-dominant	Restriction of pigment to central face of lobes and base of tubes.
delila	recessive	Block on anthocyanin synthesis in the tube.
Diluta	dominant	Intensified pigment.
Vitrix	dominant	Intensified pigment.
Venosa	dominant	Pigment limited to epidermis above veins in corolla.
rosea	recessive	Flower colour paler, and in ros^{dor} allele, restr- cted to outside of flower.
picturata	recessive	Restriction of pigment- ation in pattern similar to Eluta.

4-fold more anthocyanin in their flowers. This variation in pigment intensity is likely to have a genetic component.

Two genes have been described in *Antirrhinum* which affect pigment intensity, *Vitrix* (Doring, 1937) and *Diluta* (Baur, 1910; Sampson and Hunter, 1959). In the recessive form they both cause a reduction in intensity of flower pigmentation. In the absence of biochemical data it is difficult to know whether these genes are truly regulatory loci, or whether they are pale alleles of structural genes.

There are precedents for intensifying genes from other systems. In maize an intensifier, *In*, has been identified (Reddy and Peterson, 1978). In the dominant form *In* reduces anthocyanin production. Each copy of the dominant *In* allele relative to the recessive allele cuts pigmentation by about 30% although this effect seems to be restricted to pigmentation of the aleurone.

A similar intensifier gene, *In1*, has been described for *Petunia* (Gerats *et al.*, 1982). In the dominant form, *In1* causes a 50% reduction in the colour intensity of the flowers. When the production of anthocyanin is measured during bud development the production in *In/In* lines is similar to that in *in/in* lines until the buds are about 80% of their final length. Then the *In/In* lines make no more. The intensifier is thought to act at a late stage in the anthocyanin biosynthetic pathway affecting flavonoid conversion to anthocyanins (Gerats, 1985).

The intensifiers described in maize and *Petunia* are better candidates for regulatory genes than those reported for *Antirrhinum*. However, there does appear to be some evidence to suggest that this type of gene is active in *Antirrhinum*. It should be possible to define these genes more clearly biochemically and genetically in the near future.

Venosa. One attractive regulatory mutant is called *Venosa* (Baur, 1910; Stein, 1922)(Figure 12). The dominant allele, *Ve*, restricts anthocyanin biosynthesis to the epidermal cells above and below the vascular tissue of the corolla. This does not appear to be a mutation affecting tissue-specificity as the pigment still appears to be restricted to the epidermis (Martin, unpublished results). *Venosa* appears in many of the wild *Antirrhinum* species such as *Antirrhinum graniticum*, *A. majus* var. *majus*, var *latifolium* and var. *linkianum* (Stubbe, 1966). No physiology or biochemistry has yet been done to suggest how this gene might work. A similar gene, *An12*, has been reported in *Petunia* (Gerats, 1985).

Picturatum and Eluta. Two mutations restrict pigmentation to particular areas of the corolla. One, picturatum does so in the recessive homozygous form (Baur, 1910). The other, Eluta, is dominant (Figure 10). In both cases pigmentation is restricted to the face of the lower lobes, the tips of the upper lobes and the base of the corolla tube. Very little further information is available on picturatum. Eluta is a semi-dominant gene, more pigment being produced in the El/el heterozygote than in the El/El homozygote. The gene appears in some of the wild Antirrhinum species, including Antirrhinum majus varieties (Stubbe, 1966).

The degree of pigment reduction and restriction directed by this gene can be somewhat variable. Tetraploids have been used to test gene dosage effects (Seyffert, 1957). A single copy of Eluta in tetraploids reduces anthocyanin content to about 52% of the self-coloured flower. A second copy reduces it to 42%, a third to about 30% and when all four El alleles are present pigmentation is reduced to about 21%. The gene does not seem to be able to block anthocyanin production completely, and may work in a very complex manner.

Rosea. Another gene, rosea affects pigment intensity of the Antirrhinum flower (Baur, 1924). The recessive allele $rosea^{colorata}$, in the homozygous form, reduces pigmentation to a rose flush. The effects of the gene may not

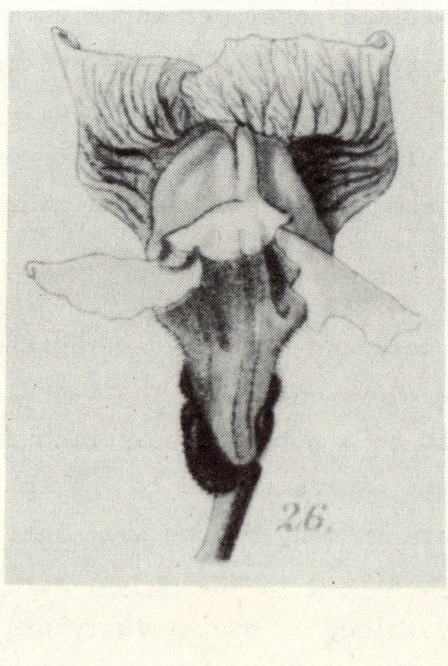

Figure 12. The Venosa mutation of Antirrhinum, from Baur (1910)

be simple however, as a second allele, rosea[dorsea], reduces pigmentation, but also restricts it, so that only the outer edges of the flower are coloured.

Delila. The best characterised regulatory mutant in Antirrhinum is delila (Vilmarin, 1896; de Vries, 1903; Wheldale, 1907; Stubbe, 1966). In the recessive homozygote, del/del, pigmentation of the corolla tube is completely blocked, although full-red pigmentation of the lobes can be achieved. In this mutant the sepals also show pigmentation at their tips, but a lack of anthyocyanin production lower down.

Delila blocks anthocyanin biosynthesis late in the pathway. Harrison and Stickland (1980) found that they could not overcome the del/del

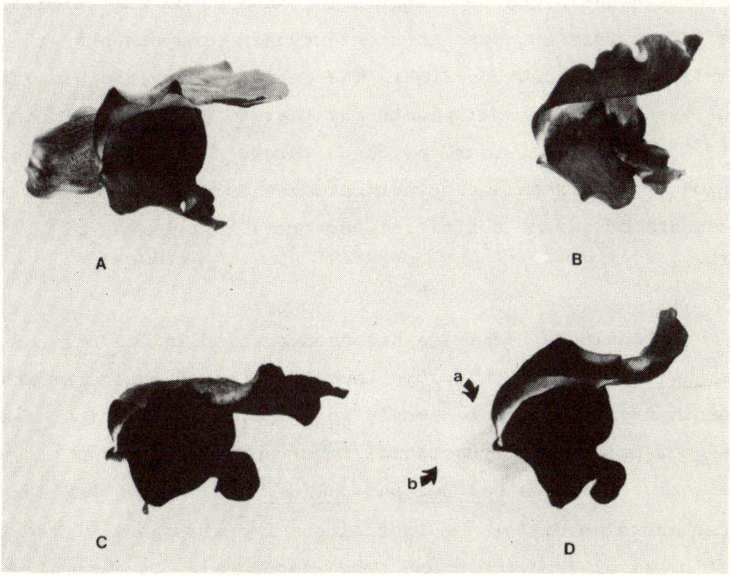

Figure 13. The delila mutation
A. Full Red ($\overline{Del^{+}}/\overline{Del^{+}}$) (JI 7)
B. Full Red del/del (JI 8) showing complete lack of pigmentation of tube
C. Commercial line with high anthocyanin intensity but the delila mutation blocking pigmentation in the tube
D. A slightly older flower from the same plant as C showing pigmentation of the tube (a) except where the calyx has shielded the tissue from the light (b)

block by feeding the flowers with known anthocyanin intermediates such as dihydroquercetin. The *delila* block can also be overcome by inoculating the tubes with *Botrytis cinerea* spores (Harrison and Hopwood, 1969). Harrison and Stickland (1980) partially purified a compound active in this process, that was heat labile and of low molecular weight, they suggested that it overcame the *delila* block in an indirect fashion. Breakdown of the *delila* block can also be seen as the flowers get older (Figure 13). Anthocyanin may then be observed, especially over the veins of the corolla tube. The breakdown of the *delila* block is more extreme in lines where pigment intensity is greater. Figure 13 shows a plant with a high anthocyanin content, that appears to carry *del*/*del*. In young flowers no pigment is observed in the corolla tube. Shortly after the full expansion, considerable pigmentation develops in the corolla tube. It is also very interesting that no anthocyanin biosynthesis can be observed in the area of the corolla tube shaded by the calyx. This suggests that anthocyanin biosynthesis at this stage may be completely dependent on light. The early anthocyanin biosynthesis at the base of the tube in normal plants may therefore be light independent.

Some preliminary evidence throws a little light onto the mode of action of *delila* on anthocyanin biosynthesis. This data is a mixture of measurements of enzyme activities and transcript levels of biosynthetic enzymes.

Chalcone isomerase has been assayed in *Del*/*Del* and *del*/*del* plants. It is possible that some isomerisation of chalconarginenin to naringenin may occur spontaneously in cells, so regulation of enzyme activity may not be of great functional importance. Comparison of chalcone isomerase activities from *Del*/*Del* and *del*/*del* plants with similar genetic backgrounds, shows no difference in chalcone isomerase activity extracted from whole flowers or from lobes and tubes separately (C. Martin and J. Bartlett, unpublished results). This confirms that *delila* is a late block that does not affect early enzymes such as chalcone isomerase.

The activity of UFGT in *Del*/*Del* and *del*/*del* plants has also been compared. There is a 90% reduction of UFGT activity in *del*/*del* corolla tubes compared to *Del*/*Del* tubes, although the lobes of *del*/*del* flowers show no loss of UFGT activity (Table 5). This is a specific effect of *delila* on this enzyme, not shown by lines homozygous for *niv*, *eos*, *inc* or *pal*. In addition two alleles of *delila* which show partial pigmentation of the tubes

TABLE 5

EFFECT OF Delila ON UFGT ACTIVITY.

GENOTYPE.		ACTIVITY nmol mg^{-1} min^{-1}.	% FULL RED LOBES	% FULL RED TUBES.
Niv^{+} Inc^{+} Eos^{+} Pal^{+} Del^{+}	LOBES	51.6	100	
	TUBES	21.3		100
Niv^{+} Inc^{+} Eos^{+} Pal^{+} del^{-}	LOBES	32.7	63	
	TUBES	2.4		12
Niv^{+} Inc^{+} eos^{-} Pal^{+} del^{-}	LOBES	49.4	96	
	TUBES	2.5		12
Niv^{+} Inc^{+} Eos^{+} pal^{-} del^{-}	LOBES	48.1	93	
	TUBES	2.6		12

TABLE 6

BREAKDOWN OF Delila BLOCK AND EFFECT ON UFGT ACTIVITY.

GENOTYPE	ACTIVITY IN TUBES nmol mg^{-1} min^{-1}	% FULL RED
Del^{+}	21.3	100
del^{-}	2.4	12
del^{a}- partial breakdown of del block (STOCK JI22)	4.4	21
del^{b}- massive breakdown of del block (STOCK JI 23)	5.1	24

show increased UFGT activity in the corolla tubes (Table 6).

The possibility that delila is a lesion in the structural gene for UFGT has also been tested. Comparison of the thermal inactivation characteristics at 50°C for UFGT from lobes of Del/Del and del/del flowers shows that both extracts have exactly the same half-life (T. Gerats and C. Martin, unpublished results). This result suggests that delila is a regulatory mutant that reduces UFGT activity in the corolla tube, probably by blocking its biosynthesis.

The levels of nivea (chalcone synthase) and pallida transcripts in whole flowers from Del/Del and del/del plants have also been measured (Table 7). There is no difference in the amounts of nivea (chalcone synthase) transcript in the two flower types suggesting that delila does not affect nivea gene expression. There is about 20% less pallida transcript of flowers of del/del plants compared to Del/Del plants. Although the amounts of each transcript in the tube have not yet been measured, a reduction of about 25% would be predicted if transcription was blocked in the corolla tubes, as the tubes contain about 25% of the pigment in the whole flower. These very preliminary data therefore suggest that pallida is blocked by delila as well as UFGT. This conclusion is currently being tested more stringently.

Thus, delila appears to block anthocyanin biosynthesis in the corolla tube by repressing activity of the later enzymes in the pathway, probably by inhibiting their gene expression. This regulation of the later steps of the biosynthetic pathway could be a general feature of genes controlling flower pigmentation. In Petunia four genes affect the activity of UFGT and methyltransferase activity, An1, An2, An6 and An9. These are thought to be regulatory mutants (Gerats et al., 1983). However, in maize, regulatory genes such as R,B,Pl and Vp appear to affect both early steps

EFFECT OF Delila ON THE EXPRESSION OF NIVEA AND PALLIDA GENES.

GENOTYPE	relative amount of transcript extracted per ug total RNA. NIVEA	PALLIDA
Niv^+ Pal^+ Del^+	1.00	1.00
Niv^+ Pal^+ del^-	1.07	0.83

such as chalcone synthase and late steps such as UFGT (Dooner and Nelson, 1977, 1979; Dooner, 1983). So there may be different regulation systems operative either within the plant or between species. The unravelling of these complex control systems is an exciting challenge for the future.

DISCUSSION

In conclusion, the regulation of anthocyanin biosynthesis is complex and multifold. However, due to the existence of a large number of mutants some ideas about the mechanisms of regulation are beginning to be formulated. These ideas not only reinforce our preconceptions, but also expand our concepts about the potential and nature of regulatory systems.

ACKNOWLEDGEMENTS

We wish to thank Mrs. M.Limberg for typing this manuscript and Mr. P.Scott and Mr. A.Davies for providing the figures. We are also very grateful to Dr. B.J. Harrison for providing references and Frau Martina Lehnert for German-to-English translations.

REFERENCES

Baur, E. (1910) Veretbungs- und Bastardierungsversuche mit Antirrhinum. Zschr. indukt. Abstamm. Vererb. lehre, 3, 34-98.

Baur, E. (1924) Untersuchungen über das wesen, die Entstehung und die Vererbung von Rossen unterschieden bei Antirrhinum majus. Bibliotecheca Genetica, 4.

Bonas, U., Sommer, H. & Saedler, H. (1986) (manuscript in preparation).

Carpenter, R., Martin, C. & Coen, E.S. (1986) (manuscript in preparation).

Coe, E.H. (1955) Anthocyanin synthesis in maize, the interaction of A_2 and Pr in leucoanthocyanin accumulation. Genetics, 40, 568.

Coe, E.H. (1957) Anthocyanin synthesis in maize. A gene sequence construction. Am. Nat. 91, 381-385.

Coe, E.H. (1985) Phenotypes in corn: control of pathways by alleles, time and place. In Plant Genetics. ed. Freeling, M. pp.509-521. New York: Alan R. Liss Inc.

Coen, E.S., Carpenter, R. & Martin, C. (1986) Transposable elements generate novel spacial patterns of gene expression in Antirrhinum majus. (in press).

Dellaporta, S.L., Wessler, S., Chen, J., Kermicle, J. & Greenblatt, I. (1985) Molecular cloning using transposon tagging and structural characterisation of the R locus of maize. Abstracts 1st Int. Cong. Plant Mol. Biol. Savannah, Georgia p.48.

Dooner, H.K. (1980) Regulation of the enzyme UFGT by the controlling element Ds in bz-m4, an unstable mutant in maize. Cold Spring Harbor Symp. Quant. Biol., 45, 457-462.

Dooner, H.K. (1983) Co-ordinate genetic regulation of flavonoid biosynthetic enzymes in maize. Mol. Gen. Genet. 189, 136-141.

Dooner, H.K. & Nelson, O.E. (1977) Genetic control of UDPglucose-flavonol 3-O-glucosyltransferase in the endosperm of maize. Biochem. Genetics, 15, 501-519.

Dooner, H.K. & Nelson, O.E. (1979) Hetrogeneous flavonoid glucosyltransferases in purple derivatives from a controlling element-suppressed bronze mutant in maize. Proc. Natl. Acad. Sci. USA, 76, 2369-2371.

Döring, H. (1937) Uber den Einfluss der Ernährung auf die mutationshäufigkeit bei Antirrhinum majus. Ber dtsch. Bot. Ges. 55, 167-182.

Emerson, R.A. (1921) The genetic relations of plant colors in maize. Cornell University Agricultural Experiment Station, Memoir, 39, 3-156.

Fincham, J.R.S. & Harrison, B.J. (1967) Instability at the Pal locus in Antirrhinum majus. Heredity, 22, 211-224.

Forkmann, G. & Dangelmayr, B. (1980) Genetic control of chalcone isomerase activity in flowers of Dianthus caryophyllus. Biochem. Genet., 18, 519-527.

Forkmann, G. & Stotz, G. (1981) Genetic control of flavanone 3-hydroxylase and flavonoid 3'-hydroxylase activity in Antirrhinum majus (Snapdragon). Z. Naturforsch., 36C, 411-416.

Gerats, A.G.M. (1985) Mutable systems; their influence on flavonoid synthesis in Petunia hybrida. Ph.D. Thesis, University of Amsterdam.

Gerats, A.G.M., Bassard, J., Coe, E.H. & Larson, R. (1984) Influence of B and Pl on UDPG:Flavonoid-3-O-Glucosytransferase in Zea mays L. Biochem. Genet., 22, 1161-1169.

Gerats, A.G.M., Cornelissen, R.T.J., Groot, S., Hogervarst, J.M.V., Schram, A.W. & Bianchi, F. (1982) A gene controlling rate of anthocyanin synthesis and mutation frequency of the gene An1 in Petunia hybrida. Theor. Appl. Genet., 62, 199-203.

Gerats, A.G.M., Wallroth, M., Donker-Koopman, W., Groot, S.P.C. & Schram, A.W. (1983) The genetic control of the enzyme UDP-G Glucose:3-O-flavonoid-glucosyltransferase in flowers of Petunia hybrida. Theor. Appl. Genet., 65, 349-352.

Hahlbrock, K. & Grisebach, H. (1975) Biosynthesis of the flavonoids. In The Flavonoids. eds. Harborne, J.B., Mabry, T.J. & Mabry, H. pp. 866-915. London: Chapman and Hall.

Hahlbrock, K., Knobloch, K-H., Kreuzaler, F., Potts, J.R. & Wellman, E. (1976) Coordinated induction and subsequent activity changes of two groups of metabolically interrelated enzymes. Eur. J. Biochem., 61, 199-206.

Hahlbrock, K., Bollman, J., Hauffe, K.D., Jahnen, W., Knogge, W., Kombrink, E., Schmelzer, E., Sonssich, I. & Walter, M. (1985) Defence reactions of plants to fungal infections. Abstracts 1st Int. Congress Plant Mol. Biol. Savannah, Georgia p. 3.

Harborne, J.B. (1967) Comparative Biochemistry of the Flavonoids. Academic Press, New York.

Harrison, B.J. & Fincham, J.R.S. (1968) Instability at the Pal locus in Antirrhinum majus. 3. A gene controlling mutation frequency. Heredity, 23, 67-72.

Harrison, B.J. & Hopwood, D.A. (1969) The release of genetically blocked anthocyanin in Antirrhinum majus by Botrytis cinerea. Report of the John Innes Institute, p.31.

Harrison, B.J. & Stickland, R.G. (1974) Precursors and genetic control of pigmentation. 2. Genotype analysis of pigment controlling genes in acyanic phenotypes in Antirrhinum majus. Heredity, 33, 114-115.

Harrison, B.J. & Stickland, R.G. (1978) Precursors and genetic control of pigmentation. 4. Hydroxylation and methoxylation stages in antyocyanidin synthesis. Heredity, 40, 127-32.

Harrison, B.J. & Stickland, R.G. (1980) Precursors and the genetic control of pigmentation. 5. Initiation of anthocyanin synthesis in Antirrhinum majus by Botrytis cinerea. Heredity, 44, 103-109.

Jonssen, L.M.V., Aarsman, M.E.G., Schram, A.W. & Bennink, G.J.H. (1982) Methylation of anthocyanins by cell-free extracts of flower buds of Petunia hybrida. Phytochemistry, 21, 2457-2459.

Jonssen, L.M.V., de Vlaming, P., Wiering, H., Aarsman, M.E.G. & Schram, A.W. (1983) Genetic control of anthocyanin-O-methyltransferase activity in flowers of Petunia hybrida. Theor. Appl. Genet. 66, 349-355.

Jonssen, L.M.V., Donker-Koopman, W.E. & Schram, A.W. (1984) Turnover of anthocyanins and tissue compartmentation of anthocyanin biosynthesis in flowers of Petunia hybrida. J. Plant Physiol., 115, 29-37.

Kamsteeg, J., van Brederode, J. & van Nigtevecht, G. (1980a) Identification properties and genetic control of UDP-L-rhamnose: anthocyanidin 3-O-glucoside,6'-O-rhamnosyltransferase isolated from petals of red campian (Silene dioica). Z. Naturforsch., 35C, 249-

Kamsteeg, J., van Brederode, J., Hammels, C.H. & van Nigtevecht, G. (1980b) Identification properties and genetic control of hydroxycinnamoyl-coenzyme A-anthocyanidin 3-rhamnosyl (1→6) glucoside, 4"-hydroxycinnamoyl transferase isolated from petals of Silene dioica. Biochem. Physiol. Pflanzen, 175, 403-411.

Kaulen, H., Schell, J. & Kreuzaler, F. (1986) Light-induced expression of the chimeric chalcone synthase-NPTII gene in tobacco cells. EMBO J., 1-8.

Kirby, L.T. & Styles, E.D. (1970) Flavonoids associated with specific gene action in maize aleurone, and the role of light in substituting for the action of a gene. Can. J. Genet. Cytol., 12, 934-940.

Kristiansen, K.N. (1984) Biosynthesis of proanthocyanidins in barley: genetic control of the conversion of dihydroquercetin to catechin and procyanidins. Carlsberg Res. Commun., 49, 503-524.

Larson, R.L. & Coe, E.H. (1968) Enzymatic action of the B_z anthocyanin factor in maize. Proc. XII Int. Cong. Genet., 1, 131.

Larson, R.L. & Coe, E.H. (1977) Gene-dependent flavonoid glucosyltransferase in maize. Biochem. Genet. 15, 153-156.

de Luca, V. & Ibrahim, R.K. (1982) Characterisation of three distinct flavonol-O-methyl transferases from Chrysosplenium americanum. Phytochemistry, 21, 1537-1540.

McClintock, B. (1948) Mutable loci in maize. Carn. Inst. Wash. Year Book, 47, 155-169.

McCormick, S. (1978) Pigment synthesis in maize aleurone from precursors fed to anthocyanin mutants. Biochem. Genet. 16, 777-785.

Martin, C., Carpenter, R., Sommer, H., Saedler, H. & Coen, E.S. (1985) Molecular analysis of instability in flower pigmentation of Antirrhinum majus, following isolation of the pallida locus by transposon tagging. EMBO J., 4, 1625-1630.

Moustafa, E. & Wong, E. (1967) Purification and properties of chalcone-flavanone isomerase from soya bean seed. Phytochemistry 6, 62-5 632.

Poulton, J.E. (1981) Transmethylation and demethylation reactions in the metabolism of secondary plant products. In The Biochemistry of Plants, 7. ed. Conn, E.E. pp. 667-723. London, New York: Academic Press.

Reddy, A.R. & Peterson, P.A. (1978) The action of the intensifier (in) gene in flavonoid production in aleurone tissue of maize. Can. J. Genet. Cytol., 20, 337-347.

Reddy, A.R. & Reddy, G.M. (1971) Chemicogenetic studies of leucoanthocyanidin in a maize mutant. Curr. Sci., 40, 335-337.

Sampson, D.R. & Hunter, A.W.S. (1959) Inheritance of shades of crimson and magenta flowers of Antirrhinum majus. Can. J. Genet. Cytol. 1, 173-181.

Schröder, J., Kreuzaler, F., Schäfer, E. & Hahlbrock, K. (1979) Concomitant induction of phenylalanine amonia-lyase and flavanone synthase mRNAs in irradiated plant cells. J. Biol. Chem., 254, 57-65.

Seyffert, W. (1957) Untersuchungen über interallele Wechselwirkungen. 1. Die unvollständige Dommanz des El-Faktors von A.majus. Zschr. indukt. Abstamm. Vererb. lehre, 88, 56-77.

Sommer, H., Carpenter, R., Harrison, B.J. & Saedler, H. (1985) The transposable element Tam3 of Antirrhinum majus generates a novel type of sequence alteration upon excision. Mol. Gen. Genet. 199, 225-231.

Sommer, H. & Saedler, H. (1986) Structure of the chalcone synthase gene of Antirrhinum majus. Mol. Gen. Genet. 202, 429-439.

Spribille, R. & Forkmann, G. (1982) Genetic control of clacone synthase activity in flowers of Antirrhinum majus. Phytochemistry, 21, 2231-2234.

Stein, E. (1922) Über den Einfluß von radiumbestraklung auf Antirrhinum. Zschr. indukt. Abstamm. Vererb. lehre, 29, 1-15.

Stickland, R.G. & Harrison, B.J. (1974) Precursors and genetic control of pigmentation. 1. Induced biosynthesis of perlargonidin, cyanidin and delphinidin in Antirrhinum majus. Heredity, 33, 108-112.

Stotz, G. & Forkmann, G. (1981) Oxidation of flavanones to flavones with flower extracts of Antirrhinum majus (Snapdragon). Z. Naturforsch., 36C, 737-741.

Stubbe, H. (1966) Genetik und Zytologie von Antirrhinum L. sec. Antirrhinum. Veb. Guystav Fischer Verlag, Jena.

Styles, E.D. & Ceska, O. (1977) The genetic control of flavonoid synthesis in maize. Can. J. Genet. Cytol., 19, 289-302.

Styles, E.D., Ceska, O. & Seah, K-T. (1973) Developmental differences in action of R and B alleles in maize. Can. J. Genet. Cytol., 15, 59-72.

Vilmarin, C. (1896) Blumengartenerei Bd. 1. Antirrhinum majus, S 756. Berlin: Paul Parey.

de Vries, H. (1903) Die Mutationstheorie, 2, Verlag von Veit Co. Leipzig.

van Weely, S., Bleumer, A., Spruyt, R. & Schram, A.W. (1983) Chalcone isomerise in flowers of mutants of Petunia hybrida. Planta, 159, 226-230.

Weinand, U., Sommer, H., Schzarz, Zs., Shepherd, N., Saedler, H., Kreuzaler, F., Ragg, H., Fautz, E., Halhbrock, K., Harrison, B.J. & Peterson, P.A. (1982) A general method to identify plant structural genes among genomic DNA clones using transposable element induced mutations. Mol. Gen. Genet., 187, 195-201.

Wheldale, M.W. (1907) The inheritance of flower colour in Antirrhinum majus. Proc. Roy. Soc. Ser. B, 79, 288-305.

Wheldale, M.W. (1916) The Anthocyanin Pigment of Plants. Cambridge: University Press.

Wiering, H. & de Vlaming, P. (1977) Glycosylation and methylation patterns of anthocyanins in Petunia hybrida. II. The genes Mf1 and Mf2. Z. Pflanzenzücht, 78, 113-123.

MITOCHONDRIAL GENOME REARRANGEMENTS ASSOCIATED WITH THE CYTOPLASMIC MALE STERILE PHENOTYPE IN MAIZE AND SORGHUM

Julia Bailey-Serres, Peter G. Isaac, Ian D. Small and Christopher J. Leaver
Department of Botany, University of Edinburgh, Edinburgh EH9 3JH

INTRODUCTION

Mitochondria and chloroplasts are the major sites of energy conversion in green plant cells and thus play an important role in determining the performance and productivity of crop plants. Both organelles possess genomes which contain only a portion of the genetic information necessary for organellar biogenesis and function. In mitochondria about 5-10% of the total organellar protein is encoded by mitochondrial DNA (mtDNA) and synthesized on mitochondrial ribosomes. The remaining organellar proteins are encoded by the nuclear genome, synthesized on cytosolic ribosomes and imported into the mitochondrion. Therefore, the biogenesis and function of mitochondria requires the coordinated and integrated expression of two separate genomes.

The major aims of our research are to obtain a detailed knowledge of the organization, information content and expression of the plant mitochondrial genome, together with an understanding of the interactions between, and coordinated expression of, nuclear and mitochondrial genomes during mitochondrial biogenesis. This information will in turn give a greater appreciation of the molecular basis and phenotypic expression of cytoplasmically inherited genetic variation which has its origins in mtDNA and may be of potential use to plant breeders. It will also allow us to investigate how genetic, environmental and physiological factors regulate mitochondrial biogenesis during the changing energy requirements of the plant cell.

To date progress has been impeded by the large and complex nature of the plant mitochondrial genome (200 Kb to 2400 Kb depending on the plant species - Bendich, 1985) and the lack of identifiable, non-lethal mutations which affect mitochondrial function. However one well characterised phenotype which is thought to be due to mutations in the mitochondrial genome is cytoplasmic male sterility.

THE CYTOPLASMIC MALE STERILE PHENOTYPE

The cytoplasmic male sterile (CMS) phenotype is recognized by the failure of the mature plant to produce or release functional pollen (Duvick, 1965; Edwardson, 1970). CMS plants occur in natural populations, and can also be generated through crosses in which a nucleus is backcrossed into a foreign cytoplasm (Figure 1), or by protoplast fusion (Hanson and Conde, 1985). CMS lines can be classified on the basis of nuclear genes which restore pollen fertility. In maize, for example, in addition to the normal (N) fertile cytoplasm, three CMS cytoplasms have been identified (S (USDA), T (Texas) and C (Charrua)) which are restored by different nuclear Rf alleles (Table 1) (for review, Laughnan and Gabay-Laughnan, 1983). Normal and CMS cytoplasms of sorghum have been similarly identified (Table 1).

The CMS phenotype is used extensively in the commercial production of F_1 hybrids of maize, sorghum, onions, sugarbeet and sunflower, thus avoiding the labour-intensive operation of hand emasculation of the female parent. Fertile F_1 hybrids are produced by crossing a male sterile female with a male parent containing nuclear genes which restore pollen fertility (Figure 1). Nuclear restorer allele(s) (termed _Rf_ alleles in maize and _Ms_ alleles in sorghum) are present in the natural nucleus of a given cytoplasm and sometimes in unrelated nuclear genotypes. Despite the obvious labour-saving advantages of using CMS lines to produce F_1 hybrids, breeding programmes which rely on a single CMS source result in a high degree of cytoplasmic genetic uniformity and as a consequence, increased genetic vulnerability. For example, the maternally inherited susceptibility of T cytoplasm maize to _Helminthosporium maydis_, Race T (_Cochliobolus heterotrophus_) (for review: Leaver and Gray, 1982) was responsible for the disastrous Southern Corn Leaf Blight which occurred in the U.S. in 1970. It is therefore important to maintain genetic diversity in the cytoplasms of crop plants by using diverse CMS lines in hybrid seed production. In addition the creation of new CMS sources, by performing specific crosses or through genetic engineering, would be aided by an understanding of the molecular and biochemical basis of the CMS phenotype.

A range of circumstantial evidence suggests that the mitochondrial genome, rather than the chloroplast genome, is the site of the mutation(s) which lead to CMS in a number of plant species (Pring and Lonsdale, 1985). The restriction endonuclease digestion patterns of mtDNA from fertile and CMS cytoplasms exhibit distinct and characteristic differences, whereas

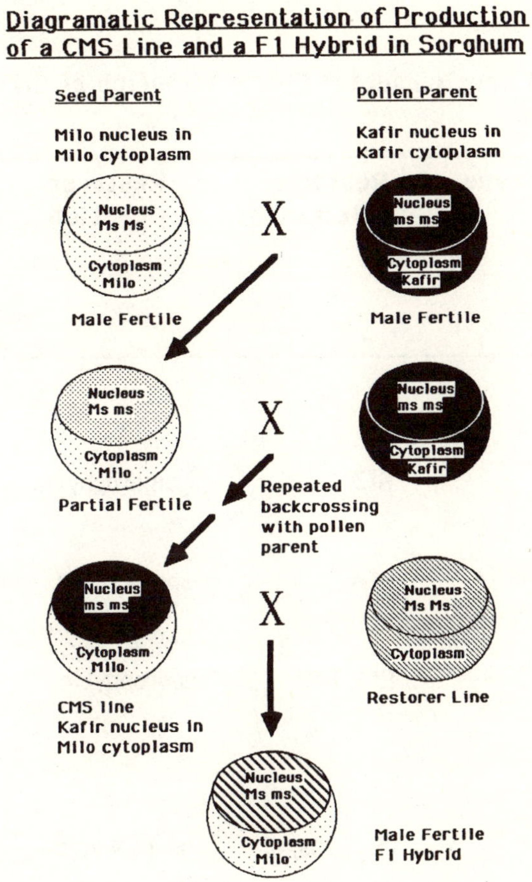

Figure 1. Production of a CMS Line and F_1 Hybrid from Two Fertile Lines of Sorghum.

The seed (female) parent (Milo nucleus in Milo cytoplasm) is hand emasculated and fertilized with pollen from the male parent (Kafir nucleus in Kafir cytoplasm) which lacks the dominant fertility allele(s) (Ms). The F_1 progeny have a hybrid nucleus (Kafir + Milo nucleus) and Milo cytoplasm, are partially or completely male fertile and therefore must be hand emasculated prior to further backcrossing with the pollen parent. Backcrossing is repeated until male sterile progeny (Kafir nucleus in Milo cytoplasm) are obtained, in practice normally about eight backcrosses are performed. The CMS line is maintained by crossing with the original pollen parent. Male fertile F_1 hybrid lines are generated by crossing the CMS line with a restorer line which contains the nuclear pollen fertility alleles.

Ms - dominant fertility allele. ms - recessive fertility allele

Table 1

Genetic and Developmental Characteristics of CMS Lines of Maize and Sorghum

Cytoplasm Type	Restorer Allele(s)	Restorer Reaction	Time of Pollen Abortion or Anther Abnormality
Maize			
S (USDA)	Rf3	gametophytic	binucleate microspore
C (Charrua)	Rf4	sporophytic	vacuolate microspore
T (Texas)	Rf1, Rf2	sporophytic	tetrad to binucleate microspore
Sorghum			
Milo	Ms_{c1},Ms_{c2} Ms_{c3}	sporophytic	vacuolate microspore
IS1112C M35-1	not known	not known	indehiscent anthers
9E	not known	not known	indehiscent anthers

Review references: Laughnan and Gabay-Laughnan, 1983; Schertz and Pring, 1980.

the corresponding chloroplast DNA patterns are virtually indistinguishable (Pring and Levings, 1978). In addition, we have observed that mitochondria isolated from CMS lines of maize and sorghum synthesize characteristic variant polypeptides not found in mitochondria from normal (male fertile) cytoplasms (Table 2) (Forde et al., 1978, 1980; Forde and Leaver, 1980; Dixon and Leaver, 1982). Based on these observations we can speculate that CMS originated as a mitochondrial mutation, perhaps as the result of changes in genome organization, with consequent effects on either replication, transcription and/or translation products. It remains to be established whether the observed variations in mitochondrial genome structure and expression in CMS lines are causally related to a failure in mitochondrial biogenesis and/or function in the anther tapetum or microspores during pollen formation.

Table 2

Variant Mitochondrial Translation Products in Maize and Sorghum

Cytoplasm Type	Phenotype MF = male fertile MS = male sterile	Molecular Weight of Variant Polypepetides (estimated from SDS-polyacrylamide gel electrophoresis)
Maize		
N	MF	-
T	MS	13,000
C	MS	17,500
S	MS	116,000
Sorghum		
Kafir	MF	-
Milo	MS	65,000
IS1112C	MS	12,000 + 82,000
M35-1	MS	82,000
9E	MS	42,000 (COI)

References: Forde et al., 1978; Forde et al, 1980; Dixon and Leaver, 1982.

PLANT MITOCHONDRIAL GENOMES: ORGANIZATION AND INFORMATION CONTENT

The mitochondrial genomes of maize, a monocot, and *Brassica*, a dicot, can be represented as a single, large 'master circle'. In both plants it has been predicted that intramolecular recombination across repeated mtDNA sequences gives rise to smaller sub-genomic circular molecules (Lonsdale *et al.*, 1984; Palmer and Shields, 1984). In N (male fertile) cytoplasm maize, six sets of repeated sequences have been mapped on the 570 Kb mitochondrial master circle (Figure 2). The observed variations in mtDNA restriction endonuclease digestion patterns between male fertile and CMS lines could result from differential replication or recombination between the sub-genomic circular molecules. These changes in mtDNA restriction endonuclease digestion patterns apparently occurred as spontaneous rare events (Laughnan *et al.*, 1981), and appear to be influenced by nuclear background (Borck and Walbot, 1982; Escote *et al.*, 1986). In addition, genome rearrangements occur following transfer to and regeneration from tissue culture, and after protoplast fusion (Belliard *et al.*, 1979; McNay, 1984; Boeshore *et al.*, 1983; Nagy *et al.*, 1981; Chetrit *et al.*, 1984; Kemble and Shepherd, 1984). In a number of instances such alterations in mitochondrial genome organization are associated with changes in mitochondrial genome expression and male fertility status (Kemble *et al.*, 1983; Dixon *et al.*, 1982; Boeshore *et al.*, 1985).

To investigate whether the observed changes in mitochondrial genome organization and expression involve specific genes, we have begun to characterize the structure and expression of selected genes in fertile and CMS lines of maize and sorghum.

Despite a large variation in size, the mitochondrial genomes of animals, fungi (yeast) and higher plants encode the same basic set of gene products which include rRNAs, tRNAs and components of the multi-subunit complexes of the inner mitochondrial membrane (Table 3). To date, the maize mitochondrial genes encoding cytochrome *c* oxidase subunits I and II (*COXI* - Isaac *et al.*, 1985b and *COXII* - Fox and Leaver, 1982), apocytochrome *b* (*COB* - Dawson *et al.*, 1984) and subunits , 6 and 9 of ATP synthase (*ATPA*- Isaac *et al.*, 1985a; Braun and Levings, 1985; *ATP6* - Dewey *et al.*, 1985a; *ATP9* - Dewey *et al.*, 1985b) have been isolated and sequenced. In addition some of these genes have been mapped on the N (normal) fertile maize mitochondrial genome (Dawson *et al.*, 1986) (Figure 2).

We are currently investigating mtDNA rearrangements within and around genes in fertile and male sterile plants to determine if variation in

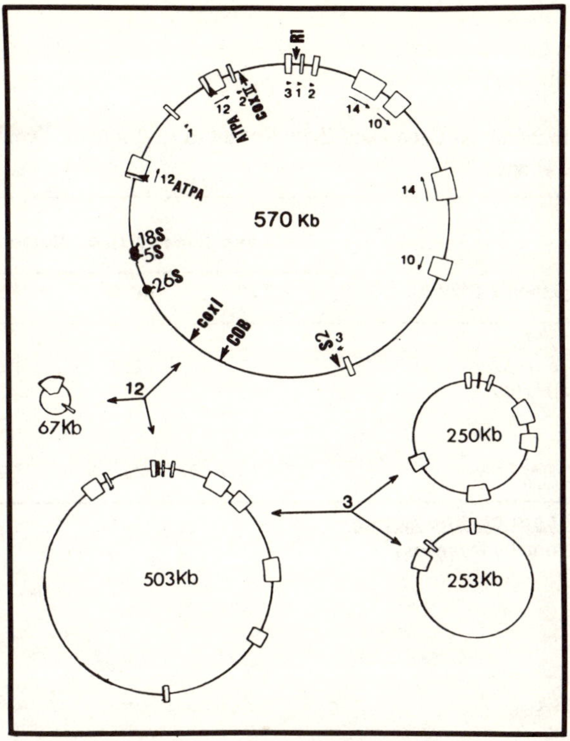

Figure 2. Map of the Fertile N Maize Mitochondrial Genome

The physical map of the normal (WF9 N) fertile maize mitochondrial genome was determined by cross hybridization of cosmid clones containing mtDNA restriction fragments. Linkage groups of cosmid clones were established and eventually a circular mitochondrial genome map of 570 Kb was completed (Lonsdale et al., 1984).

Six repeated mtDNA sequences (the '1,2,3,10,12 and 14 Kb repeats') were identified and mapped. In addition, genomic locations of mitochondrial genes (26S, 18S and 5S rRNAs, COXI, COXII, COB and ATPA) and sequences homologous to the linear mtDNAs S2 and R1 (S1) were mapped (Dawson et al., 1986).

In addition to the 570 Kb 'master circle' the existence of a number of sub-genomic circular molecules was proposed by Lonsdale et al. (1984). Recombination across the '12 kB' repeats in the 570 Kb master circle is thought to generate two sub-genomic circles of 67 and 503 Kb. Subsequent recombination across the '3 Kb' repeats in the 503 Kb circle would generate equimolar amounts of the 253 and 250 Kb sub-genomic circular molecules. The identification of sub-genomic circular molecules has led to the prediction that the maize mitochondrial genome has a complex multipartite structure which arises from homologous recombination across repeated DNA sequences.

Table 3

Mitochondrial Genes and their Products in Animals, Yeast and Plants

	Zea mays	**S.cerevisiae**	**Homo sapiens**
Mt Genome Size Kb	**570**	**80**	**16.6**
Large- rRNA	26S	21S	16S
Small- rRNA	18S	15S	12S
5S - rRNA	+	–	–
Transfer RNAs	? (ca.30)	25	22
<u>Protein coding genes</u>			
NADH dehydrogenase			
ND1	+	–	+
ND2	?	–	+
ND3	?	–	+
ND4	?	–	+
ND4L	?	–	+
ND5	?	–	+
ND6	?	–	+
Ubiquinol cytochrome c reductase			
Apocytochrome b	+	+	+
Cytochrome c oxidase			
Subunit I	+	+	+
Subunit II	+	+	+
Subunit III	+	+	+
F_1 - F_0 ATPase			
Subunit α	+	–	–
Subunit 6	+	+	+
Subunit 8	?	+	+
Subunit 9	+	+	–
Small ribosomal subunit protein	+	+	–
RNA processing enzymes			
bI2 mRNA maturases in Cob	–	+	–
bI4 introns.	–	+	–
aI1 Cox I intron.	–	+	–

mtDNA organization can be correlated with the synthesis of variant mitochondrial translation products in CMS lines. We have observed differences in the restriction fragment locations of the COXI and ATPA genes in both maize and sorghum. The most striking rearrangements are those associated with COXI which we describe in this review. In CMS-S-type cytoplasms of maize, multiple rearrangements occur 5' to the COXI gene. In the 9E-type cytoplasm of sorghum, rearrangements 5' to and within the COXI open reading frame result in alteration in gene transcription and the synthesis of a larger subunit I polypeptide.

MITOCHONDRIAL GENOME ORGANIZATION AND EXPRESSION IN S CYTOPLASM MAIZE

Mitochondria from CMS-S-type cytoplasms of maize contain two double-stranded linear mtDNA molecules in addition to the main mitochondrial genome (Pring *et al.*, 1977). The presence of these molecules, termed S1 (6.4 Kb) and S2 (5.4 Kb), can be correlated with the synthesis of several variant mitochondrial translation products (Table 2) (Forde and Leaver, 1980; Liddell and Leaver, in preparation). Sequence analysis of S1 and S2 has revealed that they have identical 208 bp terminal inverted repeats (S-TIR sequences) (Levings and Sederoff, 1983; Paillard *et al.*, 1985). Both S1 and S2 contain transcribed open reading frames, one of which may encode the protein bound to the 5' termini and may be involved in autonomous replication of these DNAs (Levings and Sederoff, 1983; Palliard *et al.*, 1985; Kemble and Thompson, 1982). Sequences homologous to the S-TIRs of S1 and S2 are found at a number of positions within the mtDNA of N and S cytoplasms (Thompson *et al.*, 1980; Lonsdale *et al.*, 1981). Schardl *et al.* (1984) have shown that the presence of these linear DNAs affects the organization of CMS-S mtDNA. Recombination between S1 and S2 and the homologous integrated S-TIR sequences of the circular mtDNA results in the formation of linear chromosomes, each bearing a S-TIR sequence at its terminus. The recombination system involved may be the same as that which catalyzes recombination across the '3 Kb' repeats which also contain the S-TIR sequence (Figure 3A).

We have recently sequenced the gene encoding subunit I of cytochrome *c* oxidase (COXI) and examined a genome rearrangement found near it in mtDNA from the S cytoplasm (Isaac *et al.*, 1985b; Leaver *et al.*, 1985). In CMS lines with S cytoplasm, a 186 bp sequence, identical to part of the S-TIR sequence is located 5' to the coding sequence of COXI (Figure 3B) (Isaac *et al.*, 1985b). Recombination between S1 and S2 and this homologous

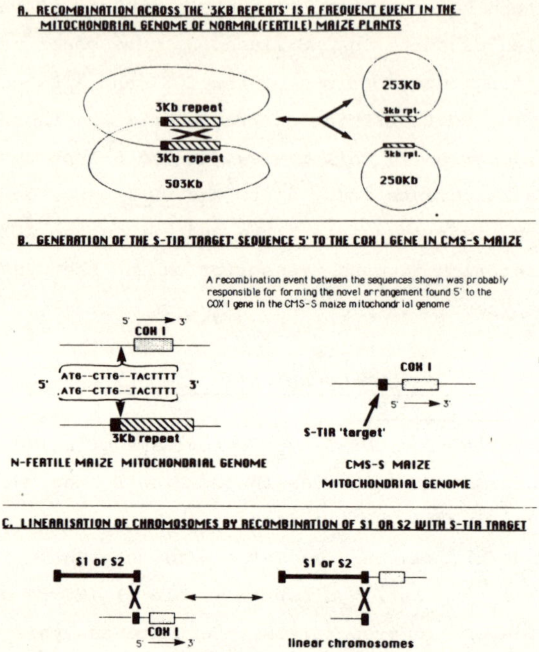

Figure 3. Recombination Associated with the COXI Gene in CMS-S Maize mtDNA.

(A) Lonsdale et al. (1984) observed that the "3 Kb" repeat can be found in four different genomic environments or linkage groups which are present in equal stoichiometries. Two copies of the "3 Kb" repeat were mapped on the 503 Kb circle and a single copy on the 253 and 250 Kb circles. This led Lonsdale et al. (1984) to predict that recombination across the '3 Kb' repeats is a frequent and reversable event. The '3 Kb' repeat (hatched region) contains a nearly complete (186 bp) copy of the terminal inverted repeat of the S and R 'plasmid-like' DNAs found in CMS-S and RU cytoplasms (S-TIR) at one end (black region).

(B) DNA sequence analysis has shown that in S cytoplasms a sequence identical to a portion of the S-TIR is found 5' to the COXI gene. An aberrant recombination event between one of the '3 Kb' repeat S-TIR sequences and a sequence 174 bp 5' to the COXI gene (23 bp 5' to the start of transcription) may have been responsible for this genome arrangement (Isaac et al., 1985b).

(C) The S-TIR sequence 5' to the COXI gene appears to act as a target for recombination with the free S1 or S2 DNAs. Such genomic rearrangements can result in the formation of linear mitochondrial chromosomes, each containing a copy of S1 or S2 at one end.

sequence (S-TIR 'target') results in the production of linear mitochondrial chromosomes with S1 or S2 attached to one end (Figure 3C).

Thus, mitochondrial genome recombination can result in the COXI gene being located in multiple genomic environments in S cytoplasm maize. Since the COXI gene is transcribed in an identical manner in mitochondria from S and N cytoplasms, these reorganization events appear to have no effect on the expression of the COXI gene (Isaac et al., 1985b). The proportion of circular mtDNAs which are linearized by such rearrangements appears to be influenced by nuclear factors (Leaver et al., 1985; Isaac et al., 1985b). For example, the COXI gene is found predominantly on linearized mtDNA molecules in S cytoplasm in the WF9 nuclear background, but as circular molecules in the B37 nuclear background (Leaver et al., 1985). Nuclear factors also influence the relative abundance of S1 and S2 and the frequency of spontaneous reversion to fertility by S cytoplasm maize (Laughnan et al., 1981; Laughnan and Gabay-Laughnan, 1983; Escote et al., 1986).

REARRANGEMENT WITHIN THE CYTOCHROME c OXIDASE SUBUNIT I GENE OF SORGHUM LEADS TO SYNTHESIS OF A LARGER POLYPEPTIDE

In sorghum, mitochondria from male sterile and male fertile lines containing 9E cytoplasm synthesize a variant form of cytochrome c oxidase subunit I (COI) with an apparent molecular weight of 42,000 (42K). This variant COI appears to replace the normal 38,000 molecular weight (38K) COI synthesized by Milo and other sorghum cytoplasms (Figure 4) (Dixon and Leaver, 1982; Bailey-Serres et al., (in press).

To establish whether synthesis of this variant COI is due to a mutation in the coding region of the COXI gene or an alteration in gene transcription, the genes encoding COI from Milo and 9E cytoplasms were compared. Analysis of the EcoRI restriction endonuclease digestion patterns of mtDNA from Milo and 9E cytoplasms revealed specific differences (Figure 5, A, B) (Bailey-Serres et al., in press). Hybridization of the EcoRI digested mtDNA with a ^{32}P-labelled portion of the maize COXI gene revealed that in cytoplasms which synthesize the normal 38K COI (e.g. Milo) the gene is located on a 4.3 Kb EcoRI fragment, whereas in 9E cytoplasm, which synthesizes the larger form of COI, the gene is located on a 10.4 Kb EcoRI fragment (Figure 5C, D).

The EcoRI fragments containing the COXI gene from Milo and 9E cytoplasm were cloned and further characterized. DNA sequence analysis of the 4.3 Kb EcoRI fragment from Milo cytoplasm and the 10.4 Kb EcoRI fragment from 9E cytoplasm revealed that both contain an open reading frame (ORF) encoding COI. The two COXI gene sequences are identical from a position -100 bp 5' to the presumed initiator codon, to position 1579 bp within the ORF. In the 4.3 Kb EcoRI fragment from Milo cytoplasm the ORF terminates 11 bp beyond this divergence point, so the normal COI polypeptide is 530 amino acids in length (58,484 mol. wt.) and is similar to COI of maize (528 amino acids) (Isaac et al., 1985b). In contrast, in the 10.4 Kb EcoRI

Figure 4. Identification of a Variant Form of Cytochrome c Oxidase Subunit I Polypeptide in Sorghum.

Mitochondrial translation products were labelled by incubating mitochondria isolated from etiolated shoots in a medium containing ^{35}S-methionine and an energy generating system. Mitochondrial polypeptides were immunoprecipitated with an antiserum raised against yeast cytochrome c oxidase subunit I protein. The immunoprecipitated protein was fractionated by electrophoresis in a 16% (w/v) SDS-polyacrylamide gel, and detected by autoradiography of the dried gel (Dixon and Leaver, 1982). In mitochondria from Milo cytoplasm a polypeptide with an apparent molecular weight of 38,000 (38K) was identified, whereas in 9E cytoplasm a 42,000 molecular weight (42K) polypeptide was identified.

fragment from 9E cytoplasm the COXI ORF continues for an additional 314 bp. It appears that a mtDNA rearrangement within the 9E COXI gene resulted in this 3' continuation of the COXI ORF, and leads to the synthesis of a COI polypeptide which has a novel 101 amino acid extension at its carboxy terminus (Figure 5). Thus, the putative COI polypeptide in 9E cytoplasm has a molecular weight of 70,358; 11,884 mol. wt. larger than the normal COI from Milo cytoplasm (Bailey-Serres et al., in press). A computer search of

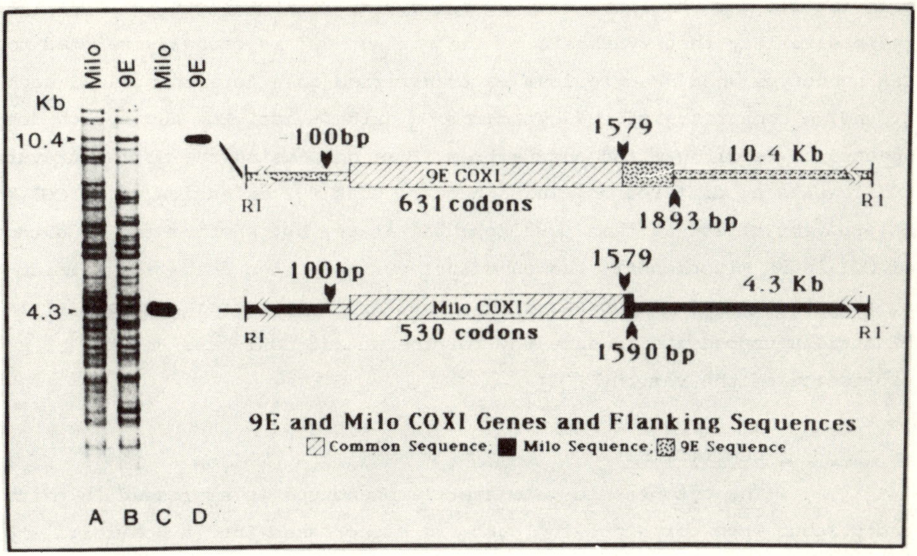

Figure 5. Location and Structure of the COXI Gene in the Mitochondrial Genome of Milo and 9E Cytoplasm Sorghum.

MtDNA from Milo (A) and 9E (B) cytoplasm was digested with EcoRI, fractionated by electrophoresis in a 0.8% (w/v) agarose gel, stained with EtBr and photographed. The mtDNA was transferred to nitrocellulose and hybridized with a ^{32}P-labelled clone containing an internal portion of the maize COXI gene (Isaac et al., 1985b). The autoradiogram of the Southern blot shows that the COXI gene is located on a 4.3 Kb EcoRI fragment in Milo (C) and a 10.4 Kb fragment in 9E (D).

DNA sequence analysis of the COXI gene of Milo and 9E cytoplasm has revealed that the Milo gene encodes a 530 amino acid (a.a) polypeptide whereas the 9E gene encodes a 631 a.a. polypeptide. The diagram shows the location of the homologous COXI DNA sequences in the 4.3 and 10.4 Kb EcoRI (RI) fragments. The points of divergence between the two genes are indicated by arrows and non-homologous sequences are shaded.

the EMBL and GenBank data bases revealed that the novel 3' coding sequence of 9E COXI is not homologous to any known DNA sequence. However, preliminary hybridization studies indicate that sequence homologous to the 3' end of 9E COXI is present at an additional genomic location in both 9E and Milo mtDNA. At present we are using an oligonucleotide probe to identify this sequence and hope to determine if the continuation of the 9E COXI gene resulted from recombination between two genes, and to elucidate the molecular events involved in this mtDNA rearrangement.

The variant form of COI is synthesized by mitochondria from both male fertile and CMS lines with 9E cytoplasm. We have begun to investigate the possibility that synthesis of the variant COI is causally related to the CMS phenotype when 9E cytoplasm is transferred to a foreign nuclear background. A comparison of the cytochrome c oxidase activity and cytochrome spectra of seedling mitochondria from lines possessing the normal or variant COI reveals no differences. This suggests that the extension of 9E COI has no apparent effect at this developmental stage, but whether the 3' extension of COI in 9E mitochondria has an effect during pollen formation remains to be investigated. It is possible that the fertility restorer allele(s) in the 9E nucleus compensate in some way for the mutant COXI gene and the expression of the variant COI.

CONCLUSIONS

The cytoplasmic male sterile phenotype is expressed in maize and sorghum when certain cytoplasms are backcrossed into a non-fertility restoring nuclear background and therefore appears to result from incompatibility between a foreign nucleus and cytoplasm. Considerable evidence now supports the view that the CMS phenotype originated as mutations in the mtDNA of male fertile plants as a result of aberrant intra- and/or intermolecular recombination events, leading to the generation of chimaeric mtDNA sequences. This would explain the considerable heterogeneity among the restriction endonuclease digestion patterns of mtDNA from normal (fertile) and mutant, male sterile cytoplasms.

Our observations described above give details of rearrangements either within or adjacent to the cytochrome c oxidase subunit I gene (COXI) in maize and sorghum mitochondrial genomes.

In the case of the COXI gene in CMS-S maize lines the recombination appears to have initially involved intermolecular recombination between the DNA sequence 5' to the COXI gene and a "3 Kb" repeat sequence, and secondly between this novel sequence arrangement and the linear S1 or S2 mtDNAs. In the 9E sorghum cytoplasm, intramolecular recombination within the mitochondrial genome appears to have generated a continuation of the COXI ORF leading to the synthesis of a larger polypeptide. These specific rearrangements do not apparently have any effect on cytochrome oxidase activity or plant growth in young seedlings. It is conceivable however that these types of genomic rearrangements may be an example of the underlying molecular events which are causally related to the CMS phenotype.

Dewey and co-workers have recently described the isolation and characterisation of a unique fragment of mtDNA from CMS-T maize which presumably arose by a series of novel recombination events between the mitochondrial and chloroplast genomes (Dewey et al., 1986). This unique fragment contains portions of the flanking and/or coding regions of the maize mitochondrial 26S ribosomal RNA gene, the ATPase subunit 6 gene, and the chloroplast tRNA-Arg gene. It also contains two open reading frames (ORFs) which could encode proteins of 12,961 M_r and 24,675 M_r, the former ORF is only transcribed in the T cytoplasm, while the latter is expressed in normal and all three CMS lines. It is interesting to note that our earlier studies showed that isolated mitochondria from T cytoplasms synthesise a unique 13,000 M_r polypeptide (Forde and Leaver, 1980).

Based on the results reviewed above we can speculate that novel mitochondrial genome rearrangements result in the formation of chimaeric DNA sequences. Occasionally such rearrangements may involve blocks of sequence from normal genes and/or non-coding sequences which could generate a unique 'mosaic gene'. Providing this 'gene' contains a suitable promoter and translational control sequences, upstream of an open reading frame, it may be expressed as a variant mitochondrial polypeptide. The resulting variant polypeptide may contain blocks of amino acid sequence which are derived from existing polypeptides, and these domains may endow the polypeptide with some of the structural or enzymatic properties of its progenitor. The properties of the variant polypeptide (e.g. hydrophobicity) may target it to the inner mitochondrial membrane where it may compete with related native polypeptides and become associated with specific enzyme complexes. The presence of a complex containing a variant polypeptide may be tolerated if the activity of the complex is sufficient to sustain the normal cellular demand for energy. During microsporogenesis, a developmental stage associated with increased rates of respiration and mitochondrial biogenesis and where the first visible abnormality in CMS lines is mitochondrial degeneration (Warmke and Lee, 1978), the mitochondrial enzyme complexes may be unable to provide sufficient energy to sustain further cellular development.

The CMS phenotype is only expressed when the cytoplasm containing the mutation is transferred to a foreign nuclear background which cannot suppress or compensate for the mitochondrial mutation. When a male sterile cytoplasm is backcrossed into a nuclear background containing fertility restorer genes mitochondrial biogenesis and function is normal. Therefore, we can speculate that the mode of action of nuclear fertility restorer genes is either to suppress synthesis of the variant polypeptide below a certain threshold level (as seen for example in the case of the 13,000 M_r polypeptide synthesised by CMS-T maize lines, Forde and Leaver, 1980), or to compensate for the mitochondrial dysfunction in some way, thus restoring their potential for replication and/or respiratory function.

To date no evidence of a direct causal relationship has been established between synthesis of variant mitochondrial polypeptides and the CMS phenotype. We and others have, however, demonstrated that mitochondrial genome rearrangements result in altered genome organisation and expression and it seems likely that these types of molecular events will eventually be shown to underlie the male sterility trait.

ACKNOWLEDGEMENTS

The work described in this review has been supported by research grants from the AFRC, SERC and the Gatsby Foundation to C.J.L.

REFERENCES

Belliard, G., Vedel, F. & Pelletier, G. (1979) Mitochondrial recombination in cytoplasmic hybrids of Nicotiana by protoplast fusion. Nature, 281, 401-403.

Bendich, A.J. (1985) Plant mitochondrial DNA: Unusual variation on a common theme. In: Genetic Flux in Plants, ed. B. Hohn and E.S. Dennis. Berlin: Springer-Verlag.

Boeshore, M., Liftshitz, I., Hanson, M. & Izhar, S. (1983) Novel composition of mitochondrial genomes in Petunia somatic hybrids derived from cytoplasmic male sterile and fertile plants. Mol. Gen. Genet., 190, 459-467.

Boeshore, M.L., Hanson, M.R. and Izhar, S. (1985) A variant mitochondrial DNA arrangement specific to Petunia stable sterile somatic hybrids. Plant Mol. Biol., 4, 125-132.

Borck, K.S. & Walbot, V. (1982) Comparisons of restriction endonuclease patterns of mitochondrial DNA from normal and male sterile cytoplasms of Z. mays L. Genetics, 102, 109-128.

Braun, C.J. & Levings, C.S., III (1985) Nucleotide sequence of the F_1 ATPase α-subunit gene from maize mitochondria. Plant Physiol., 79, 571-577.

Chetrit, P., Mathieu, C., Vedel, F., Pelletier, G. & Primard, C. (1985) Mitochondrial DNA polymorphism induced by protoplast fusion in cruciferae. Theor. Appl. Genet., 69, 361-366.

Dawson, A., Jones, V.P. & Leaver, C.J. (1984) The apocytochrome b gene in maize mitochondria does not contain introns and is preceded by a potential ribosome binding site. EMBO Journal, 3, 2107-2113.

Dawson, A.J., Hodge, T.P., Isaac, P.G., Leaver, C.J. & Lonsdale, M.D. (1986) Location of the genes for cytochrome oxidase subunits I and II, apocytochrome b, α-subunit of the F_1ATPase and the ribosomal RNA genes on the mitochondrial genome of maize (Zea mays L.). Curr. Genet., 10, 561-564.

Dewey, R.E., Levings, C.S., III & Timothy, D.H. (1986) Novel recombinations in the maize mitochondrial genome produce a unique transcriptional unit in Texas male-sterile cytoplasm. Cell, 44, 439-449.

Dewey, R.E., Levings, C.S., III & Timothy, D.H. (1985a) Nucleotide sequence of the ATPase subunit 6 gene of maize mitochondria. Plant Physiol, 79, 914-919.

Dewey, R.E., Schuster, A.M., Levings, C.S., III & Timothy, D.H. (1985b) Nucleotide sequence of F_oATPase proteolipid (subunit 9) gene of maize. Proc. Nat. Acad. Sci. USA., 82, 1015-1019.

Dixon, L.K. & Leaver, C.J. (1982) Mitochondrial gene expression and CMS in Sorghum. Plant Mol. Biol., 1, 89-102.

Dixon, L.K., Leaver, C.J., Brettell, R.I.S. & Gengenbach, B.Y. (1982) Mitochondrial sensitivity to Dreschlera maydis T-toxin and the synthesis of a variant mitochondrial polypeptide in plants derived from maize tissue cultures with Texas male-sterile cytoplasm. Theor. Appl. Genet., 63, 75-80.

Duvick, D.N. (1965) Cytoplasmic pollen sterility in corn. Adv. Genet. 13, 1-56.

Edwardson, J.R. (1970) Cytoplasmic male sterility. Bot. Rev., 36, 341-420.

Escote, L.J., Gabay-Laughnan, S.J. & Laughnan, J.R. (1986) Cytoplasmic reversions to fertility in CMS-S maize need not involve loss of linear mitochondrial plasmids. Plasmid, 14, 264-267.

Forde, B.G. & Leaver, C.J. (1980) Nuclear and cytoplasmic genes controlling synthesis of variant mitochondrial polypeptides in male sterile maize. Proc. Nat. Acad. Sci. USA., 77, 418-422.

Forde, B.G., Oliver, R.J. & Leaver, C.J. (1978) Variation in mitochondrial translation products associated with male sterile cytoplasms in maize. Proc. Nat. Acad. Sci. USA., 75, 3841-3845.

Forde, B.G., Oliver, R.J., Leaver, C.J., Gunn, R.E. & Kemble, R.J. (1980). Classification of normal and male sterile cytoplasms in maize. I. Electrophoretic analysis of variation in mitochondrially synthesized proteins. Genetics, 95, 443-450.

Fox, T.D. & Leaver, C.J. (1981) The Zea mays mitochondrial gene coding cytochrome oxidase subunit II has an intervening sequence and does not contain TGA codons. Cell, 26, 315-323.

Hanson, M. & Condé, M. (1985) Functioning and variation of cytoplasmic genomes: lessons from cytoplasmic-nuclear interactions affecting male fertility in plants. Int. Rev. Cytol., 94, 213-265.

Isaac, P.G., Brennicke, A., Dunbar, S.M. & Leaver, C.J. (1985a) The mitochondrial genome of fertile maize (Zea mays L.) contains two copies of the gene encoding the α-subunit of the F_1ATPase. Curr. Genet., 10, 321-328.

Isaac, P.G., Jones, V.P. & Leaver, C.J. (1985b) The maize cytochrome c oxidase subunit I gene: sequence, expression and rearrangement in cytoplasmic male sterile plants. EMBO Journal, 4, 1617-1623.

Kemble, R.J., Flavell, R.B. & Brettell, R.I.S. (1983) Mitochondrial DNA analysis of fertile and sterile maize plants derived from tissue culture with the Texas male sterile cytoplasm. Theoretical Applied Genetics, 62, 213-217.

Kemble, R.J. & Shephard, J.F. (1984) Cytoplasmic variation in a potato protoclonal population. Theoretical and Applied Genetics, 69, 211-216.

Kemble, R.J. & Thompson, R.D. (1982) S1 and S2, the linear mitochondrial DNAs present in a male sterile line of maize, possess terminally attached proteins. Nucleic Acids Res., 10, 8181-8186.

Laughnan, J. & Gabay-Laughnan, S. (1983) Cytoplasmic male sterility in maize. Ann. Rev. Genet., 17, 27-48.

Laughnan, J.R., Gabay-Laughnan, S.J. & Carlson, J.E. (1981) Characteristics of cms-S reversion to male fertility in maize. Stadler Symp., 13, 93-114.

Leaver, C.J. & Gray, M. (1982) Mitochondrial genome organization and expression in higher plants. Ann. Rev. Plant Physiol., 33, 373-402.

Leaver, C.J., Isaac, P.G., Bailey-Serres, J., Small, I.D., Hanson, D.K. & Fox, T.D. (1985) Recombination events associated with the cytochrome c oxidase subunit I gene in fertile and cytoplasmic m male sterile maize and sorghum. In: E. Quagiarello., E.C. Slater,, F. Palmieri., C. Saccone, and A.M. Kroon., pp. 111-122. Amsterdam: Elsevier.

Levings, C.S., III & Sederoff, R.R. (1983) Nucleotide sequence of the S2 mitochondrial DNA from the S cytoplasm of maize. Proc. Nat. Acad. Sci. USA, 80, 4055-4059.

Lonsdale, D.M., Hodge, T.P. & Fauron, C.M.-R. (1984) The physical map and organization of the mitochondrial genome from fertile cytoplasm of maize. Nucleic Acids Res., 12, 9249-9261.

Lonsdale, D.M., Thompson, R.D. & Hodge, T.P. (1981) The integrated forms of the S1 and S2 DNA elements of maize male sterile mitochondrial DNA are flanked by a large repeated sequence. Nucleic Acids Res. 9, 3657-3669.

McNay, J.W., Choury, P.S. & Pring, D.R. (1984) Molecular analysis of genomic stability of mitochondrial DNA in tissue culture cells of maize. Theor. Appl. Genet., 67, 4333-4337.

Nagy, F., Torok, I. & Maliga, P. (1981) Extensive rearrangements in the mtDNA in somatic hybrids of Nicotiana tabacum and Nicotiana knightiana. Mol. Gen. Genet., 183, 437-439.

Paillard, M., Sederoff, R.R. & Levings, C.S., III (1985) Nucleotide sequence of the S-1 mitochondrial DNA from the S cytoplasm of maize. EMBO Journal, 4, 1125-1128.

Palmer, J.D. & Shields, C.R. (1984) Tripartite structure of the Brassica campestris mitochondrial genome. Nature, 307, 437-440.

Pring, D.R. & Levings, C.S., III (1978) Heterogeneity of maize cytoplasmic genomes among male-sterile cytoplasms. Genetics, 89, 121-136.

Pring, D.R., Levings, C.S., III, Hu, W.W. & Timothy, D.H. (1977) Unique DNA associated with mitochondria in the 'S' type cytoplasm of male-sterile lines. Proc. Nat. Acad. Sci. USA, 74, 2904-2908.

Pring, D.R. & Lonsdale, D.R. (1985) Molecular biology of higher plant mitochondrial DNA. Int. Rev. Cytol., 97, 1-46.

Schardl, C.L., Lonsdale, D.M., Pring, D.R. & Rose, K.R. (1984) Linearization of maize mitochondrial chromosomes by recombination with linear episomes. Nature, 310, 292-296.

Schertz, K.F. & Pring, D.R. (1982) Cytoplasmic sterility systems in Sorghum. In: Sorghum in the eighties. eds. L.R. House, L.K. Mughogho & J.M. Peacock. Vol. 1, pp. 373-383. International Crops Research Institute for Semi-Arid Tropics, Patancheru, India.

Thompson, R.D., Kemble, R.J. & Flavell, R.B. (1980) Variations in mitochondrial DNA organization between normal and male-sterile cytoplasms of maize. Nucleic Acids Res., 8, 1999-2008.

Warmke, H.E. & Lee, S.L. (1978) Pollen abortion in T-cytoplasmic male-sterile corn (Zea mays): a suggested mechanism. Science, 200, 561-563.

TOMATO RIPENING MUTANTS

D. Grierson, M.E. Purton, J.E. Knapp and B. Bathgate
Department of Physiology and Environmental Science,
University of Nottingham School of Agriculture,
Sutton Bonington, Loughborough, LE12 5RD, U.K.

INTRODUCTION

The main body of a fleshy fruit can be derived from a number of different parts of a flower. At the beginning of fruit development specific groups of cells derived from the mesocarp, pericarp or receptacle undergo cell division and expansion (Coombe, 1976). During the growth phase, when the fruit protects or supports the developing seeds, the cells are generally hard and unpalatable. At maturity, the fruit cells undergo a series of changes in ultrastructure, chemical composition and physiological activity that makes them attractive to animals and birds and they are eaten. The seeds survive the digestive process and are dispersed some distance from the parent plant. The changes in colour, texture, flavour and aroma that serve to attract potential consumers (Table 1) constitute ripening. This is not a uniform process. Different fruits are derived from different cell types. There is a range of patterns of cell division and expansion (Coombe, 1976; Rhodes, 1980), the cells accumulate different types of storage materials, and the biochemical conversions that occur during ripening also vary (Tucker and Grierson, 1987). For example, unripe bananas store large amounts of starch, which is converted to sugars during ripening, whereas grapes accumulate sugars before ripening. In some fruits, such as tomatoes, dramatic changes in colour are due to the production of β-carotene and lycopene in plastids, in others, such as apples and strawberries, colour production is brought about by the accumulation of anthocyanins in the vacuoles. There are even some fruits such as avocado and kiwi fruit that show little or no colour change during ripening. Similarly, there appear to be several different mechanisms of softening and specific fruits also have quite different flavours and aromas. This variation in ripening patterns in different types of fruit means that it is impossible to give a definition of the process in terms of biochemical pathways. The most appropriate definition is to consider ripening as a series of processes rendering the cells attractive to eat. In any specific fruit the changes may be defined precisely (Table 1). They invariably involve all the compartments of a cell and are highly coordinated. Although it is quite clear that the cells are alive and metabolically active when they are consumed, ripening may be considered as a specialised case of senescence. The developmental programme that leads to ripening is only just beginning to be understood at the molecular level.

In climacteric fruits, such as apples, tomatoes and bananas, ripening is associated with an increase in ethylene synthesis and respiration (Rhodes, 1980; Biale and Young, 1981). Ethylene appears to stimulate ripening and is considered by some researchers to be a ripening hormone, despite the fact that it is produced in the same organ where it has its effect. Ethylene appears not to be involved in ripening of non-climacteric fruits (Rhodes, 1980; Biale and Young, 1981) and there must be some other regulatory factors involved in the initiation and progression of ripening in all fruits. As part of the ripening programme many alterations occur in the regulation of existing enzymes and biochemical pathways. RNA and protein synthesis continues and there is now good evidence for specific changes in

Table 1 Ripening Changes in Tomato

Metabolism	50-100 fold increase in synthesis and evolution of ethylene Increase in respiration Solubilization of starch Changes in metabolism of organic acids
Colour	Loss of chlorophyll, dismantling of the photosynthetic apparatus and accumulation of carotenoids such as β-carotene and lycopene
Texture	Partial solubilization and modification of pectins and cellulose in cell walls
Flavour and aroma	Accumulation of sugars and changes in organic acid metabolism Destruction of alkaloids such as α-tomatine Production of a complex mixture of volatiles
Gene expression	Disappearance of some mRNAs and proteins synthesised before ripening, such as those for chloroplast photosynthetic enzymes Appearance of new ripening-specific mRNAs directing the synthesis of enzymes catalysing specific ripening changes

gene expression that are necessary for ripening to take place (Grierson, 1986; Grierson et al., 1986).

Tomatoes (Lycopersicon esculentum Mill) are an obvious choice for biochemical and developmental studies on fruit ripening. Following fertilization of the flower, fruit growth and maturation lasts for about seven weeks. The ripening process itself extends over a period of seven to ten days and occurs in the absence of cell division or expansion. It is stimulated by ethylene, and synthesis of the gas by tomato fruit is one of the earliest signs of ripening. Using controlled environment facilities plants can be grown throughout the year, thus ensuring a continuous supply of material for experiments. Furthermore, tomato has a good genetic map (Rick, 1980) and a number of ripening mutants exist that are proving useful in dissecting the mechanism of ripening and its regulation.

CHANGES DURING NORMAL RIPENING

Some of the major changes occurring during tomato ripening are shown in Figure 1a. They begin with ethylene synthesis, which is followed by the appearance of polygalacturonase activity and the production of lycopene. The increase in ethylene synthesis is also accompanied by a rise in respiration, called the respiratory climacteric, which may actually be caused by ethylene (see Rhodes, 1980). The appearance of polygalacturonase activity has been shown to be due to de novo enzyme synthesis (Tucker and Grierson, 1982). This enzyme accumulates in the cell walls where it plays a major part in solubilization of the pectin fraction (Themmen et al., 1982; Crookes and Grierson, 1983) leading to softening. Lycopene and β-carotene accumulate in the plastids as they are converted from chloroplasts to chromoplasts. This is a complex process. Initial changes in the chloroplasts, such as the dissolution of starch, may begin before ripening (Figure 2). This is followed by the loss of chlorophyll, which is associated with the disappearance of thylakoids. The β-carotene is believed to accumulate in the osmiophyllic globules seen in the electron microscope. Lycopene is deposited as crystalloid bodies surrounded by a membrane. During preparation for electron microscopy the lycopene tends to be dissolved, leaving behind the characteristic undulating internal membrane system (Figure 2). Although the carotenoid-synthesising enzymes are located within the plastids (Kreutz et al., 1982; Lutke-Brinkhaus et al., 1982) the genes regulating their production are inherited in a Mendelian fashion (Goodwin, 1980) and thus are present in the nucleus.

Throughout the ripening process protein synthesis and turnover continues. Changes in the labelling pattern of total proteins from fruit supplied in vivo with radioactive amino acid have recently been catalogued by Baker et al. (1985). Increases and decreases in the synthesis of specific proteins have been noted in the cell wall (Figure 3) and chromoplast fractions (Figure 4) (Grierson et al., 1985; Bathgate et al., 1986). Many of the chloroplast proteins disappear during chromoplast formation, and new but unidentified proteins accumulate. There are differences when total proteins stained with Coomassie Blue are compared with newly synthesised proteins. These are presumably due to variation in the turnover rates of individual polypeptides (Bathgate et al., 1985).

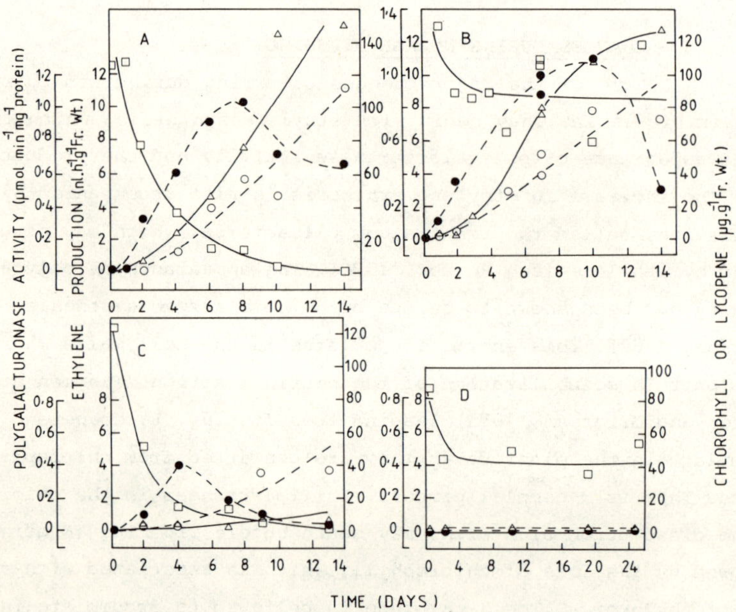

Figure 1. Ripening changes in normal and mutant tomatoes. Plants were grown in controlled environment chambers (16h light, 23°C; 8h dark, 15°C) and flowers tagged at anthesis. Time 0 = 49 days after anthesis. Fruit were sampled at intervals after this time and measurements were made of ethylene synthesis (●), PG synthesis (O), chlorophyll (□) and lycopene (Δ). A = normal fruit, B,C,D are fruit homozygous for the gf, Nr and rin mutations respectively.

MUTATIONS AFFECTING RIPENING

Many mutations that affect tomato ripening have been noted by plant breeders. Their chromosomal location and map position have been determined by standard genetic procedures (Rick, 1980). Often these mutations have little or no effect on growth and development of fruit or the parent plant but exert a profound influence on the ripening process (Table 2). Some of them have been incorporated into isogenic lines of standard

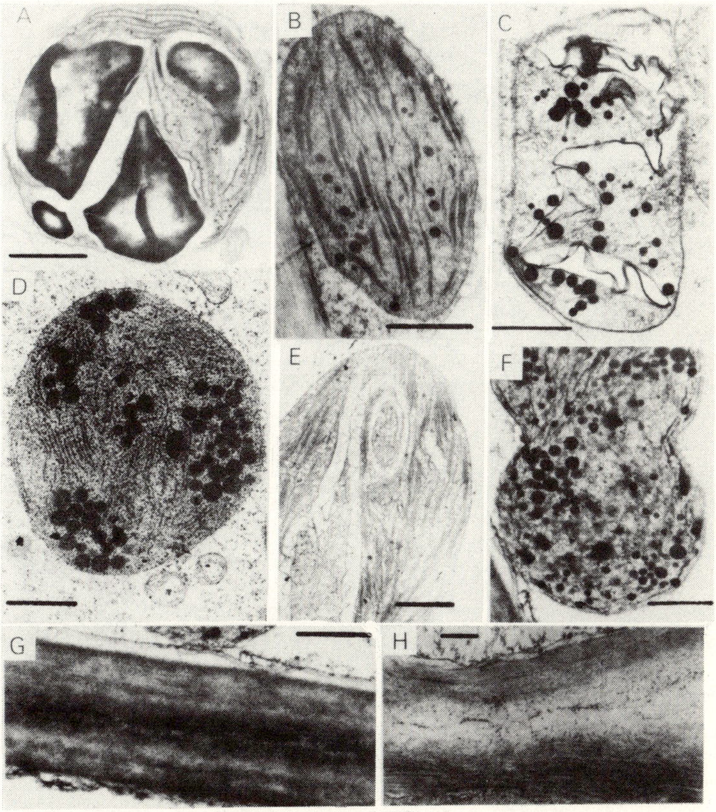

Figure 2. Electron micrographs of plastids and cell walls from normal and mutant fruit. Chloroplast from immature fruit (A), mature green fruit (B), normal chromoplast from ripe fruit (C), chromoplast for Nr fruit (D), chromoplast from gf fruit (E), chromoplast from rin fruit (F), cell wall from "ripe" Nr fruit (H), cell wall from rin fruit at a comparable age (G). Cell walls from normal ripe fruit show very extensive solubilisation of the middle lamella region at the same age (Crookes and Grierson, 1983). The bar represents 1 μm.

varieties such as Ailsa Craig (Darby, 1978) and Rutgers (Tigchelaar *et al.*, 1978). Generally there is an effect on fruit colour. This is probably because colour changes are easy to select for and should not be taken to imply that colour is the only aspect of ripening that is affected by mutation.

The *yellowflesh* and *tangerine* mutations (Table 2) affect the process of carotenoid production but other aspects of ripening such as ethylene production and softening continue normally. The mutations presumably affect the genes specifying enzymes in the carotenoid biosynthesis pathway

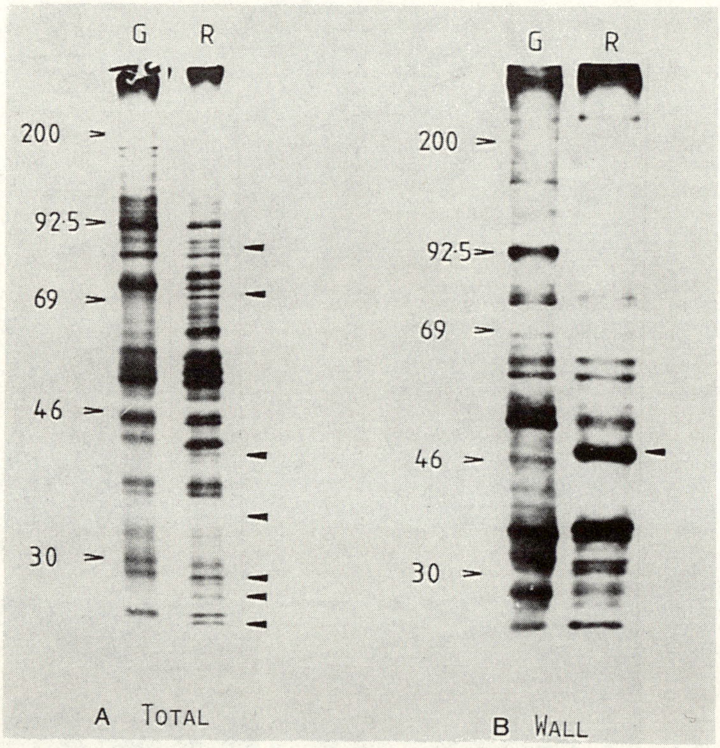

Figure 3. Comparison of proteins synthesised in green and ripening tomato fruit. Discs of pericarp tissue from green (G) or ripening (R) fruit were incubated with ^{35}S-methionine for 16h. The radioactive proteins were extracted, fractionated by sodium dodecyl sulphate-polyacrylamide gel electrophoresis and detected by fluorography. The positions of marker proteins (molecular weight $x10^{-3}$) are shown on the left. Proteins labelled in ripening but not green fruit are shown by arrows on the right. A) Total SDS-soluble proteins. B) Cell wall bound proteins. (See Grierson *et al.*, 1985).

(Goodwin, 1980). The finding that these and other mutations affecting the same pathway map to several different chromosomes indicates that genes for carotenoid biosynthesis are not clustered. Other mutations have more complex effects on ripening. Fruit of the greenflesh mutant, for example, appear a dirty brown colour when ripe (Ramirez and Tomes, 1964). This is caused by the retention of much of the chlorophyll in the chromoplasts. Ethylene production and the synthesis of lycopene and polygalacturonase continue as normal (Figure 1). Examination of the greenflesh chromoplasts in the electron

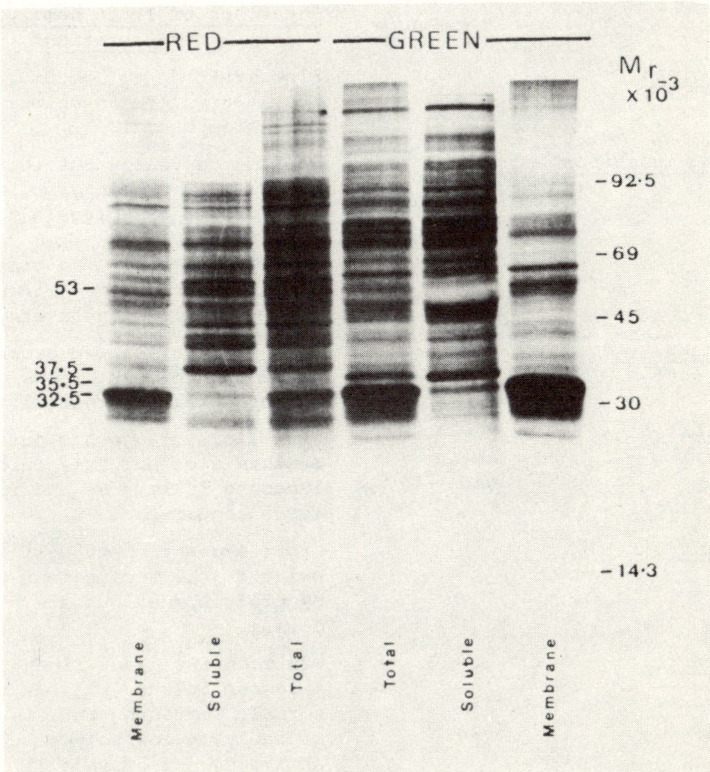

Figure 4. Synthesis of plastid proteins during conversion of chloroplasts to chromoplasts. Fruit were cut in half, seeds and jelly were removed and ^{35}S-methionine was added for 6h. Plastids were then isolated and total, soluble and membrane proteins prepared from chloroplasts ("green") and chromoplasts ("red").

Equal quantities were separated by sodium-dodecyl sulphate-polyacrylamide gel electrophoresis and detected by autoradiography. The positions of radioactive marker proteins are shown (molecular weight x10^{-3}). Major differences in labelled proteins (molecular weight x10^{-3}) are indicated.

microscope shows that they retain not only chlorophyll but a substantial amount of thylakoid material during ripening (Figure 2). It seems likely that the failure to degrade chlorophyll is related to the retention of the thylakoids, since a number of chloroplast membrane proteins also persist for longer during ripening in this mutant. Ultrastructural studies also show that thylakoids and the membrane system associated with lycopene crystalloids can occur in the same plastid (Figure 2). This indicates that the

Table 2 Some Ripening Mutants of Tomato

Name	Chromosome	Phenotype of fruit homozygous for the mutation[1]
Yellow flesh (r)	3	Ripe fruit do not accumulate lycopene. They are yellow in colour but otherwise normal.
Ripening inhibitor (rin)	5	Fruit turn yellow but there is no respiratory climacteric, ethylene production, or polygalacturonase synthesis. Fruit soften very little, store for a very long time but have poor flavour. Normal ripening is not restored by exogenous ethylene.
Greenflesh (gf)	8	Ripe fruit are a red-brown colour because chlorophyll loss is incomplete but are otherwise normal.
Neverripe (Nr)	9	Fruit turn orange and soften slowly. Synthesis of polygalacturonase and lycopene is reduced and fruit store for a long time.
Tangerine (t)	10	Fruit are a rich tangerine colour owing to the replacement of lycopene by prolycopene, but are otherwise normal.
Alcobaca (alc)	10	Fruit on the plant ripen slowly to a pale red colour. Flavour is almost normal. Softening rate is reduced, due probably to low polygalacturonase activity, and storage life is thus increased. Fruit picked mature green are reported to show reduced respiration and ethylene production and only to ripen to a yellow colour.
Non ripening (Nor)	10	Similar to rin except that the final fruit colour is pale orange.

[1] Descriptions based on information in: Darby (1978), Tigchelaar et al, (1978), Rick (1980), Kopeliovitch et al. (1980), Mutschler (1984), Grierson (1985) and personal observations by the authors.

destruction of the photosynthetic apparatus is not necessary for chromoplast formation. It remains to be seen whether genes for both these developmental stages of tomato plastids can be expressed concurrently.

There are also mutations that have pleiotropic effects on ripening, such as _Neverripe_, _ripening inhibitor_, _alcobaca_ and _nonripening_ (Table 2). _Neverripe_ fruit show a reduction in ethylene production and synthesise only small amounts of lycopene and polygalacturonase (Figure 1). In normal fruit three isoforms of polygalacturonase (PG) are produced. PG-1 is synthesised early during ripening when it is the dominant form (Tucker _et al._, 1980). Later, PG-2, which can be resolved into two components PG-2a and PG-2b (Brady _et al._, 1982) becomes the major form (Tucker _et al._, 1980). Generally, the total amount of PG in _Neverripe_ (_Nr_) fruit is about 15% of normal. This consists predominantly of PG-1 (Tucker _et al._, 1980) although PG-2 does appear after some months (Crookes and Grierson, 1983). The fruit never ripen fully and remain perfectly sound for 12 months or more. This is probably due, at least in part, to the reduced cell wall solubilization resulting from low levels of PG (Figures 1 & 2).

The _ripening inhibitor_ (_rin_) mutation has an even more extreme effect on ripening behaviour (Figure 1). _rin_ fruit lose chlorophyll very slowly. They show no increase in ethylene synthesis and do not accumulate lycopene or PG. After six months the fruit are a lemon yellow colour and a trace of PG may be present. Fruit store for a year or more, during which time the seeds may germinate and produce seedlings inside the fruit. _rin_ tomatoes of the same chronological age as soft red tomatoes show little evidence of cell wall solubilization (Figure 2), although some softening occurs in older fruit. Chromoplasts in _Nr_ and _rin_ fruit lose their thylakoids and accumulate osmiophyllic droplets and some internal membranes (Figure 2). There is, however, no evidence for the characteristic internal membrane system associated with the deposition of lycopene.

The protein composition of the plastids from _rin_ and normal plants changes as the chloroplasts are transformed to chromoplasts (Figure 5). Characteristic chloroplast proteins of molecular weight 50,000 (probably the large subunit of rubisco) and a group from 21-23,000 decline. Membrane proteins of 52,000, 54,000 and 27,000 accumulate, as does a soluble protein of 15,000. The main difference between _rin_ and normal chromoplasts is the presence of a prominent soluble protein of molecular weight 17,000 in _rin_. This protein is present in small amounts in chloroplasts but normally disappears in chromoplasts. In contrast, in rin it accumulates. In an early report on rin proteins (Mizrahi et al., 1976) a prominent protein was noted in mutant fruit that was suggested as an inhibitor of ripening. We do not know if the 17,000 molecular weight chromoplast protein is the same as that detected previously. There is no evidence that it is causally related to the rin phenotype but its existence is of interest considering the important role that altered organelle proteins can play in plant development (see Chapter 3).

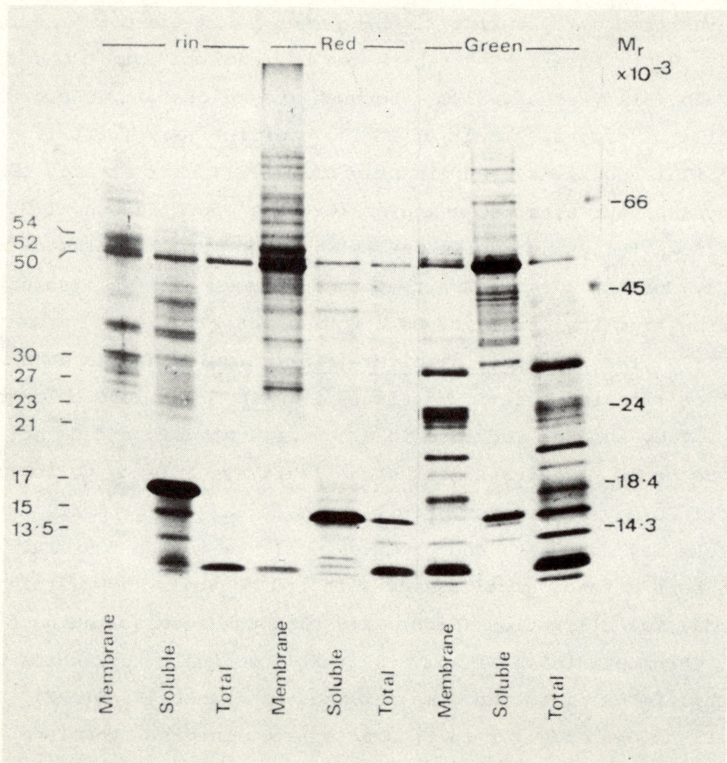

Figure 5. Comparison of plastid proteins from green and ripe fruit and from "ripe" rin mutant fruit. Plastids were isolated and total, soluble and membrane proteins prepared. Equal quantities of protein were separated by sodium dodecyl sulphate-polyacrylamide gel electrophoresis and stained with Coomassie blue. Proteins were extracted from normal fruit chloroplasts ("green"), chromoplasts ("red") or rin chromoplasts from yellow fruit ("rin"). The positions of marker proteins are shown (molecular weight $\times 10^{-3}$), as are the major proteins that change in amount.

It has been reported that supplying ethylene to rin fruit can stimulate synthesis of some PG and lycopene (Mizrahi, et al., 1975). We have not been able to reproduce this observation.

RIPENING mRNAs

Changes in mRNA content during tomato ripening were first demonstrated by Rattanapanone et al. (1978). These initial observations have been confirmed and extended by more recent studies with tomato (Speirs et al., 1984; Grierson et al., 1985) and avocado (Christoffersen et al., 1982, 1984). The current view is that there are at least three classes of fruit mRNA and the relative proportions of these change during fruit development and ripening. Thus some mRNAs disappear during fruit maturation, before ripening, others persist throughout development and ripening, and further 'ripening-specific' mRNAs also get switched on (Grierson et al., 1985). The appearance of these mRNAs in mature green tomatoes is stimulated by ethylene (Figure 6) (Grierson et al., 1986). A number of them are genuine ripening-specific mRNAs but others are present before ripening and increase greatly in concentration in response to ethylene (Maunders et al., 1986). cDNAs for nineteen mRNA sequences that accumulate during ripening have been cloned (Slater et al., 1985). Seven of these, corresponding to most of the major mRNAs shown in Figure 6, have been characterized in detail. The mRNA sizes have been measured by Northern hybridization, the molecular weights of the corresponding polypeptides determined by hybrid-release translation, and expression of the mRNAs during natural ripening and in response to ethylene have been studied by hybridization (Maunders et al., 1986). These results are summarised in Table 3. Five of the mRNAs appear specific to ripening in that they increase in concentration at least 300-fold as ripening occurs and they are not expressed in leaves or roots. Three of the cloned cDNAs (pTOM5, 6, 13) have been completely sequenced, and from this information the amino acid sequence of the proteins they encode can be predicted. The N-terminal amino acid sequence for tomato polygalacturonase-2 has also been determined. A 26 amino acid sequence at the N-terminus coincides exactly with the sequence predicted from a DNA region near the 5' end of the pTOM6 cDNA (Grierson et al., 1986). This indicates that pTOM6 encodes polygalacturonase-2 or a closely related protein. The expression of PG-2 mRNA during natural ripening, measured by dot hybridization of RNA samples to ^{32}P-labelled pTOM6 cDNA, is shown in Figure 7. Maximum expression of the mRNA occurs when tomatoes reach the fully orange stage.

Hybrid-release translation experiments show that the mRNA that hybridizes to pTOM6 cDNA directs the synthesis of a polypeptide with a molecular weight of approximately 55,000 (Slater et al., 1985) (Table 3). This is substantially larger than the estimate of 46,000 for the molecular weight of mature PG-2 (Tucker et al., 1980; Tucker and Grierson, 1982). This apparent discrepancy is accounted for by the information provided by the DNA sequence of pTOM6, which predicts the presence of the PG-2 sequence plus an additional 71 amino acids (mol. wt. = 8,500) at the N-terminus of the mRNA translation product. This pre-sequence is presumed to be removed post-translationally. Part or all of it may serve to direct the transport of PG-2 through the cell endomembrane system and across the plasmalemma into the cell wall. Polygalacturonase-2 is glycosylated (Ali and Brady, 1982; Moshrefi and Luh, 1983) and carbohydrate residues are presumably added to the polypeptide after translation.

The translation product of the PG-2 mRNA is shown in Figure 6.

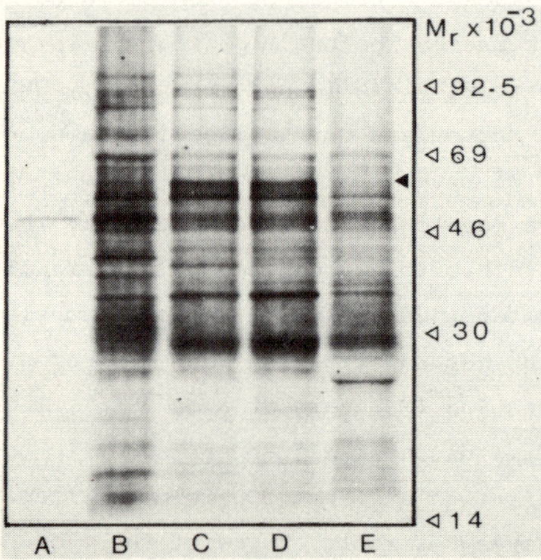

Figure 6. Effect of ethylene on changes in mRNA content of normal and rin mutant tomatoes. Total RNA was extracted and translated in vitro in a rabbit reticulocyte lysate in the presence of ^{35}S-methionine. Radioactive proteins were fractionated by sodium dodecyl sulphate - polyacrylamide gel electrophoresis and detected by fluorography. A sample of the translation mixture incubated without added RNA is shown in track A. Translation products of RNA from mature green tomatoes (B), 30h or 48h after incubation in $10 \mu l.l^{-1}$ ethylene (C,D) or rin fruit incubated in ethylene (E). The positions of radioactive marker proteins (molecular weight x 10^{-3}) are shown on the right. The position of the translation product of polygalacturonase mRNA is shown by the black arrowhead.

Little or no mRNA for this protein is present in mature green tomatoes but it accumulates within 30h of ethylene treatment. At this stage the fruit are still green. Comparison with the mRNA translation products from yellow rin tomatoes treated with ethylene indicates that this mRNA is present in very small amounts in rin (Figure 6). Confirmation of this has been obtained by hybridization of mRNA from rin fruit to ^{32}P-labelled pTOM6 DNA (Figure 8). Thus the failure of rin tomatoes to produce polygalacturonase is explained by the absence of its mRNA. This cannot be the sole basis for the rin phenotype however, since other mRNAs are also reduced or absent in this mutant (Figure 6) and, of course, rin fails to synthesise ethylene and lycopene. In similar experiments with Nr fruit, which contain very reduced quantities of PG-2 (Tucker et al., 1980; Crookes and Grierson, 1983), we found approximately 4% of the normal level of PG-2 mRNA in the mutant.

The only ripening-related mRNAs identified to date are those for tomato polygalacturonase (Grierson, et al., 1986) and avocado cellulase (Christoffersen et al., 1984). It seems probable that some of the others (Slater et al., 1985; Maunders et al., 1986) may code for enzymes involved in carotenoid biosynthesis, production of flavour volatiles and possibly also the increases in respiration and the metabolism of starch, sugars and organic acids. The possibility that the production of ethylene also requires the appearance of new mRNAs is discussed below.

Table 3. Properties and Expression of Cloned Ripening-Related mRNAs from Tomato

Clone (pTOM)	mRNA (approx. no. of nucleotides)	Protein (approx.mol.wt.)	Ripening Specificity
5	1960	48,000	+
6	1600	55,000	+
13	1400	35,000	+
36	1470	52,000	+
75	1200	28,000	-
99	1520	44,000	+
137	1680	57,000	-

ETHYLENE AND THE CONTROL OF RIPENING

There is no doubt that ethylene plays an important role in stimulating ripening. Adding ethylene to unripe fruit frequently hastens the onset of ripening (McGlasson et al., 1978; Rhodes, 1980; Procter and Caygill, 1985). Inhibiting ethylene synthesis also retards or prevents ripening. Removing ethylene from fruit stores is recognised commercially as an important method of retarding ripening. The pathway for ethylene biosynthesis (Adams and Yang, 1979) is shown in Figure 9. The production of

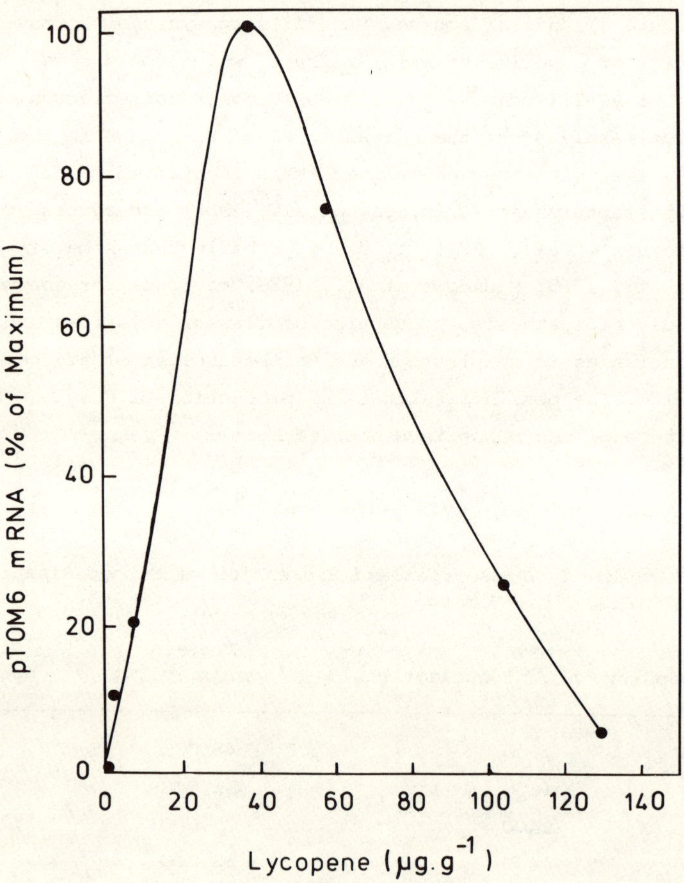

Figure 7. Change in concentration of polygalacturonase mRNA during the ripening of tomato fruit. Total polyA+ mRNA was purified from fruit at different stages of ripening (indicated by the lycopene content) and 1 μg hybridized to ^{32}P-labelled pTOM 6 DNA (the clone for tomato PG).

ethylene is one of the earliest detectable events during initiation of tomato ripening (Sawamura *et al*., 1978) and the increase in ethylene occurs before polygalacturonase synthesis (Grierson and Tucker, 1983). The enhanced synthesis of ethylene is due to increases in the activities of two enzymes: ACC (1-amino-cyclopropane-1-carboxylic acid) synthase and EFE (the 'ethylene forming enzyme', see Yang and Hoffman, 1984). The control of activity of these enzymes is thus of critical importance for ripening. Ethylene synthesis is autocatalytic, that is the presence of the gas induces increases in the activity of ACC synthase and EFE and results in enhanced ethylene production. It is possible that the process requires mRNA accumulation and enzyme synthesis but this is not certain (Yang and Hoffman, 1984). Ethylene stimulates the respiration rate and enhances the concentration of existing mRNAs. Once autocatalytic ethylene production is under way the ripening-specific mRNAs begin to accumulate and the changes in synthesis of enzymes required for ripening proceed. It is not yet clear whether the continuous presence of ethylene is required for the expression of ripening genes but circumstantial evidence from experiments with inhibitors of ethylene synthesis and action suggest that this may be the case. We still do not understand what factors control the *initiation* of ethylene production, whether by control of mRNA and enzyme synthesis or by regulation of existing enzymes.

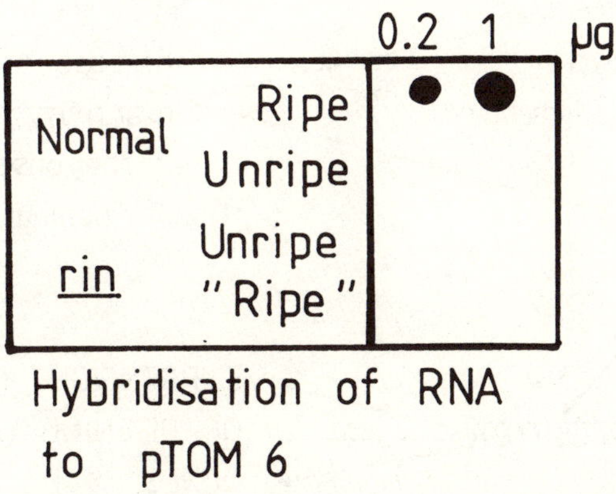

Figure 8. Hybridization of mRNA from normal and *rin* mutant tomatoes to cloned cDNA for polygalacturonase. PolyA+ mRNA was prepared from unripe and ripe normal and *rin* mutant fruit. The RNA was applied as dots to membranes, immobilised, and hybridized to ^{32}P-labelled pTOM 6 cDNA. Hybridization was detected by exposure of the membrane to X-ray film.

The mechanism for perceiving and responding to ethylene is also poorly understood.

It is assumed that ethylene perception by plant cells involves a receptor and that binding of ethylene to this site(s) sets in train a sequence of biochemical events leading to the response (see Smith and Hall, 1985). Analogue and inhibitor studies have suggested several properties for the receptor (see Smith and Hall, 1985; Beyer, 1985; Sisler, 1984). It is possible that metabolism of ethylene to ethylene oxide or ethylene glycol may be involved in the response mechanism (Beyer, 1985). However, although there have been several studies of ethylene binding sites these have not yet been shown to be involved in mediating a physiologically significant response. Ethylene action can, however, be inhibited by low concentrations of silver

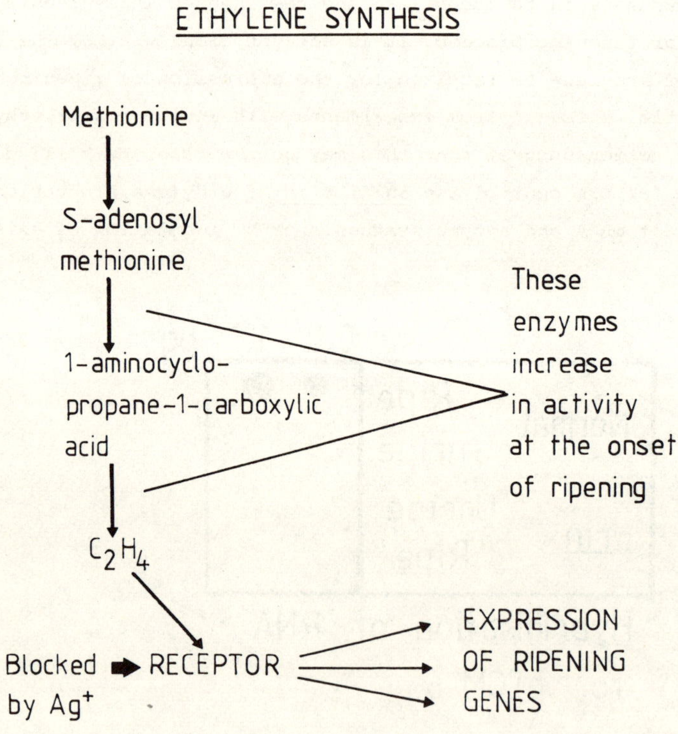

Figure 9. Ethylene synthesis and action. The two enzymes specific to the pathway that increase in activity during ripening are ACC(1-amino-cyclopropane-1-carboxylic acid) synthase and the ethylene forming enzyme.

ions (Beyer, 1976) and treatment of plant tissues with silver as the thiosulphate complex inhibits flower petal senescence (Veen, 1983) and fruit ripening (Saltveit et al., 1978; Hobson et al., 1984). It seems likely that Ag^+ interferes either with the perception of ethylene or with the response mechanism.

DISCUSSION AND CONCLUSIONS

The biochemical and genetic evidence outlined above suggests that the expression of specific genes is required for normal ripening. One of the 'ripening genes' codes for the enzyme PG, which plays a major part in cell wall solubilization which is responsible for the changes in texture during tomato ripening. A model for the regulation of gene expression during ripening is shown in Figure 10. Ethylene seems to stimulate the process but

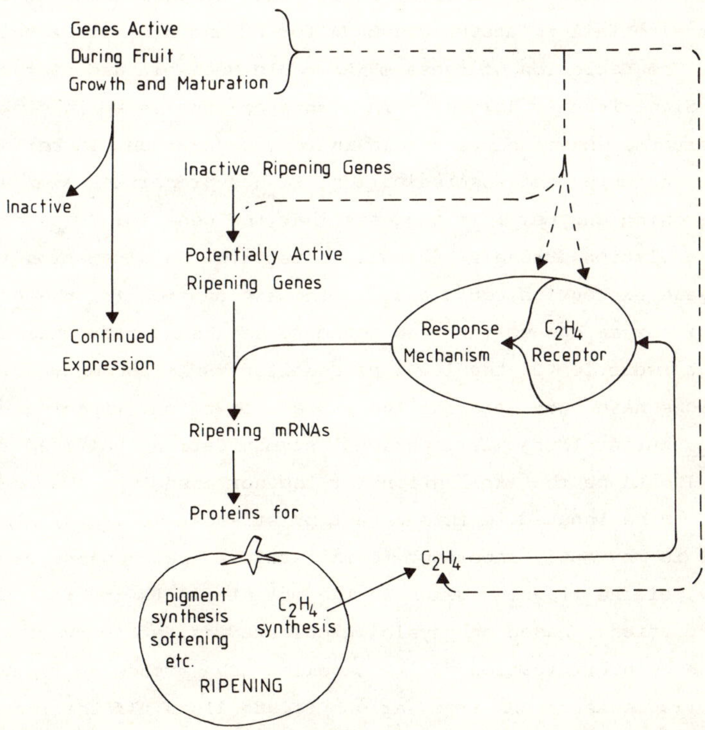

Figure 10. The control of gene expression during tomato ripening.

we do not know how its synthesis is regulated, how it is perceived, or how it brings about its effects. Clearly, we need more information about these aspects of the process. How do the mutants fit into the picture?

It has been suggested that the rin mutation affects the ethylene receptor (McGlasson, 1985). This would explain why rin fruit do not ripen. However, one objection to this hypothesis is that rin fruit do respond to ethylene by showing a rise in respiration (Herner and Sink, 1973). Ethylene also induces a rise in activity of the ethylene forming enzyme in rin fruit (Liu et al., 1985). They must, therefore, be able to perceive the gas. Thus, either rin fruit are lacking a specific ripening receptor or they carry a lesion affecting the response rather than the perception mechanism. Although rin tomatoes neither synthesise ethylene nor ripen in response to added ethylene, they are capable of ethylene synthesis in response to wounding (McGlasson et al., 1975). This suggests that they are not deficient in the genes for ethylene biosynthetic enzymes but rather they lack the biochemical machinery for turning on ethylene synthesis during ripening. rin fruit also fail to accumulate mRNA for PG and some other unidentified proteins. The depletion of these mRNAs could be explained if the genes were deleted. Significantly however, rin plants do express PG in flower abscission zones where the enzyme plays a role in cell separation (Tucker and Roberts, 1984). Furthermore, the abscission zone PG has properties similar to those of the ripening enzyme. This suggests that the gene for PG is present in rin but its regulation during ripening is defective. An alteration in the regulation of gene expression could also be the explanation for the Nr phenotype. No mRNA or enzyme has so far been found to be absent from this mutant but there is a reduction in the level of specific mRNAs and enzymes. Similar observations have been made for the expression of our ripening-related cDNAs in the alc mutant (Mutschler, personal communication). Altered regulation appears also to be the explanation for the nor phenotype (Table 2), since ripening can be induced in this mutant by stress treatments (Mizrahi et al., 1982). It is obviously important to investigate the presence or absence and integrity, of the ripening genes in the mutant plants in order to test whether these predictions, based on physiological studies, are correct. This is now being done by hybridization of ^{32}P-labelled cDNA probes to restriction endonuclease fragments of DNA from various tomato lines. If the results support the idea that the genes are present, then this would favour the altered regulation hypothesis, since several different genes appear to be affected by one mutation.

As discussed before, altered expression of the ripening genes could be due to a defective receptor for ethylene. Alternatively, it could be explained by the absence of a metabolite required for expression of some of the genes, or to the altered activity of a regulator gene. However the genes are regulated during normal ripening, it is reasonable to expect that there will be specific factors, probably proteins, that interact with specific DNA sequences associated with the ripening genes and switch them on. The chromosomal DNA sequences for ripening genes can now be isolated, using the cDNA clones as probes. This will permit a direct test of the hypothesis that ripening fruit contain proteins that bind specifically to regulatory DNA sequences adjacent to ripening genes. The possibility exists that one or more of the ripening mutations with pleiotropic effects, such as rin or Nr, results from an alteration in such a regulatory protein. This proposal can be tested experimentally.

REFERENCES

Adams, D.O. & Yang, S.F. (1979) Ethylene biosynthesis. Identification of 1-aminocyclopropane-1-carboxylic acid as an intermediate in the conversion of methionine to ethylene. Proc. Nat. Acad. Sci. U.S.A. 76, 170-174.

Ali, Z.M. & Brady, C.J. (1982) Purification and characterisation of the polygalacturonases of tomato fruits. Aust. J. Plant Physiol. 9, 155-169.

Baker, J.E., Anderson, J.D. & Hruschka, W.R. (1985) Protein synthesis in tomato pericarp tissue during ripening characteristics of amino acid incorporation. J. Plant Physiol. 120, 167-179.

Bathgate, B., Purton, M.E., Grierson, D. & Goodenough, P.W. (1985) Plastid changes during the conversion of chloroplasts to chromoplasts in ripening tomatoes. Planta 165, 197-204.

Bathgate, B., Goodenough, P.W. & Grierson, D. (1986) Regulation of the expression of the psbA gene in tomato fruit chloroplasts and chromoplasts. J. Plant Physiol. 124, 223-233.

Beyer, E.M. Jr. (1976) A potent inhibitor of ethylene action in plants. Plant Physiol. 58, 268-271.

Beyer, E.M. Jr. (1985) Ethylene metabolism. In: Ethylene and Plant Development. eds. J.A. Roberts & G.A. Tucker, pp. 125-137, London, Butterworths.

Biale, J.B. & Young, R.E. (1981) Respiration and ripening in fruits - retrospect and prospect. In: Recent Advances in the Biochemistry of Fruit and Vegetables. eds. J. Friend & M.J.C. Rhodes. pp.1-39, London, Academic Press.

Brady, C.J., MacAlpine, G., McGlasson, W.B. & Ueda, Y. (1982) Polygalacturonase in tomato fruits and the induction of ripening. Aust. J. Plant Physiol. 9, 171-178.

Christoffersen, R.E., Warm, E. & Laties, G.G. (1982) Gene expression during fruit ripening in avocado. Planta 155, 52-57.

Christoffersen, R.E., Tucker, M.L. & Laties, G.G. (1984) Cellulase gene expression in ripening avocado fruit: The accumulation of cellulose mRNA and protein as demonstrated by cDNA hybridization and immunodetection. Plant. Molec. Biol. 3, 385-391.

Coombe, B.G. (1976) The development of fleshy fruits. Ann. Rev. Plant Physiol. 27, 507-528.

Crookes, P.R. & Grierson, D. (1983) Ultrastructure of tomato fruit ripening and the role of polygalacturonase isoenzymes in cell wall degradation. Plant Physiol, 72, 1088-1093.

Darby, L.A. (1978) Isogenic lines of tomato fruit colour mutants. Hort. Res. 18, 73-84.

Goodwin, T.W. (1980). The biochemistry of the carotenoids. Vol. 1. Plants, 2nd edition, London, Chapman & Hall.

Grierson, D. (1985) Gene expression in ripening tomato fruit. CRC Crit. Rev. Plant Sci., 3, 113-132.

Grierson, D. (1986) Molecular biology of fruit ripening. In Oxford Surveys of Plant Cell and Molecular Biology. ed. B.J. Miflin, in press.

Grierson, D., Maunders, M.J., Slater, A., Ray, J., Bird, C.R., Schuch, W., Holdsworth, M.J., Tucker, G.A. & Knapp, J.E. (1986) Philosophical Transactions of the Royal Society, in press.

Grierson, D., Slater, A., Speirs, J. & Tucker, G.A. (1985) The appearance of polygalacturonase in tomatoes: one of a series of changes in gene expression during development and ripening. Planta, 163, 263-271.

Grierson, D. & Tucker, G.A. (1983) Timing of ethylene and polygalacturonase synthesis in relation to the control of tomato fruit ripening. Planta, 157, 174-179.

Herner, R.C. & Sink, K.C. (1973) Ethylene production and respiratory behaviour of the rin tomato mutant. Plant Physiol. 52, 38-42.

Hobson, G.E., Nichols, R., Davies, J.N. & Atkey, P. (1984) The inhibition of tomato fruit ripening by silver. J. Plant Physiol. 116, 21-29.

Kopeliovitch, E., Mizrahi, Y., Rabinowitch, H.D. & Kedar, N. (1980). Physiology of the tomato mutant alcobaca. Physiol. Plant., 48, 307-311.

Kreutz, K., Beyer, P. & Kleinig, H. (1982) The site of carotenogenic enzymes in chromoplasts from Narcissus pseudonarcissus L., Planta, 154, 66-69.

Liu, Y., Hoffman, N.E. & Yang, S.F. (1985) Promotion by ethylene of the capability to convert 1-aminocyclopropane-1-carboxylic acid to ethylene in preclimacteric tomato and cantaloupe fruits. Plant Physiol., 72, 407-411.

Lutke-Brinkhaus, F., Liedvogel, B., Kreutz, K. & Kleinig, H. (1982) Phytoene synthase and phytoene dehydrogenase associated with envelope membranes from spinach chloroplasts. Planta, 156, 176-180.

Maunders, M.J., Holdsworth, M.J., Slater, A., Knapp, J.E., Bird, C.R., Schuch, W. & Grierson, D. (1986) Ethylene stimulates the accumulation of ripening-specific mRNAs in tomatoes. Submitted for publication.

McGlasson, W.B. (1985) Ethylene and fruit ripening. HortScience 20, 51-54.

McGlasson, W.B., Poovaiah, B.W. & Dostal, H.C. (1975) Ethylene production and respiration in ageing leaf segments and in disks of fruit tissue of normal and mutant tomatoes. Plant Physiol. 56, 547-549.

McGlasson, W.B., Wade, N.L. & Adato, I. (1978). Phytohormones and fruit ripening. In Phytohormones and Related Compounds - A Comprehensive Treatise, Volume II, eds. D.S. Letham, P.B. Goodwin & T.J.V. Higgins, pp. 447-493. Amsterdam, Elsevier/North-Holland.

Mizrahi, Y., Dostal, H.C. & Cherry, J.H. (1976) Protein differences between fruits of rin, a non-ripening tomato mutant, and a normal variety. Planta, 130, 223-224.

Mizrahi, Y., Dostal, H.C. & Cherry, J.H. (1975) Ethylene induced ripening in attached rin fruits, a non-ripening mutant of tomato. Hortscience 10, 414-415.

Mizrahi, Y., Zohar, R. & Malis-Arad, S. (1982) Effect of sodium chloride on fruit ripening of the non-ripening tomato mutants nor and rin. Plant Physiol. 69, 497-501.

Moshrefi, M. & Luh, B.S. (1983) Carbohydrate composition and electrophoretic properties of tomato polygalacturonase isoenzymes. Eur. J. Biochem. 135, 511-514.

Mutschler, M.A. (1984) Ripening and storage characteristics of the alcobaca ripening mutant in tomato. J. Am. Soc. Hortic. Sci. 109, 504-507.

Proctor, F.J. & Caygill, J.C. (1985) Ethylene in commercial post-harvest handling of tropical fruit. In Ethylene and Plant Development. eds. J.A. Roberts & G.A. Tucker, pp. 317-332, London: Butterworths.

Ramirez, D.A. & Tomes, M.L. (1964) Relationship between chlorophyll and carotenoid biosynthesis in dirty-red (green flesh) mutant in tomato. Bot. Gaz. 125, 221-226.

Rattanapanone, N., Speirs, J. & Grierson, D. (1978) Evidence for changes in mRNA content related to tomato fruit ripening. Phytochemistry, 17, 311-312.

Rhodes, M.J.C. (1980) The maturation and ripening of fruits. In Senescence in Plants. ed. K.V. Thimann, pp. 157-205, Boca Raton, CRC Press.

Rick, C.M. (1980) Tomato linkage survey. Tomato Genet. Coop. Report 30, 2-17.

Salveit, M.E., Bradford, K.J. & Dilley, D.R. (1978) Silver ion inhibits ethylene synthesis and action in ripening fruits. J. Amer. Soc. Hortic. Sci., 103, 472-475.

Sawamura, M., Knegt, E. & Bruinsma, J. (1978) Levels of endogenous ethylene, carbon dioxide, and soluble pectin, and activities of pectin methylesterase and polygalacturonase in ripening tomato fruits. Plant Cell Physiol. 19, 1061-1069.

Sisler, E.C. (1984) Distribution and properties of ethylene-binding component from plant tissue. In Ethylene-Biochemical, Physiological and Applied Aspects. ed. Y. Fuchs, pp. 45-54. The Hague: Martinus Nijhoff/Dr. W. Junk.

Slater, A., Maunders, M.J., Edwards, K., Schuch, W. & Grierson, D. (1985) Isolation and characterisation of cDNA clones for tomato polygalacturonase and other ripening-related proteins. Plant Molec. Biol. 5, 137-147.

Smith, A.R. & Hall, M.A. (1985) Ethylene binding. In Ethylene and Plant Development. eds. J.A. Roberts & G.A. Tucker, pp. 101-116. London: Butterworths.

Speirs, J., Brady, C.J., Grierson, D. & Lee, E. (1984) Changes in ribosome organisation and messenger RNA abundance in ripening tomato fruits. Aust. J. Plant. Physiol. 11, 225-233.

Themmen, A.P.N., Tucker, G.A. & Grierson, D. (1982) Degradation of isolated tomato cell walls by purified polygalacturonase in vitro. Plant Physiol. 69, 122-124.

Tigchelaar, E.C., McGlasson, W.B. & Buescher, R.W. (1978) Genetic regulation of tomato fruit ripening. HortScience 13, 508-513.

Tucker, G.A. & Grierson, D. (1982) Synthesis of polygalacturonase during tomato fruit ripening. Planta, 155, 64-67.

Tucker, G.A. & Grierson, D. (1987) Fruit ripening. In: Biochemistry of Plants: A Comprehensive Treatise. Volume 9, ed. D.D. Davies, Academic Press. In press.

Tucker, G.A. & Roberts, J.A. (1984) Flower abscission in mutant tomato plants. Planta, 160, 164-167.

Tucker, G.A., Robertson, N.G. & Grierson, D. (1980) Changes in polygalacturonase isoenzymes during the 'ripening' of normal and mutant tomato fruit. Eur. J. Biochem. 112, 119-124.

Veen, H. (1983) Silver thiosulphate: an experimental tool in plant science. Scientia Horticulturae, 20, 211-224.

Yang, S.F. & Hoffman, N.E. (1984) Ethylene biosynthesis and its regulation in higher plants. Ann. Rev. Plant Physiol. 35, 155-189.

EFFECTS OF MUTANT GENES ON THE SYNTHESIS OF STORAGE COMPONENTS IN DEVELOPING BARLEY ENDOSPERMS

P.R.Shewry, M.S. Williamson and M. Kreis
Biochemistry Department, Rothamsted Experimental Station, Harpenden, Herts. AL5 2JQ, U.K.

INTRODUCTION

The major storage components of cereal endosperms, starch and alcohol-soluble storage proteins (prolamins), together account for over 50% of the mature tissue. Any changes in their amounts or compositions can have important consequences for cereal production and utilization. Starch is a major component of yield, and any decrease in amount inevitably results in a yield deficit. Changes in the composition of starch also affect its suitability as a substrate for industrial processing. Although prolamins are quantitatively less important (accounting for only about 5-10% of total dry matter), they have important effects on the nutritional quality and technological properties of the grain. The nutritional quality of all the major cereals, except oats and rice, as feed for humans and monogastric livestock such as pigs and chickens is limited by the low levels of certain essential amino acids in the prolamin fraction (see Bright and Shewry, 1983). In barley the first and second limiting amino acids are lysine and threonine respectively, in maize lysine and tryptophan. In wheat the prolamins also form gluten, the visco-elastic mass that is largely responsible for the ability of wheat flour to be baked into leavened bread (see Miflin et al., 1983).

Mutant genes affecting the synthesis of starch and prolamins have been described in the diploid cereals maize, barley, sorghum and rice (Nelson, 1980). In maize an extensive series of genes affecting starch synthesis has been identified by screening gene banks for mutant endosperm phenotype. These have been given names which describe either the appearance of the endosperm or the biochemical effect: sugary, shrunken, waxy, dull, brittle, opaque, floury, amylose-extender and others. Some of them, e.g. opaque-2, also affect prolamin synthesis resulting in a "high lysine" phenotype.

Mutant genes affecting storage components are also known in

sorghum. These include *waxy* and *sugary* genes affecting only starch synthesis, and two "high lysine" genes which also affect prolamins. The latter were again selected on mutant endosperm phenotype, one being induced and one spontaneous. Although *waxy* mutants have also been described in rice, no genes are known to affect starch *and* prolamin synthesis.

A number of "high lysine" genes have also been identified in barley by direct screening for an increased proportion of basic amino acids in the grain. With one exception they have been induced using a range of chemical and physical mutagens. All result in decreased total amounts of starch and prolamins and some also affect their compositions. They may also affect the levels of other endosperm components including specific proteins and lipids. Other mutant genes affect only starch synthesis, resulting in *waxy* and *amylose-extender* phenotypes.

The present discussion will be restricted to high lysine genes of barley, and particularly to studies of three contrasting genes which have been carried out in our own laboratory. Further information on these and other genes can be found in a series of review articles (Miflin & Shewry, 1979; Nelson, 1980; Tallberg, 1982; Bright & Shewry, 1983).

THE STRUCTURE AND GENETICS OF HORDEIN

Before discussing the effects of high lysine genes on prolamin synthesis it is necessary to describe briefly the properties and genetics of barley prolamins (hordeins).

Hordein is a complex mixture of polypeptides which vary in number and electrophoretic mobility between genotypes (Shewry *et al.*, 1978). The total number of polypeptides resolved by two-dimensional (2-D) iso-electric focusing (IEF)/sodium dodecylsulphate polyacrylamide gel electro-phoresis (SDS-PAGE) is usually in the range 20-40. Despite this polymorphism the polypeptides can be classified into three major groups. Related groups of prolamins can also be recognised in other members of the Triticeae including the cultivated cereals wheat and rye. These three groups have been called the high molecular weight (HMW) prolamins, sulphur-rich (S-rich) pro-lamins and S-poor prolamins (Shewry *et al.*, 1984). In all cases each group of prolamins is controlled by co-dominant alleles at one or more complex multigenic loci. Although the major loci encoding the groups have now been mapped in barley and wheat (see Shewry *et al.*, 1984), minor loci undoubtedly exist in at least some genotypes and remain to be identified.

The characteristics of the major groups of hordein polypeptides are summarised in Figure 1. Whereas the S-rich (B hordein) and S-poor (C hordein) groups are separated into a number of bands by SDS-PAGE and show great genotypic variation, the HMW D hordein gives only a single band in most genotypes. B, C and D hordeins account for about 80-90%, 10-20% and 2-4% of the total fraction respectively (Shewry _et al._, 1983). Whereas D hordein is only present in aggregates stabilised by disulphide bonds, C hordein contains no cysteine and is monomeric. B hordein is at least partially aggregative (see below), and may form mixed aggregates with D hordein.

Genetic studies show that B, C and D hordeins are encoded by structural genes at single loci designated _Hor 2_, _Hor 1_ and _Hor 3_ respectively (Shewry _et al._, 1984). All three loci are located on chromosome 5, with _Hor 1_ and _Hor 2_ on the short arm and _Hor 3_ on the long arm. "Southern

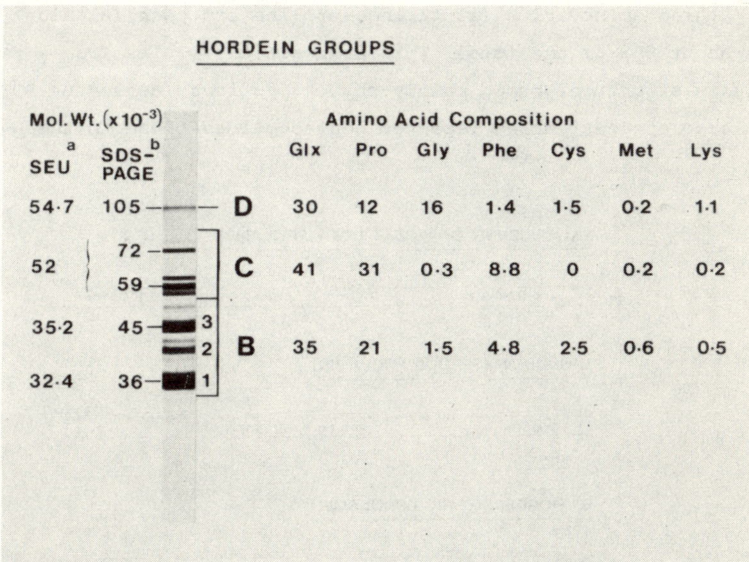

Figure 1. Characteristics of the groups of hordein polypeptides separated by SDS-PAGE.

a. mol. wt. determined by sedimentation equilibrium ultracentrifugation.

b. mol. wt. determined by SDS-PAGE.

Data are from Shewry _et al._ (1980b), Field _et al._ (1982) and Kreis _et al._ (1984).

Blot" analyses show that whereas the Hor 1 and Hor 2 loci are highly polymorphic with copy numbers in the range 20-30 genes per haploid genome (Shewry et al., 1985a; Bunce et al., 1986), the Hor 3 locus shows little polymorphism and has a low copy number (Forde, 1985; Bunce et al., 1986).

D hordein differs from the other groups in its high content of glycine (about 14 mol %) and relatively low content of proline (11 mol %), Although no amino acid sequences of D hordein have been reported, the cross-hybridization of cDNAs for D hordein and the HMW prolamins of wheat (Forde, J. et al., 1983) suggest that the structure is similar to that of the wheat protein depicted in Figure 2 (Forde, J., et al., 1985). This has non-repetitive N- and C-terminal domains of about 100 and 40 residues respectively, which contain all the cysteines and most of the charged residues. These are separated by a repetitive domain, about 440 residues in total consisting of tandem and interspersed nonapeptides (consensus GYYPTSLQQ) and hexapeptides (consensus PGQGQQ).

C hordeins have the most unusual amino acid composition, with only three amino acids (glutamine, proline and phenylalanine) accounting for about 80% of the total. This composition results from a repetitive primary structure, based mainly on octapeptides (consensus PQQPFPQQ). There are also at least three repeated pentapeptides close to the N-terminus, and

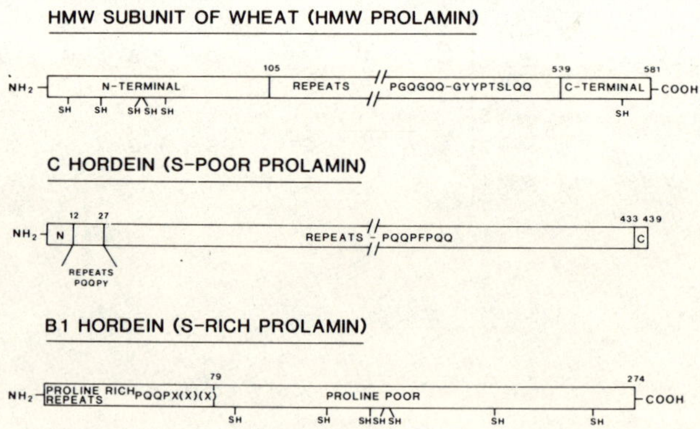

Figure 2. Diagrammatic representations of the structural organisation of an HMW prolamin of wheat and C and B1 hordeins of barley.

The figure is based on the data of Forde, J. et al. (1985), Forde, B.G. et al. (1985a,b) and Tatham et al. (1985).

short non-repetitive N- and C-terminal domains of 12 and 6 residues respectively. A summary of the structural organisation of C hordein based on partial sequences from whole protein, peptides and cDNAs (Forde, B.G. et al., 1985b, Tatham et al., 1985) is shown in Figure 2.

B hordein constitutes the major, but not the only, group of S-rich prolamins in barley. It is highly polymorphic, and the polypeptides were initially classified into three classes on the basis of cyanogen bromide (CNBr) peptide maps (Faulks et al., 1981). More recently Kreis et al. (1983a) showed that these classes corresponded to two sub-families of cDNAs, and suggested that these represented the products of two sub-families of genes at the Hor 2 locus. These sub-families encode the B1 and B2 + B3 hordein bands of cv. Sundance (Figure 1), and correspond to the CNBr classes I + II and III respectively.

Although all the major B hordein polypeptides appear to be encoded by the two sub-families of genes at the Hor 2 locus described by Kreis et al. (1983a), minor sub-families of genes probably also exist at the locus. Also, in some genotypes additional loci have been reported to encode minor B hordein-like polypeptides (Netsvetaev & Sozinov, 1984; Shewry et al., 1986).

The structure of a B1 hordein polypeptide of cv. Sundance is summarized in Figure 2. This was the first complete B hordein sequence determined (Forde, B.G. et al., 1985a). Comparison with the more recently reported genomic sequence of Brandt et al. (1985) and with partial cDNA sequences (Forde, B.G. et al., 1985b; Rasmussen et al., 1983) indicate that the structure is typical of the B1 hordein sub-family, and of B hordeins in general. There are two structural domains. A proline-rich domain of about 80 residues extends to the N-terminus, and contains a mixture of pentapeptides, hexapeptides and heptapeptides, all based on a PQQPX motif. The C-terminal proline-poor domain of about 190 residues is non-repetitive, rich in cysteine but with a low content of proline. The B3 type hordeins of cv. Sundance have a similar structure to the B1 type, but show some differences in the C-terminal domain (notably in a region close to the C-terminus) and more extensive variation at the N-terminal end of the proline-rich repetitive domain (Forde, B.G. et al., 1985b).

Detailed comparisons of the sequences of the S-poor, S-rich and HMW prolamins of barley, wheat and rye show that they are all structurally related. The repeated peptides in the S-rich and S-poor prolamins are based on the same PQQPX motif, while the S-rich and HMW prolamins have homologous sequences present in their non-repetitive regions. These relationships have been discussed in detail by Kreis et al. (1985a,b), and suggest that the genes encoding the three groups of prolamins have evolved from a single ancestral gene. If this is so, it is reasonable to suppose that they may have retained some common sequences regulating developmental and tissue-specific expression. This is relevant when considering the effects of "regulatory" high lysine genes on their expression (see below).

THE lys GENE OF HIPROLY, A SPONTANEOUS "REGULATORY" MUTATION

This was the first high lysine gene of barley to be discovered, and is still one of the most intriguing in its range of biochemical effects. It was found by Munck and his co-workers at Svalof, Sweden, who were stimulated by the discovery of mutant high lysine genes of maize to screen 2000 lines from the world barley collection. A single line, CI3947, was found to have an unusually high dye binding capacity (a measurement of total basic amino acids) relative to grain nitrogen (Munck et al., 1970). The line was called Hiproly to reflect its elevated levels of protein and lysine, and contains about 30% more lysine but has only about 30% of the yield of normal cultivars (Munck et al., 1970). The high lysine character was found to be controlled by a single recessive gene, designated lys, which maps to chromosome 7 (Karlsson, 1976). Although the lys gene has since been incorporated into lines with yields approaching those of normal cultivars (Persson & Karlsson, 1977), they have not proved to be sufficiently competitive for commercial release.

The reduced yield results from a smaller seed size (Persson & Karlsson, 1977), which may result from decreases in the amounts of starch and prolamins. There have, however, been no detailed studies of the amounts of starch in Hiproly and derived lines, or its composition. The amounts of hordein in Hiproly and high yielding lys lines are only decreased by about 10-20%, and there are no apparent effects on hordein polypeptide composition (Miflin and Shewry, 1979). This decrease is not sufficient to account for the increased lysine content of the grain, and the answer lies in the presence of increased amounts of specific lysine-rich salt-soluble proteins.

Hejgaard and Boisen (1980) showed that four such proteins accounted for about 7% of the grain lysine in normal cultivars, but 17% in Hiproly. Together they accounted for about half of the increased lysine content of Hiproly. The four proteins are β-amylase (5.0 g % lysine), Protein Z (7.1 g % lysine), Chymotrypsin Inhibitor CI-1 (9.5 g % lysine) and

Chymotrypsin Inhibitor CI-2 (11.5g% lysine). The amounts present in normal lines and Hiproly-derived lines with the lys gene are given in Table 1.

The mechanism for these specific increases is not known, but information may come from current studies of the molecular biology of these proteins. We have isolated cDNA clones for the Chymotrypsin Inhibitors CI-1 and CI-2 from libraries prepared from membrane bound poly $(A)^+$ RNA fractions isolated from developing endosperms of Hiproly (Williamson et al., 1986). "Southern blots" show several EcoRI and BamHI restriction fragments of barley genomic DNA hybridizing to the CI-1 and CI-2 cDNA inserts. About 4 to 6 copies of CI-2 and CI-1 genes per haploid genome were estimated for Hiproly and other cultivars. Northern hybridization experiments show that the abundance of CI-2 mRNAs is proportional to the accumulation of CI-2 in normal and high-lysine barley.

THE RISØ SERIES OF INDUCED "REGULATORY" GENES

The discovery of only one naturally occurring high-lysine line of barley was disappointing when compared with the success achieved with maize, and led a number of research groups to use mutagenesis. The most well-known and successful programme was carried out by Dr. H. Doll and co-workers at the Risø National Laboratory, Roskilde, Denmark (see Doll, 1983). They initially treated seed of two cultivars (Carlsberg II and Bomi) with a range of chemical and physical mutagens (Table 2) and screened a total of about 15,000 lines for the high lysine character using combined analyses for dye binding capacity (DBC) and % N. A total of 10 different mutants were isolated,

Table 1. β-Amylase, Protein Z and Chymotrypsin Inhibitors in Normal and High Lysine Barleys

	Protein mg/g				
	Total Protein	β-amylase	Protein Z	CI-1	CI-2
Mona	110	0.98	2.2	0.24	0.08
Hiproly x Mona^5	127	4.3	4.3	1.8	0.67
Bomi	98	0.85	2.5	0.23	0.06
Hiproly x Bomi^2	113	3.2	4.5	1.7	0.38

Data of Hejgaard and Boisen (1980).

of which only one (Risø 56) was later found to be a structural mutation at one of the Hor loci. This is discussed in detail in a later section. The nine remaining lines all appeared to have mutations at "regulatory" genes, and include the well-known Risø mutant 1508.

Later attempts to induce high lysine genes at Risø have been less successful. Analysis of 2000 lines by the DBC and % N method and 3000 lines by SDS-PAGE of hordein yielded only two mutants (Risø 18 and 19), both of which proved to be allelic to Risø 1508. A further screen of about 50,000 seeds for low hordein content also yielded only mutants resembling Risø 1508 (see Doll, 1983, for a detailed discussion of the selection of the Risø mutants).

The characteristics of the Risø "regulatory" mutants are summarized in Table 2. The increase in lysine varies from less than 10% to about 35% but this is always associated with a yield deficit. With the exception of Risø 8, which is semi-dominant, all the mutant genes are recessive. Eight of the mutant genes have been mapped. Mutants 1508, 18 and 19 are alleles at the lys 3 locus, which is present on chromosome 7 but not linked to the lys locus of Hiproly (Karlssen, 1976). Three other mutant genes (29, 86, 13) are alleles at a second locus (lys 5) on chromosome 6, while Risø 527 maps to a further locus (lys 6) on the same chromosome. Finally, Risø 8 maps to a locus (lys 4) on chromosome 5, which is not closely linked to the hordein structural loci (see Jensen, 1984).

Table 2

Line	Parent Cultivar	Mutagen	Gene	Chromosome	Yield (% of parent cv.)	Lysine content (% protein)	1000 seed wt (% of parent cv.)
Riso 1508	Bomi	EI	lys 3a	7	77	136	88
18	..	NaN_3	lys 3b	7	-	-	-
19	..	NaN_3	lys 3c	7	-	-	-
8	..	EMS	lys 4d	5	79	109	84
13	..	EMS	lys 5f	6	72	123	79
29	Carlsberg II	EMS	lys 5g	6	85	109	88
86	..	EMS	lys 5h	6	79	105	79
527	Bomi	γ-rays	lys 6i	6	69	112	80
7	..	n_{th}	-	-	89	111	100
9	..	EMS	-	-	54	103	65
16	..	n_f	-	-	84	108	86
17	..	n_f	-	-	57	110	85

EI, ethylinimine; EMS, ethyl methane sulfonate; n_{th}, thermal neutrons; n_f, fast neutrons; NaN_3, sodium azide.
From Doll et al., 1974; Doll, 1976, 1980, 1983; Jensen, 1979a,b; Karlsson, 1977.

The pleiotropic effects of the mutant genes of Risø 29 and Risø 1508 on the amounts of hordein, lysine and starch were elegantly demonstrated by Doll and co-workers (Køie et al., 1976; Doll & Køie, 1978). They produced homozygous chromosome-doubled monoploid lines from crosses between the mutants and normal cultivars using the Hordeum bulbosum technique. No recombination between the high lysine, low hordein and low starch characters was observed in a total of 135 lines.

BIOCHEMICAL ANALYSIS OF THE MUTANT LINES

Risø 1508

Because of its greatly increased lysine content and only modest yield deficit, Risø 1508 was introduced into a number of breeding programmes shortly after its release in 1973. The results were disappointing to say the least. It proved impossible to break the linkage between the lysine increase and yield decrease, and most of the breeding programmes were soon discontinued. An exception was at the Carlsberg Institute, Denmark, where they recently reported the development of modified Risø 1508 types with yields comparable with those of some currently grown cultivars (Munck, 1985).

The lys 3a gene affects a wide range of aspects of endosperm development and composition, including the synthesis of proteins, carbohydrates and lipids. A number of comparisons of the gross protein compositions of Bomi and Risø 1508 have been reported, and all show a greatly decreased content of hordein, and increases in other protein fractions and in free amino acids. For example, the results of Shewry et al. (1980a) showed that only 15.6% of the total grain N was present in the hordein fraction compared with about 45% in Bomi. Salt-soluble proteins and glutelins (insoluble proteins) together accounted for about 52% compared with 33%, and non-protein nitrogen (including free amino acids) for 19.9% compared with 11.6%.

Analysis of developing grain showed that the increases in non-protein nitrogen and salt-soluble proteins were associated with a prolonged duration of accumulation, while the reduced hordein level resulted from a lower rate of accumulation between two and four weeks after anthesis, after which accumulation was arrested in Risø 1508 but continued till 5 weeks after anthesis in Bomi (Figure 3a) (Shewry et al., 1979).

Changes have also been reported in the levels of specific metabolic proteins (including enzymes), and these may reflect changes in endosperm metabolism. For example the levels of RNA-ase (Brandt and Ingversen,

1976; Miflin and Shewry, 1979) and malate dehydrogenase (Miflin and Shewry, 1979) are increased in the developing endosperm, and of α-amylase in the malted grain (Munck, 1976; Allison, 1978). In contrast the level of β-amylase, a protein which may have a storage function in the mature grain, is decreased (Allison, 1978; Giese and Hejgaard, 1984) (Figure 3b). A further group of proteins which is affected and which may or may not have a storage function is the chloroform/methanol-soluble proteins (CM proteins). The amount of CMe, a trypsin inhibitor, is decreased, and the amounts of CMa and CMb increased (Lazaro et al., 1985).

The decreased hordein content is associated with changes in the morphology of the protein bodies in the developing starchy endosperm (Miflin and Shewry, 1979). Transmission electron microscopy shows that these are biphasic in Bomi, with one phase being densely stained and granular. Whereas the granular phase is also present in Risø 1508 and may be greater in extent, the rest of the protein body has a spongy appearance.

Although a number of studies of the effects of the lys 3a gene on the hordein fraction have been reported, the following account is based on the most recent and detailed study by Kreis et al. (1984). As well as the total amount of hordein being reduced to about 30% of that in Bomi, there

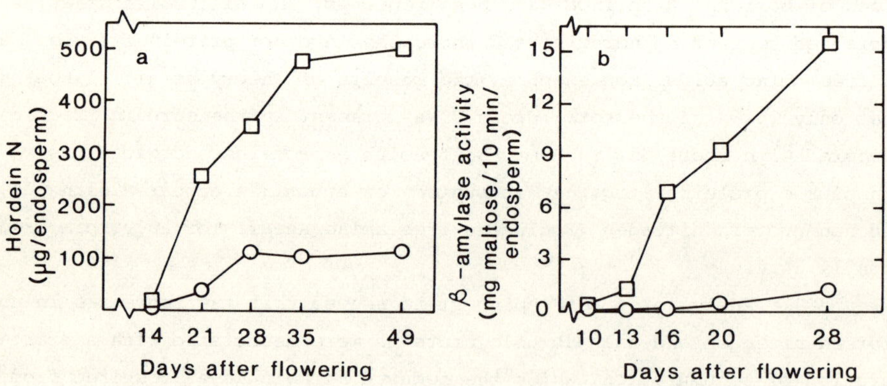

Figure 3. The total amount of hordein (a) and the activity of β-amylase (b) in developing endosperms of Bomi (□-□) and Risø (O-O).

The data are from Shewry et al. (1979)(a) and Kreis (1979)(b). Because the plants were grown under different conditions, the stages of development in the two experiments are not directly comparable.

are also drastic changes in the polypeptide composition. The C hordein polypeptides present in Bomi are absent, but an additional minor band is present in this region of the SDS-PAGE gel. Assuming that the latter band is C hordein-related, the total amount is reduced from about 73 to 5 μg N/seed. The total amount of B hordein is reduced from about 380 to 75 μg N/seed and comparison of the effects of the gene on the band patterns of B hordeins in cultivars with different structural alleles at the Hor 2 locus indicates that class I polypeptides are affected more than class III polypeptides. In contrast, the amount of D hordein is increased from about 2 to 8 μg N/seed, and its mobility on SDS-PAGE is unaffected.

An attempt in the same study to explain the effects of the lys 3a gene at the molecular level was unsuccessful. Populations of mRNAs in developing endosperms were studied by colony hybridization, hybrid-selected translation, "Northern hybridization" and "hybrid dot" analysis using cloned cDNAs related to B, C and D hordeins. The populations of mRNAs for the different groups of hordeins in Risø 1508 and Bomi were generally consistent with the proportions of the proteins in the mature grain. Finally, "Southern Blot" analysis showed no differences in the number or organisation of B hordein-related genes in the two lines. It was concluded that although the lys 3a gene has been shown to have differential and specific effects on the expression of different Hor genes, there is insufficient evidence to explain the exact molecular mechanism by which this is achieved.

Whereas the accumulation of hordein is arrested at an earlier stage of endosperm development in Risø 1508 than in Bomi, the accumulation of starch extends over the same time period in both lines (Kreis, 1978). However, a lower rate of accumulation in Risø 1508 results in about 20% less starch in the mature grain. Risø 1508 also contains about twice the level of free sugars, which probably results from their decreased utilization for starch synthesis.

Kreis (1979, 1980) showed that Risø 1508 had a lower level of starch synthetase throughout endosperm development. He also purified the primer independent starch synthetase (glucosyltransferase II) from Risø 1508 and Bomi, and showed that the primed reaction was similar in both lines, and was increased about 3-fold on the addition of 0.4M sodium citrate (Figure 4). The unprimed reaction was also increased by the addition of sodium citrate, but was higher in Bomi at all concentrations. The significance of these observations for the different rates of starch accumulation in the two lines is not known. He also showed that the proportions of amylose (about 28%) and amylopectin (about 72%) were not affected by the mutation (Kreis, 1979; Kreis and Doll, 1980). Differences in starch accumulation are also apparent at the subcellular level; Risø 1508 having a decreased number of small starch granules (Burgess et al., 1982).

Munck (1976) and Tallberg (1977) reported increased total lipid in whole grains of Risø 1508, and assumed that this was due to the larger embryo. However, Shewry et al. (1979) showed that isolated endosperms of Risø 1508 contained about twice as much total oil as those of Bomi (3.6% dry wt. compared with 1.8%), and that this resulted from increases in neutral lipids (triglycerides) and polar lipids. These observations were confirmed by Bhatty and co-workers (Bhatty and Rosnagel, 1979, 1980; Batty, 1982).

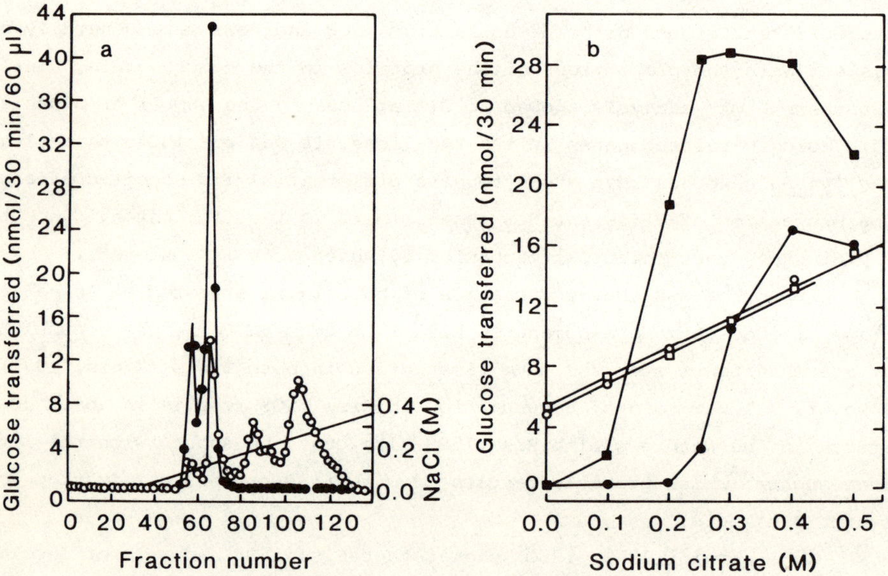

Figure 4.

a. Chromatography of soluble starch synthetase from endosperms of Risø 1508 on DEAE cellulose. Aliquots were assays with amylopectin (primed) (O - O) and without amylopectin (unprimed) (● - ●) as described by Kreis (1979).
b. The effect of sodium citrate on unprimed (□ ●) and primed (□ O) glucosyltransferase II activity from Bomi (□ □) and Risø 1508 (O ●). Assay conditions are described by Kreis (1979).

Data are taken from Kreis (1979).

OTHER MUTANTS

All the other Risø "regulatory" mutants have a decreased proportion of hordein (Table 3), with compensatory increases in other proteins and often also in non-protein nitrogen (including free amino acids). Table 3 also shows that less starch and more soluble sugars are present in all the mutants except Risø 7. In fact, Doll (1983) recently reported that high lysine doubled monoploid lines from the cross Risø 7 x Sultan had a reduction of 7% in seed weight (presumably due to decreased starch), but of 39% in hordein.

Mutants 7, 527, 13, 29, and 86 also have altered proportions of B hordein polypeptides (Køie and Doll, 1979; Doll, 1983; Bright and Shewry, 1983; Shewry et al., 1985c) (see Figure 5). Mutants 7, 527 and 13 are in Bomi, and have different effects on class I (B1) and Class III (B3) polypeptides. Mutants 7 and 527 have increased proportions of B1 hordein, and mutant 13 of B3 hordein. Mutants 29 and 86 are in Carlsberg II, and it is not possible to interpret the changes in B hordein pattern in terms of effects on specific sub-families of polypeptides and their genes. Risø 527 also has a specific decrease in D hordein.

Although the starch compositions of all mutants have not been determined, only Risø 16 has so far been shown to have an altered ratio of amylose to amylopectin (Torp, 1979) (Table 3).

Table 3 The effects of the Risø "regulatory" high lysine genes on the carbohydrate and nitrogenous fractions of the mature grain

	Starch (% dry wt.)	Amylose (% starch)	Soluble sugars (% dry wt.)	N (% dry wt.)	Hordein (% of N)	Non-protein N (% of N)	Non-prolamin proteins (% of N)	Free Amino Acids (mg/g)	Lysine (mg/g)
Carlsberg II	53	28.3	1.9	1.37	40	5	55	0.9	3.5
29	45	28.4	2.6	1.54	37	5	58	1.0	4.3
86	45		2.6	1.60	38	5	57	1.0	4.4
Bomi	52	28.9	2.0	1.36	41	5	55	0.9	3.5
527	41	29.3	5.1	1.56	31	8	62	1.4	4.2
1508	44	28.4	3.6	1.49	15	8	77	1.7	5.1
7	51		1.9	1.39	33	6	61	0.9	3.8
8	45		3.2	1.53	29	7	64	1.4	4.3
13	35		6.7	1.69	27	10	62	2.9	5.0
Bomi	50		2.0	1.45	44	6	50	0.9	3.5
16	37	16.7	5.7	1.75	38	5	58	1.3	4.6
17	33		8.3	1.85	37	6	57	1.5	5.1

Data from Køie and Doll (1979) except amylose contents, which are from Kreis and Doll (1980), and Torp (1979).

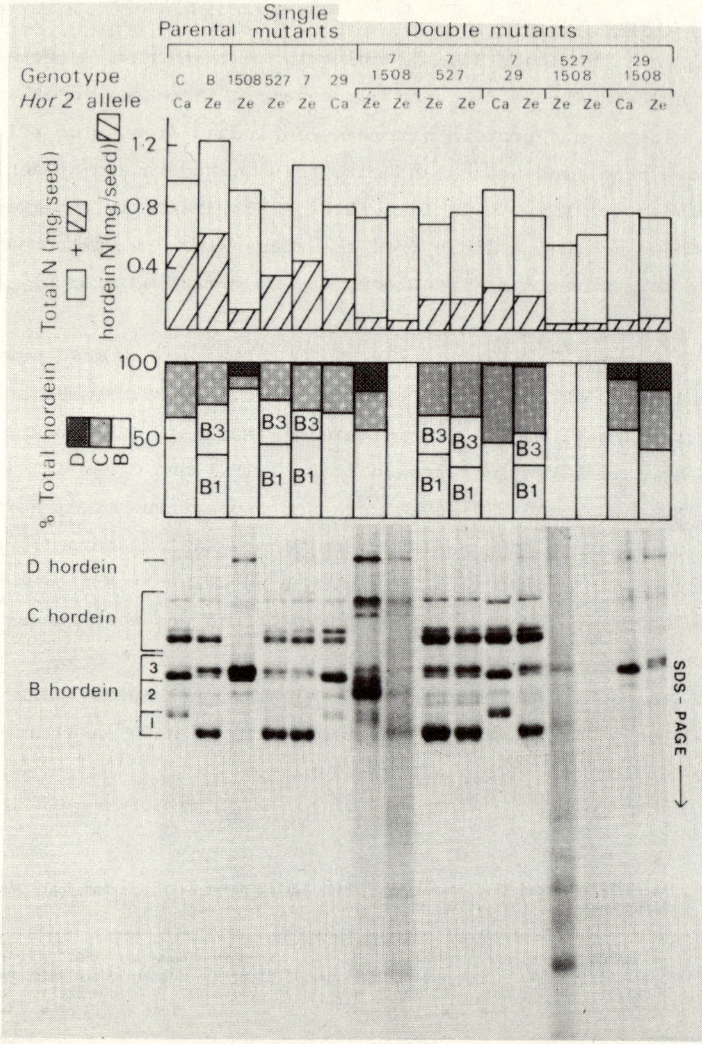

Figure 5. The effects of mutant high lysine genes on the hordein fraction of barley grain.

The lower part of the figure shows SDS-PAGE separations of the fractions. B1, B2 and B3 are bands present in Bomi and other lines with the Hor 2 Ze allele. B1 and B2 + B3 correspond to the products of two sub-families of mRNAs and presumably genes at the Hor 2 locus (Kreis et al., 1983b). Quantitative determinations of the proportions of the groups of polypeptides are shown in the central part of the figure. In this B3 includes the B2 + B3 bands. The Hor 2 alleles in the Risø 29/Risø 1508 lines were determined by back-crossing.

Taken from Shewry et al. (1985c).

Double mutant combinations

Kreis and Doll (1980) compared the accumulation of dry matter, starch, soluble sugars, total grain nitrogen and hordein in Bomi, Risø 1508 and double mutants combining Risø 1508 with Risø 29 and Risø 527 (Figure 6). The Risø 29/1508 line and Risø 1508 itself had similar rates of accumulation of dry matter, total N and hordein. However Risø 29/1508 had less starch and more soluble sugars than Risø 1508, and also a higher content of water throughout endosperm development. In contrast the Risø 527 1508 combination had drastically reduced rates of accumulation of total dry matter (about 30 compared to 45 mg/kernal at 40 days after flowering), starch, total N and hordein compared with Risø 1508. Although the content of soluble sugars was increased considerably during grain filling, this was only slightly greater at maturity.

More recently we have compared the hordein fractions of a number of double mutant combinations including those discussed above (Shewry et al., 1985c) (Figure 5). Although the results are not easy to interpret, some limited conclusions can be drawn.

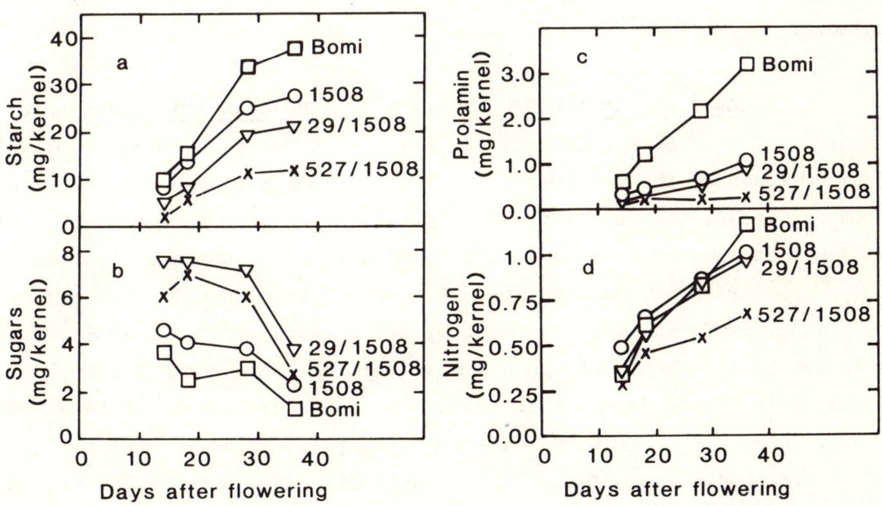

Figure 6. The accumulation of starch (a), soluble sugars (b), prolamins (c) and nitrogen (d) in developing endosperms of Bomi, Risø 1508 and the double mutants 29/1508 and 527/1508. Taken from Kreis and Doll (1980).

1. Risø 527 causes a specific decrease in D hordein, when alone or in combination.
2. Although Risø 7 and Risø 527 both have an increased proportion of B1 hordein, this is not observed in the Risø 7/527 double mutant lines. An increased proportion of B1 hordein is present in the Risø 7/Risø 29 line with the Hor 2Ze allele.
3. All double mutants with Risø 1508 have less hordein than Risø 1508 itself. The Risø 29/1508 lines have similar hordein patterns to Risø 1508, allowing for minor differences due to the presence of the Hor2Ca allele in one line. The Risø 7/1508 lines have different patterns to each other and to Risø 1508. The former is surprising as these should be isogenic. Although Risø 527 has limited effects on hordein when alone or with Risø 7, the Risø 527/1508 combination has almost no detectable hordeins. This observation is in agreement with the results of Kreis and Doll (1980).

The mutant genes of Risø 29 (lys 5g) and Risø 527 (lys 6i) clearly differ in their interactions with the lys 3a gene. Whereas the effects of the lys 5g and lys 3a are approximately additive, lys 6i and lys 3a appear to be synergistic in their effects on the accumulation of starch and hordein. The molecular basis of these interactions is, of course, unknown.

RISØ 56, AN INDUCED MUTATION AT A HORDEIN STRUCTURAL LOCUS

Risø 56 was selected by Doll and co-workers (Doll et al., 1974) from 10 lines of the cultivar Carlsberg II which had been treated with 20 krad of γ-rays. The mutation was initially associated with a translocation between chromosomes 2 and 4 (Doll, 1980) which was subsequently eliminated.

Risø 56 contains about 17% more lysine than Carlsberg II (Doll, 1976), but produces fewer and smaller seeds (Køie and Doll, 1979). The lower seed weight (by about 10%) is at least partially accounted for by a decrease in starch from 53% to 49% of the grain dry weight (Køie and Doll, 1979). There is, however, no effect on the ratio of amylose to amylopectin (M. Kreis, unpublished results). In contrast to a number of the regulatory high lysine genes, Risø 56 has a normal level of non-protein nitrogen (Køie and Doll, 1979; Miflin and Shewry, 1979) and only a small increase in soluble sugars (2.2 compared to 1.9% dry wt.) (Køie and Doll, 1979).

The main effect of the mutation is on the hordein fraction, which is decreased by about 20-30% (Køie and Doll, 1979; Miflin and Shewry, 1979; Kreis et al., 1983b). There are also increases, which are probably compensatory, in total salt-soluble proteins (Køie and Doll, 1979; Miflin and Shewry, 1979) and in specific components of this fraction such as protein Z, chymotrypsin inhibitors (Giese and Andersen, 1984), β-amylase (diastase) (Giese and Andersen, 1984; Allison, 1978) and the chloroform/ methanol-soluble proteins CMa and CMb (Lazaro et al., 1985).

Electrophoretic analysis of the hordein fraction showed that the major effect of the mutation is on B hordein, which is reduced in amount and has an altered polypeptide composition (Køie and Doll, 1979; Miflin and Shewry, 1979). Doll (1980) subsequently crossed Risø 56 with normal cultivars and showed that the mutation was located at or close to the Hor 2 (B hordein) structural locus. Although Doll (1980) designated the mutant locus as a recessive form (Hor 2ca) of that present in the parental cultivar (Hor 2Ca), it is clear from his work that it is co-dominant, but its products are not observed in heterozygotes due to the presence of large amounts of polypeptides specified by the wild type allele. It is also co-dominant with respect to the amount of hordein synthesized.

We have recently made a detailed study of the nature of the mutation and its effects in Risø 56 (Kreis et al., 1983b; Forde, B.G. et al., 1985b; Shewry et al., 1985b), and the following description is taken from this.

Quantitative analyses of the hordein fractions from Risø 56 and Carlsberg II showed that the B hordein fraction was reduced from 324 to 80 μg N/seed, and the amount of C hordein increased from 81 to 192 μg N/seed. The amount of D hordein was constant at about 6-7 μg N/seed. Analysis of the hordein fraction showed that all the major B hordein polypeptides were absent (Figure 7), but that several minor polypeptides which gave three bands on SDS-PAGE remained. Amino acid analysis of the purified B hordein bands from Risø 56 indicated that they were related to the major B hordein polypeptides present in Carlsberg II, although the lower band contained less proline (16.5 compared to 22.8 mol %). They also differ in their N-termini. Whereas the major B hordeins are blocked to Edman degradation with proline-rich repeats extending to the N-terminus, the B hordein-like polypeptides of Risø 56 have short non-repetitive N-terminal regions which are homologous with those of γ-type gliadins of wheat and γ-secalins of rye. This is the only report of γ-type prolamins in barley.

Analysis of RNA fractions indicated that transcripts for the major B hordeins were also absent from Risø 56. Products corresponding in mobility on SDS-PAGE to these proteins were not observed when total cellular

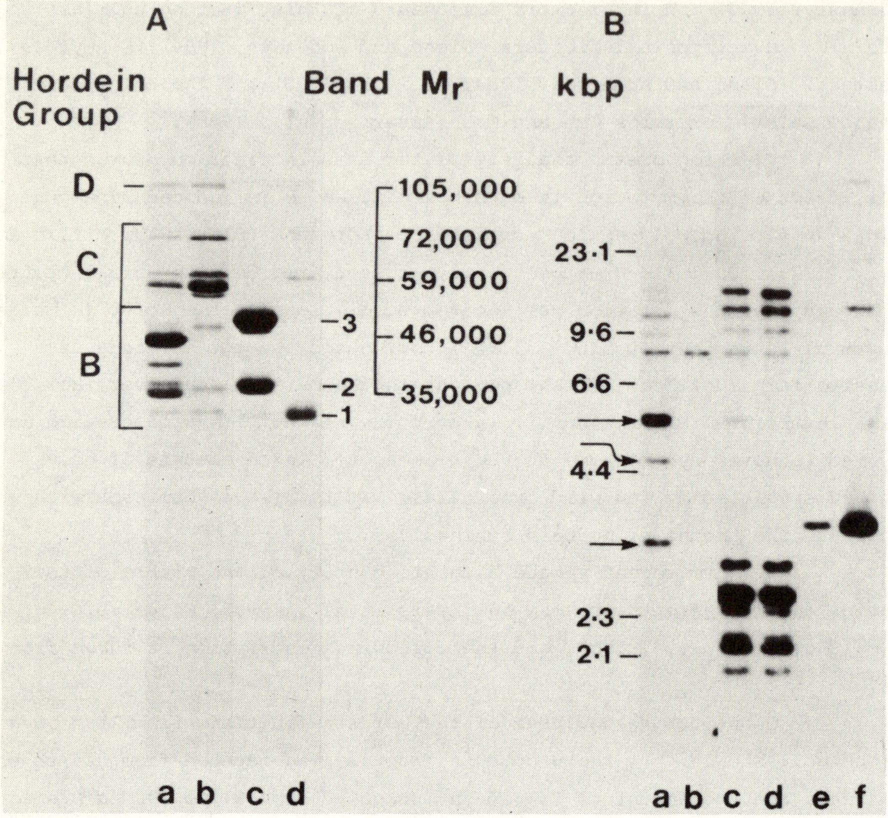

Figure 7.

Part A. SDS-PAGE of total hordein fractions from Carlsberg II(a) (a) and Risøγ(b), and purified B hordein fractions from Risø 56 consisting of bands 2 + 3 (c) and band 1 (d).

Taken from Shewry et al. (1985b).

Part B. "Southern blot" analysis of total genomic DNA from Carlsberg II (a, c) and Risø 56 (b,d). The DNA was digested with Hind III, separated by electrophoresis, transferred to a nylon membrane and hybridized sequentially with ^{32}P cDNAs related to C hordein (c,d) and B3 hordein (a,b). Tracks e and f are reconstructions representing 1 and 10 copies respectively of the C hordein sequence per haploid genome. The arrows indicate restriction fragments which hybridize to both probes, but are absent from the mutant. Full details are given in Forde, B. G. et al. (1985b), from which the figure is taken.

RNA or a poly (A)$^+$ RNA fraction prepared from membrane-bound polysomes were translated in vitro using the wheat germ system. These RNA fractions were also analysed by dot hybridization with cDNAs related to the two major sub-families of B hordein polypeptides, and the poly (A)$^+$ RNA fraction only by colony hybridization and hybrid selection translation with the same cDNAs. The results were again negative. These studies established beyond reasonable doubt that neither normal nor defective RNA transcripts related to the major B hordeins were present in developing endosperms of Risø 56.

The numbers of B hordein genes in Carlsberg II and Risø 56 were compared by "Southern Blot" analysis (Figure 7). Total genomic DNA was digested with HindIII and the fragments separated by electrophoresis and transferred to a membrane. Sequences related to B hordeins were detected by hybridization to ^{32}P-labelled cDNAs for B3 type (Class III) hordein polypeptides, under conditions where they should hybridize to genes for all the major B hordeins. This showed 11 B hordein-related fragments in Carlsberg II, ranging in size from about 2 to 24 kb, and with a total copy number of about 13 genes per haploid genome. A similar examination of the Hor 1 (C hordein) locus showed no difference in the number or organisation of genes in the two lines.

These results indicate that a deletion of genes at the Hor 2 locus has occurred in Risø 56. The total size of the fragments which are absent from Risø 56 is about 85 kb, and this can be taken as a minimum value for the extent of the deletion. The actual extent may be considerably greater due to the presence of other DNA interspersed with the B hordein genes.

It is not known whether the single B hordein related gene in Risø 56 is expressed. Although it could encode a γ-type hordein, this is unlikely in that at least six γ-type hordeins can be resolved by 2-D electrophoresis, and these have two types of N-terminal amino acid sequences indicating that they are encoded by at least two structural genes. Also, we have not observed cross-hybridization between DNAs for B hordeins and those for γ-type prolamins of wheat and rye, and so would not expect the B3 type hordein probe to hybridize to genes for γ-type hordeins.

The presence of a single B hordein-related gene in Risø 56 may indicate that the deletion of the Hor 2 locus is incomplete. However, there is no evidence that this fragment is present at Hor 2. In at least one other case B hordein-like polypeptides have been shown to be encoded by genes not present at Hor 2. Similarly, the chromosomal locations of the genes for the γ-type hordeins are also not known. In wheat the genes for the γ-type gliadins and the ω-gliadins (which are homologous with C hordeins) are located at a single multigenic locus. A similar location of the genes for the γ-type hordeins at the Hor 1 locus would explain why they have not been affected by the deletion in Risø 56.

If, as our studies indicate, the "high lysine" mutation in Risø 56 is a deletion at the Hor 2 locus, it is necessary to explain its effects on the amounts of other grain proteins and starch. In some, if not all, cases the proteins which are increased in Risø 56 are also increased under conditions of high N availability, indicating a storage role. Their increase in Risø 56 can therefore be regarded as a general increase in other storage proteins to compensate for the reduction in B hordeins. The reduced amount of starch is more difficult to explain. Although it is possible that genes encoding proteins important in starch synthesis are interspersed with those for B hordeins at the Hor 2 locus and therefore also affected by the mutations, there is no evidence in support of this. It is more likely that, as in "regulatory" mutants, the maintenance of high levels of synthesis of storage proteins and starch are in some way interdependent, with adverse effects on one almost inevitably affecting the other.

CONCLUSIONS

It is clear from the results discussed above that the high lysine genes vary greatly in their effects on endosperm metabolism. This may indicate that they also vary in their molecular mechanisms, but this remains to be established.

The most striking feature is that effects on hordein synthesis are almost always associated with decreased starch accumulation. The only possible exception to this is Risø mutant 7 and this is consequently being studied as a source of increased lysine in breeding programmes (Tallberg, 1981). The correlation between the levels of starch and hordein synthesis even holds for Risø 56, in which the mutation has been shown to involve the deletion of B hordein genes at the Hor 2 locus.

Doll and Kreis (1979) suggested that prolamin synthesis is essential for normal grain filling, possibly because surplus nitrogen compounds interfere with starch synthesis. Although there is no evidence for this mechanism, it implies that the effects on grain filling are secondary, and that future attempts to elucidate the primary mechanism should be concentrated on the regulation of synthesis of storage proteins.

REFERENCES

Allison, M.J. (1978) Amylase activity and endosperm hardness of high lysine barleys. J. Inst. Brew., 84, 231-232.

Bhatty, R.S. (1982) Distribution of lipids in embryo and bran-endosperm fractions of Risø 1508 and Hiproly barley grains. Cer. Chem., 59, 154-156.

Bhatty, R.S. & Rossnagel, B.G. (1979) Oil content of Risø 1508 barley. Cer. Chem., 56, 586.

Bhatty, R.S. & Rossnagel, B.G. (1980) Lipid and fatty acid composition of Risø 1508 and normal barley. Cer. Chem., 57, 382-386.

Brandt, A. & Ingversen, J. (1976) In vitro synthesis of barley endosperm proteins on wild type and mutant templates. Carlsberg Res. Commun., 41, 312-320.

Brandt, A., Montembault, A., Cameron-Mills, V., & Rasmussen, S.K. (1985) Primary structure of a Bl hordein gene from barley. Carlsberg Res. Commun., 50, 333-345.

Bright, S.W.J. & Shewry, P.R. (1983) Improvement of protein quality in cereals. CRC Critical Reviews in Plant Sciences 1, 49-93.

Bunce, N.A.C., Forde, B.G., Kreis, M. & Shewry, P.R. (1986) DNA restriction fragment length polymorphism at hordein loci: application to identifying and fingerprinting barley cultivars. Seed Sci. Technol. 14, 419-429.

Burgess, S.R., Turner, R.H., Shewry, P.R. & Miflin, B.J. (1982) The structure of normal and high-lysine barley grains. J. Exp. Bot., 33, 1-11.

Doll, H. (1976) Genetic studies of high-lysine barley mutants. In: Barley Genetics III, ed. H. Gaul. pp. 542-546, Munich: Verlag Karl Thiemig.

Doll, H. (1980) A nearly non-functional mutant allele of the storage protein locus Hor 2 in barley. Hereditas, 93, 217-222.

Doll, H. (1983) Barley seed proteins and possibilities for their improvement. In: Seed Proteins: Biochemistry, Genetics, Nutritive Value. ed. W. Gottschalk & H.P. Muller pp. 205-223. The Hague: Martinus Nijhoff.

Doll, H., Køie, B. & Eggum, B.O. (1974) Induced high lysine mutants in barley. Radiation Bot., 14, 73-80.

Doll, H. & Køie, B. (1978) Influence of the high-lysine gene from barley mutant 1508 on grain, carbohydrate and protein yield. In: Seed Protein Improvement by Nuclear Techniques. pp. 107-114. Vienna: IAEA.

Doll, H. & Kreis, M. (1979) Significance of storage proteins for grain filling in barley. In: Crop Physiology and Cereal Breeding. Proceedings Eucarpia Workshop, Wageningen, 14-16 November 1978, pp. 173-175.

Faulks, A.J., Shewry, P.R. & Miflin, B.J. (1981) The polymorphism and structural homology of storage polypeptides (hordein) coded by the Hor-2 locus in barley (Hordeum vulgare L.). Biochem. Genet., 19, 841-858.

Field, J.M., Shewry, P.R., Miflin, B.J. & March, J.F. (1982) The purification and characterization of homologous high molecular weight storage proteins from grain of wheat, rye and barley. Theor. Appl. Genet., 62, 329-336.

Forde, B.G., Heyworth, A., Pywell, J. & Kreis, M. (1985a) Nucleotide sequence of a Bl hordein gene and the identification of possible upstream regulatory elements in endosperm storage protein genes from barley, wheat and maize. Nucleic Acids Res., 13, 7327-7339.

Forde, B.G., Kreis, M., Williamson, M.S., Fry, R., Pywell, J., Shewry, P.R., Bunce, N. & Miflin, B.J. (1985b) Short tandem repeats shared by B- and C-hordein cDNAs suggest a common evolutionary origin for two groups of cereal storage protein genes. EMBO Journal 4, 9-15.

Forde, J. (1985) The Molecular Biology of the High Molecular Weight Glutenin Subunits of Wheat. Ph.D. Thesis. CNAA, U.K.

Forde, J., Forde, B.G., Fry, R., Kreis, M., Shewry, P.R. & Miflin, B.J. (1983) Identification of barley and wheat cDNA clones related to the high M_r polypeptides of wheat gluten. FEBS Lett., 162, 360-366.

Forde, J., Malpica, J-M., Halford, N., Shewry, P.R., Anderson, O.D., Greene, F.C. & Miflin, B.J. (1985) The nucleotide sequence of a HMW subunit gene located on chromosome 1A of wheat (Triticum aestivum L.). Nucleic Acids Res., 13, 6817-6832.

Giese, H. & Hejgaard, J. (1984) Synthesis of salt-soluble proteins in barley. Pulse-labelling study of grain filling in liquid-cultured detached spikes. Planta, 161, 172-177.

Giese, H. & Andersen, B. (1984) The course of protein synthesis during grain filling in normal and high-lysine barley. In: Cereal Grain Protein Improvement. pp. 217-226, Vienna: IAEA.

Hejgaard, J. & Boisen, S. (1980) High lysine proteins in Hiproly barley breeding: Identification, nutritional significance and new screening methods. Hereditas 93, 311-320.

Jensen, J. (1979a) Location of a high-lysine gene and the DDT-resistance gene on barley chromosome 7. Euphytica 28, 47-56.

Jensen, J. (1979b) Chromosomal location of one dominant and four recessive high lysine genes in barley mutants. In: Seed Protein Improvement in Cereals and Grain Legumes Vol. 1. pp. 89-96, Vienna: IAEA.

Jensen, J. (1984) Co-ordinators Report. Chromosome 5. Barley Genet. Newslett., 14, 62-65.

Karlsson, K.E. (1976) Linkage studies on the lys gene in relation to some marker genes and translocations. In: Barley Genetics III, ed. H. Gaul, pp. 536-541. Munich, FRG: Verlag Karl Thiemig.

Karlsson, K.E. (1977) Linkage studies in a gene for high lysine content in Risø barley mutant 1508. Barley Genet. Newslett., 7, 40-43.

Køie, B., Doll, H. & Kreis, M. (1976) Evaluation of a high lysine barley gene using chromosome doubled monoploids. Genetika, 8, 177-182.

Køie, B. & Doll, H. (1979) Protein and carbohydrate components in the Risø high lysine barley mutants. In: Seed Protein Improvement in Cereals and Grain Legumes, vol. 1. pp. 205-215. Vienna: IAEA.

Kreis, M. (1978) Starch and free sugars during kernal development of Bomi barley and its high lysine mutant 1508. In: Seed Protein Improvement by Nuclear Techniques, pp. 115-120, Vienna: IAEA.

Kreis, M. (1979) Starch Synthesis in Barley Mutants deficient in Storage Proteins. Ph.D. Thesis, Universite Catholique de Louvain, Belgium.

Kreis, M. (1980) Primer dependent and independent forms of soluble starch synthetase from developing barley endosperms. Planta 148, 412-416.

Kreis, M. & Doll, H. (1980) Starch and prolamin level in single and double high-lysine barley mutants. Physiol. Plant., 48, 139-143.

Kreis, M., Rahman, S., Forde, B.G., Pywell, J., Shewry, P.R. & Miflin, B.J. (1983a) Sub-families of hordein mRNA encoded at the Hor 2 locus of barley. Mol. Gen. Genet., 191, 194-200.

Kreis, M., Shewry, P.R., Forde, B.G., Rahman, S. & Miflin, B.J. (1983b) Molecular analysis of a mutation conferring the high lysine phenotype on the grain of barley (Hordeum vulgare). Cell, 34, 161-167.

Kreis, M., Shewry, P.R., Forde, B.G., Rahman, S., Bahramian, M.B. & Miflin, B.J. (1984) Molecular analysis of the effects of the mutant lys 3a gene on the expression of Hor loci in developing endosperms of barley (Hordeum vulgare). Biochem. Genet., 62, 231-255.

Kreis, M., Forde, B.G., Rahman, S., Miflin, B.J. & Shewry, P.R. (1985a) Molecular evolution of the seed storage proteins of barley, rye and wheat. J. Mol. Biol., 183, 499-502.

Kreis, M., Shewry, P.R., Forde, B.G., Forde, J. & Miflin, B.J. (1985b) Structure and evolution of seed storage proteins and their genes, with particular reference to those of wheat, barley and rye. In: Oxford Surveys of Plant Cell and Molecular Biology vol. 2, pp. 253-317. OUP: Oxford.

Lazaro, A., Barker, D., Salcedo, G., Mendez, E. & Garcia-Olmedo, F. (1985) Differential effects of high lysine mutations on the accumulation of individual members of a group of proteins encoded by a disperse multigene family in the endosperm of barley (Hordeum vulgare L.). Eur. J. Biochem., 149, 617-623.

Miflin, B.J. & Shewry, P.R. (1979) The synthesis of proteins in normal and high lysine barley seeds. In: Recent Advances in the Biochemistry of Cereals ed. D. Laidman & R.G. Wyn Jones, pp. 239-273. London: Academic Press.

Miflin, B.J., Field, J.M. & Shewry, P.R. (1983) Cereal storage proteins and their effects on technological properties. In: Seed Proteins, ed. J. Daussant, J. Mosse & J. Vaughan, pp. 255-319. London: Academic Press.

Munck, L. (1976) Genetics and breeding techniques for improving feed and malt quality. In: Barley Genetics III, ed. H. Gaul, pp. 526-535. Munich, FRG: Verlag Karl Thiemig.

Munck, L. (1985) The impact of industry on development of cereals through genetic biotechnology. Proc. Biotech. '85 Asia Conference. Singapore, Nov. 1985.

Munck, L., Karlsson, K-E., Hagburg, A. & Eggum, B.O. (1970) Gene for improved nutritional value in barley seed protein. Science, 168, 985-987.

Nelson, O.E. (1980) Genetic control of polysaccharide and storage protein synthesis in the endosperm of barley, maize and sorghum. In: Advances in Cereal Science and Technology, Vol. 3. Y. Pomeranz ed. American Association of Cereal Chemists. St. Paul. Minnesota, USA, pp. 41-71.

Netsvetaev, V.P. & Sozinov, A.A. (1984) Location of a hordein G locus, Hrd G, on chromosome 5 of barley. Barley Genet. Newslett., 14, 4-6.

Persson, G. & Karlsson, K.E. (1977) Progress in breeding for improved nutritive value in barley. Cer. Res. Commun., 5, 53-58.

Rasmussen, S., Hopp, H.E. & Brandt, A. (1983) Nucleotide sequences of cDNA clones for B1 hordein polypeptides. Carlsberg Res. Commun., 48, 187-199.

Shewry, P.R., Hill, J.M., Pratt, H.M., Leggatt, M.M. & Miflin, B.J. (1978) An evaluation of techniques for the extraction and separation of hordein and glutelin from barley seed and a comparison of the protein composition of Bomi and Risø 1508. J. Exp. Bot., 29, 677-692.

Shewry, P.R., Pratt, H.M., Leggatt, M.M. & Miflin, B.J. (1979) Protein metabolism in developing endosperms of high-lysine and normal barley. Cer. Chem., 56, 110-117.

Shewry, P.R., Faulks, A.R. & Miflin, B.J. (1980a) The effect of high-lysine mutations on the protein fractions of barley grain. Biochem. Genet., 18, 133-151.

Shewry, P.R., Field, J.M., Kirkman, M.A., Faulks, A.J. & Miflin, B.J. (1980b) The extraction, solubility and characterization of two groups of barley storage polypeptides. J. Exp. Bot., 31, 393-407.

Shewry, P.R., Franklin, J., Parmar, S., Smith, S.J. & Miflin, B.J. (1983) The effects of sulphur starvation on the amino acid and protein compositions of barley grain. J. Cer. Sci., 1, 21-31.

Shewry, P.R., Miflin, B.J. & Kasarda, D.D. (1984) The structural and evolutionary relationships of the prolamin storage proteins of barley, rye and wheat. Phil. Trans. Roy. Soc., London B. 304, 297-308.

Shewry, P.R., Bunce, N.A.C., Kreis, M. & Forde, B.J. (1985a) Polymorphism at the Hor 1 locus of barley (Hordeum vulgare L.). Biochem. Genet., 23, 389-402.

Shewry, P.R., Kreis, M., Parmar, S., Lew, E.J-L. & Kasarda, D.D. (1985b) Identification of γ-type hordeins in barley. FEBS Lett., 190, 61-64.

Shewry, P.R., Parmar, S. & Franklin, J. (1985c) The effects of mutant high lysine genes on the amount and composition of hordein. Barley Genet. Newslett., 15, 48-51.

Shewry, P.R., Parmar, S., White, R.P. & Franklin, J. (1986) Mapping and biochemical analysis of the Hor 4 (Hrd G) locus encoding B hordein-like polypeptides. Barley Genet. Newslett. (in press).

Tallberg, A. (1977) The amino acid composition in endosperm and embryo of a barley variety and its high lysine mutant. Hereditas, 87, 43-46.

Tallberg, A. (1981) Protein and lysine content in high-lysine double recessives of barley II. Combinations between mutant 7 and a Hiproly backcross. Hereditas 94, 261-268.

Tallberg, A. (1982) Characterization of high-lysine barley genotypes. Hereditas, 96, 229-245.

Tatham, A.S., Drake, A.F. & Shewry, P.R. (1985) A conformational study of 'C' hordein, a glutamine and proline-rich cereal seed protein. Biochem. J., 226, 557-562.

Torp, J. (1979) Patterns of variation in major chemical constituents of the barley grain. Ph.D. Thesis, Royal Veterinary and Agricultural University, Copenhagen, Denmark.

Williamson, M.S., Forde, J., Buxton, B.M. & Kreis, M. (1986). Nucleotide sequence of barley chymotrypsin inhibitor 2 (CI.2) and its expression in normal and high lysine barley (submitted).

HORMONE MUTANTS AND SEED DORMANCY IN ARABIDOPSIS AND TOMATO

C.M. Karssen,[1] S.P.C. Groot[1] and M. Koornneef[2]
[1]Department of Plant Physiology
[2]Department of Genetics, Agricultural University, Nl-6703 BD Wageningen, The Netherlands

INTRODUCTION

The isolation of single gene mutants showing more or less reduced levels of a certain hormone or a group of hormones has added powerful new tools to the study of plant hormones. They particularly provide means to determine which aspect of growth and development depend on the presence of a hormone and, of equal importance, which do not. Hormone mutants may also help to answer the question whether environmental factors influence morphogenesis by means of changes in hormone levels or in the sensitivity to hormones. In this context, also the recently isolated hormone-insensitive mutants are important.

It is the aim of the present paper to review and discuss the contributions that have been made to our knowledge of the hormonal control of seed dormancy by the study of GA- an- ABA-deficient and/or -insensitive mutants.

CHARACTERISTICS OF GA- AND ABA-MUTANTS

GA-deficient mutants

Mostly GA-deficiency is looked for among mutants with dwarfed shoot growth. Not all dwarfs are GA deficient, however. Therefore, complete phenotypic reversion by GA applications and the demonstration of lesions for all or for certain specific GAs are essential additional criteria to identify dwarfs as GA-deficient mutants. The absence of all GAs like e.g. in the d5 mutant of maize (Hedden and Phinney, 1979) and the na mutant of pea (Potts and Reid, 1983) indicates a block in the GA biosynthetic pathway prior to GA_{12} aldehyde. Biochemical specification led to the conclusion that the d5 gene in maize controls the activity of ent-kaurene synthetase (Hedden and Phinney, 1979).

Dwarfed shoot growth may also result from blockades that occur much later, like e.g. between GA_{12} aldehyde and GA_{53} aldehyde (d3 in maize),

GA_{53} aldehyde and GA_{53} (d2 in maize) or the ultimate block between GA_{20} and GA_1 (dl in maize, le in pea) (Phinney and Spray, 1982; Sponsel, 1985).

In Arabidopsis thaliana and tomato, GA-deficient mutants have been isolated by selecting for seeds that do not germinate without an application of GAs (Koornneef and van der Veen, 1980; Koornneef et al., 1981). The seedlings that develop from GA-treated seeds show dwarf growth which can be reverted to normal growth by GA sprays. Deficiency for all GAs was demonstrated in the recessive mutants ga-1, ga-2 and ga-3 of Arabidopsis and ga-1 and ga-2 of tomato (Zeevaart, 1983, 1984). The ga-1 mutant of Arabidopsis has very low ent-kaurene synthesizing capacity (Barendse and Koornneef, 1982; Barendse et al., 1986). In tomato, feeding studies of young seedlings indicate that the GA pathway in the ga-1 mutant is blocked prior to ent-kaurene and in ga-2 between ent-kaurenoic acid and GA_{12} (Zeevaart, 1983).

In contrast to Arabidopsis and tomato, reduced seed germination has not been reported for GA-deficient mutants of e.g. pea (Sponsel, 1985), maize (Phinney and Spray, 1982) or barley (Hopp et al., 1981). A first explanation for the difference between germinating and non-germinating dwarfs may be that GA deficiency is not evenly distributed over all plant organs. Seeds of na mutant of pea contain for instance the normal complement of seed GAs, whereas shoots are devoid of all GAs (Potts and Reid, 1983; Ingram et al., 1984). If it is supposed that seed germination requires lower levels of endogenous GAs than seedling growth, a slight leakiness of the mutated gene may provide an alternative explanation for the germination of dwarf-mutant seeds. Indeed, in Arabidopsis several germinating dwarfs have been found that are allelic to non-germinating dwarfs (Koornneef and van der Veen, 1980). The supposed difference in GA content between the two allelic lines could not be demonstrated because GA levels in both lines were below the detection level (Koornneef et al., 1985a). It will be discussed below that also the absence of dormancy due to an ABA-deficiency may cause germination of GA-deficient seeds.

Besides a strong reduction of elongation growth and the absence of seed germination GA-deficient mutants of Arabidopsis and tomato also show a reduction of apical dominance, a dark green colour of the leaves and an abnormal flower development, petals and stamens develop poorly, whereas sepals and pistils are almost normal (Koornneef and van der Veen, 1980; Koornneef et al., 1985a).

ABA-deficient mutants

ABA-deficient mutants have been looked for among mutants with more or less serious wilting phenomena. The mutants were primarily identified as ABA-deficients when ABA sprays caused complete reversion to wild type (see Koornneef, 1986 for references). Wilting results in these mutants from the failure to respond to increasing water stress by stomatal closure. The mutants are not totally ABA-deficient. In the non-stressed wilty tomato mutants *notabilis* (*not*), *flacca* (*flc*) and *sitiens* (*sit*) ABA levels reach to 49, 26 and 15% of the value in wild type, respectively (Neill and Horgan, 1985). In response to rapid stress the ABA level in detached leaves of wild type tomato increases 30% in 2 h and 3-fold in 24 h, whereas leaves of *flc* and *sit* fail to respond and leaves of *not* only react during the first 3 h with little further increase. Both the failure to synthesize ABA in response to water stress and the ABA level in non-stressed leaves correspond very well with the lack of stomatal response. Such a correlation may also be found in the different alleles of the *aba* locus in *Arabidopsis*. The ABA level in immature seeds of the least wilty mutant *aba*^{3} is 10% of the comparable wild type value, in mature seeds it amounts to 26% of wild type. In the very weak *aba*^{1} mutant the latter value lies below 4% (Koornneef *et al.*, 1982). Still an even less vigorous allele *aba*^{4} is known that is almost lethal due to extreme wilting.

Most probably all wilty ABA-deficient mutants also show reduced seed dormancy (Koornneef, 1986). In *Arabidopsis* the absence of seed dormancy was taken as primary selection criterion (Koornneef *et al.*, 1982). Non-germinating GA-dependent *ga-1* mutants were subjected to a mutagenic treatment and in their progeny seeds were selected that germinated independently of GA, but still developed into dwarfs. Genetic analysis showed that this partial reversion to wild type was due to a mutation in a suppressor gene segregating independently from the *ga-1* gene. Subsequently, this suppressor gene was obtained as a homozygote recessive in a *Ga-1/Ga-1* background. It is denoted *aba* since it shows all the characteristics of an ABA-deficient mutant (see Koornneef *et al.*, 1985a; Koorneef, 1986, for a detailed presentation of the isolation procedure). In tomato the double recessive mutant *sit*^{w}/*sit*^{w} , *ga-1/ga-1* does not germinate without GA, indicating that the revertant procedure used in *Arabidopsis* for the isolation of ABA-deficient mutants is not applicable to every species.

Processes like fruit development, senescence and root geotropism which were often associated with endogenous ABA are not clearly affected in

ABA mutants (Koornneef, 1986). Thus, these processes are either not regulated by endogenous ABA or the residual levels of ABA in the mutants are saturating for their ABA demand.

ABA mutants have not yet been extensively used to unravel ABA biochemistry. Studies are in progress, however (Taylor and Linforth, 1985; Cornish and Zeevaart, 1985).

Mutants with reduced sensitivity to GA or ABA

In a few species mutants have been isolated that phenotypically strongly resemble GA- or ABA-deficient mutants but which either do not respond at all to applications of GAs or ABA or only to much higher concentrations than the deficient lines.

In *Arabidopsis* a partially dominant gene (denoted *Gai*) controls the GA insensitivity (Koornneef *et al.*, 1985b). The *Gai* mutant resembles the GA-deficient *ga-1* mutant not only in its dwarfy growth habit but also in reduced apical dominance and seed germination. Endogenous GAs in *Gai* rozettes do not differ from similar wild-type material, neither in quantity nor in composition.

Mutants with a reduced sensitivity to ABA have also been isolated in *Arabidopsis* (Koornneef *et al.*, 1984). Three mutants were isolated at 2 different loci (*Abi-1* and *abi-2*) which show both features of ABA deficiency, reduced seed dormancy and wilting phenomena. A third locus (*abi-3*) shows only reduced seed dormancy. Nevertheless, all these mutants have a normal or even slightly higher endogenous ABA content. Inhibition of seed germination and seedling growth requires much higher ABA concentrations than in wild type or *aba* mutant.

The nature of these reduced sensitivities is still a matter of debate. Reduced uptake mechanisms are not likely since apart from a reduced sensitivity to exogenous ABA or GAs also the responsiveness to endogenous hormones is affected. Since the internal concentrations in the mutants equal those in wild type, an enhanced rate of hormone metabolism is not a suitable explanation either. It is more likely that the mutations involve a step in the action of the hormones. The fact that both GA- and ABA-insensitive genotypes show most of the symptoms that are found in the deficient mutants suggests that a relatively early step in the hormone action chain is involved, possibly the binding capacity of receptor sites, or the reactivity of the hormone receptor complex.

HORMONAL CONTROL OF SEED DORMANCY

Numerous reports have shown that exogenous growth regulators applied singly or in combination maintain, impose or release dormancy in seeds. Convincing evidence that such effects reflect the action of naturally occurring hormones within the seeds is hard to produce. Attempts to correlate changes in dormancy with changes in endogenous hormone levels mostly fail to present reliable evidence for a causal relationship. Improvement is expected from recent developments in hormone determinations, in particular the use of immunological techniques. Studies of hormone compartmentation and distribution in seed tissues now seem to be within the reach of technical possibilities. Nevertheless hormone mutants still offer unique possibilities to study the role of hormones in seed dormancy without the disturbance of the intact system.

Induction of dormancy in developing seeds

The study of dormancy induction in developing seeds is far less accessible than dormancy studies with mature and harvested seeds. The experiments are more time-consuming and costly because plants have to be raised at controlled conditions. Therefore, our present knowledge on the regulation of the induction process is rather limited. It is restricted to some studies on the effects of daylength, light quality, temperature and soil nitrate on dormancy induction (Bewley and Black, 1982). No reports have been published about effects of environmental factors on endogenous hormone levels in developing seeds or about influences of exogenous hormone applications on dormancy development. Studies on the hormonal control of dormancy induction have been restricted to a search for correlative relations between ABA levels and the cessation of embryo development and the start of dehydration.

Role of ABA. Studies with the ABA-deficient *aba* mutant of *Arabidopsis thaliana* have clearly shown for the first time that ABA is absolutely required for the induction of seed dormancy (Karssen *et al.*, 1983). Dormancy induction is also prevented by a blockade of ABA action. It is shown in Figure 1 that immature seeds of both an ABA-deficient (aba^1) mutant and an ABA-insensitive (*abi-3*) mutant developed a full germination capacity, whereas wild-type seeds only showed some precocious germination around day 14. Note that wild type and *abi-3* seeds contain high levels of ABA half way through seed development, whereas *aba* seeds contain practically no ABA during all stages of seed development (Table 1). The aba plants develop more serious wilting phenomena than abi plants and, therefore, seeds develop at a somewhat slower rate.

Reciprocal crosses between wild-type and aba mutants have shown that dormancy induction only depends on ABA formed in embryo and endosperm (Karssen et al., 1983). The presence or absence of maternal ABA is irrelevant to dormancy induction. ABA sprays during seed development did not influence dormancy induction either. Nevertheless, both maternal and applied ABA evidently reach the periphery of the developing seeds because they stimulate in aba seeds the normal development of the mucilage producing testa cells. Thus, to induce dormancy ABA has to be synthesized close to its site of action or has to be localized in a specific subcellular compartment not accessible to ABA transported from outside the embryo. These conclusions interfere with the classic hormone theory that considers hormone transport an essential step in hormone action.

Tomato seeds are less suitable for dormancy studies than Arabidopsis. Dormancy develops in wild-type seeds only at certain cultivation

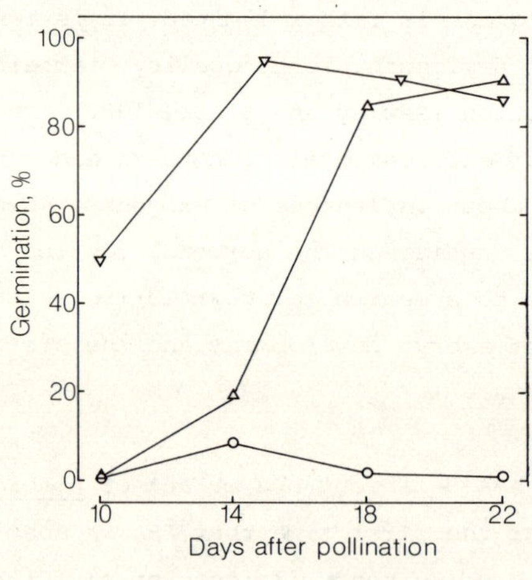

Figure 1. Precocious seed germination of different Arabidopsis thaliana genotypes in light at 24°C. Seeds were isolated from the siliquae at the indicated time after pollination. Germination was recorded after 7 days. Wild type (o), aba[1] mutant (Δ) and abi[3] mutant (∇) (original data) (see Karssen et al. (1983) for details on experimental approach).

conditions (no early defoliation, higher temperatures) and persists for short after-ripening periods only. Seeds of the sit^w mutant do not even possess this feeble form of dormancy. The results in Table 2 indicate that the development of dormancy in tomato seeds is also dependent on ABA synthesized in embryo and endosperm. As a result of selfing of Sit/sit^w, seeds with and without dominant alleles developed in heterozygous fruits, apparently in a ratio close to 3:1. All sit^w/sit^w seeds germinated in water within 3 days. About 50% of the seeds with at least one dominant allele germinated in water, the remaining 50% required an addition of GA_{4+7}.

Overripe fruits of the sit^w mutant regularly contain germinated seeds, in contrast to wild-type fruits: an indication that endogenous ABA also prevents viviparous germination. In *Arabidopsis* vivipary of *aba* seeds does not occur (Karssen *et al.*, 1983). The dehydrated state of the seeds at the end of maturation suffices to prevent viviparous germination.

The role of GA during seed development. In *Arabidopsis* studies have also been made on the influence of a GA deficiency on the induction of dormancy (Karssen and Laçka, 1986). Precocious germination of wild type, *ga-1*, *aba* and double mutants could only be compared in GA_{4+7} solutions because the germination of immature *ga-1* seeds is also absolutely dependent on addition of GA. Half way through development, 10 μM GA_{4+7} strongly stimulated precocious germination of all genotypes. However, during maturation ABA-containing seeds became insensitive to GA_{4+7} whereas ABA-deficient seeds were still fully responsive. It is concluded that the presence or absence of GAs during seeds development is irrelevant for the induction of dormancy by endogenous ABA.

Table 1. Endogenous ABA levels in unripe siliquae containing seeds of different genotypes of *Arabidopsis thaliana* at two stages of seed development. Seeds contained about 95% of total ABA. DAP: days after pollination; total development 20 days. LSD 5%: 34 $ng.g^{-1}$ (adapted from Koornneef *et al.*, 1984).

Genotype	ABA, ng (g fr.wt.)$^{-1}$	
	8-11 DAP	14-16 DAP
wild type	169	98
aba mutant	10	10
abi-3 mutant	279	144

In contrast, endogenous GAs are clearly required for normal fruit growth (Barendse et al., 1986). In the ga-1 mutant of Arabidopsis the rate of siliquae growth and the final length are strongly inhibited. However, seed set is only slightly reduced, and the average seed weight is not affected. Therefore, the ga-1 mutant develops small siliquae "crowded" with seeds of normal volume and weight. Reciprocal crosses between wild type and ga-1 demonstrate that siliquae growth is primarily dependent on endogenous GAs originating from maternal tissue, but also embryo GAs have a stimulatory effect.

Dormancy in mature seeds

In our studies with developing Arabidopsis seeds germination was only tested in light. Germination tests in darkness were not relevant because the seeds were prepared from the siliquae in light. Mature, dehydrated seeds were tested both in light and darkness, however (Karssen and Laçka, 1986). Germination tests of mature seeds of the aba mutant in darkness revealed that our previous conclusion about the absence of dormancy in ABA-deficient seeds is slightly incorrect. Shortly after harvest seeds of the aba mutant do not germinate in darkness, which is a form of dormancy. Nevertheless, freshly harvested wild-type seeds are in a much deeper state of dormancy because germination does not occur in light nor in darkness. It is not likely

Table 2. The influence of the genotype of the embryo of tomato seeds cv. Moneymaker on the germination of freshly harvested mature seeds from selfed Sit/sit^{W} plants. Seeds were collected from ripe fruits and germinated in water during 5 days. All seeds that had not germinated were thereafter soaked in 10 μM GA_{4+7}. Germinated seeds were removed from the Petri dishes, transplanted in soil and checked for wilting phenomena under water stress. A few seedlings did not survive (Groot, original data).

Seedling phenotype and genotype	Germination, number						
	Incubation, days						
	1	2	3	4	5	7 (+ GA)	Total
wilting (sit^{W}/sit^{W})	12	49	59	-	-	-	59
non wilting (Sit/Sit, Sit/sit^{W})	1	22	78	79	79	173	173
not classified	0	1	14	14	14	18	18

that the dormant state of the wild-type seeds depends on ABA that is still present in the seeds at the time of harvest. Experiments with ^{14}C-labelled ABA solutions indicate that at the minimal inhibitory ABA concentration about 7 ng ABA enters a wild-type seed (Karssen et al., 1983). The ABA content of these seeds is a tenth of that. Thus, the dormant state of mature wild-type seeds is the result of ABA action during seed development and does not require the permanent presence of endogenous ABA.

Relief of dormancy

During 6 to 8 weeks of dry storage at room temperature or 4 to 8 days of cold stratification at 2°C wild-type seeds of Arabidopsis develop the capacity to germinate in light, dark germination is not affected (Karssen and Laçka, 1986). Both dormancy breaking treatments do not promote the germination of ga-1 seeds in water. However, when germination is tested in GA_{4+7} solutions it is seen that during cold stratification and dry storage the capacity to germinate in 10 μM GA_{4+7} rises in a similar way in seeds of wild type and ga-1 mutant. A test in darkness showed that cold stratification

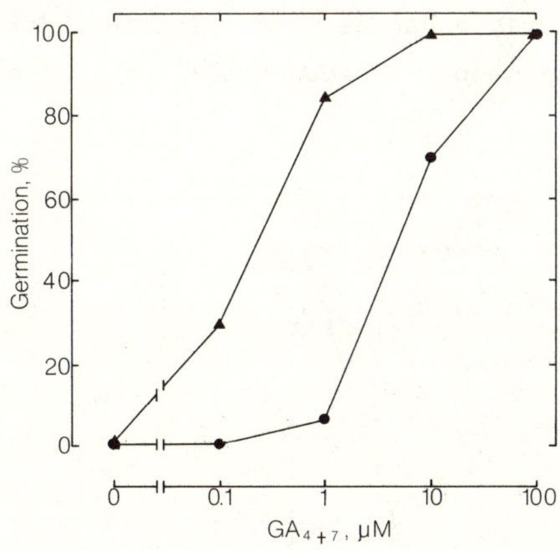

Figure 2. Germination of ga-1 mutant seeds of Arabidopsis thali thaliana at 24°C in darkness in a range of GA_{4+7} concentrations with (▲) or without (●) a preceding 7 days incubation at 2°C. (Zagórski and Karssen, unpublished results).

increased the sensitivity of ga-1 seeds to GA_{4+7} about 100 fold (Figure 2). Thus the relief of dormancy in Arabidopsis seeds involves a sensitization to GAs that occurs independently of GA biosynthesis. In the Gai-mutant seeds the sensitivity to GA_{4+7} is not restored by 7 days at 2°C (Koornneef et al., 1985b).

Germination

GA requirement. The absence of seed germination in GA-deficient mutants of Arabidopsis and tomato has clearly shown that endogenous GAs are indispensible factors for germination. The GA-requirement is strongly influenced by environmental factors. Besides a cold stratification, light also enhances the GA-sensitivity of ga-1 seeds. It is shown in Figure 3 that in light ga-1 seeds require about 50 times less GA_{4+7} than in darkness. Light also enhances the sensitivity of Gai seeds, the increase is particularly striking in a Ga-deficient ga-1/ga-1 background (Koornneef et al., 1985b).

In light wild-type seeds are independent of exogenous GA. Such a change has never been observed in ga-1 seeds (Figure 3). These results have been interpreted before as an indication that light stimulates GA-biosynthesis (Karssen and Laçka, 1986). However, it cannot be excluded that also in wild-type seeds light only enhances GA-sensitivity. Germination in water and light will then occur when the seeds have been made sensitive to

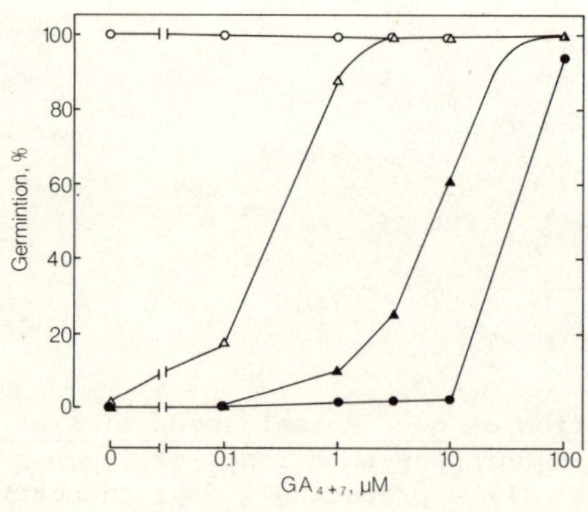

Figure 3. Germination of wild-type (circles) and ga-1 mutant (triangles) seeds of Arabidopsis thaliana at 24°C in a range of GA_{4+7} concentration in constant white fluorescent light (open symbols) or darkness (closed symbols). (Kepçzyński and Karssen, unpublished results).

the level of GA that is present in the seeds in darkness. Further kinetic studies are required to solve this dilemma.

Another interesting aspect of the results presented in Figure 3 is the lower GA-sensitivity of wild-type seeds in darkness as compared to ga-1 seeds. We suppose that the presence of GAs during seed development of wild-type seeds has caused a so-called "receptor-down" regulation of the GA-receptor sensitivity in a way comparable to animal systems.

Revision of the hormone balance theory

We noticed above that freshly harvested mature aba seeds do not germinate in darkness. A few days of cold stratification suffices to remove this feeble form of dormancy (Karssen and Laçka, 1986). In seeds of the double mutants (ga-1/ga-1, aba/aba) the dark inhibition is more persistent, indicating a dependency on endogenous GA of ABA-deficient seeds.

This was confirmed when the GA-requirement of non-afterripened seeds of the aba, ga-1 and double mutants and wild type was compared in darkness. ABA-deficient seeds required indeed a little less GA_{4+7} than seeds of the double mutant, but both genotypes required about 100 times less GA_{4+7} than seeds of the genotypes with dominant Aba alleles. Thus, ABA action during seed development increases the GA-requirement during germination (Figure 4). Therefore, the classic theory of dormancy control which supposes a balance between simultaneously acting promotive and inhibiting hormones has to be substituted by a kind of "remote control" of the GA-requirement during germination by the ABA of seed development through the intermediate dormant condition.

Localization of hormone action in tomato seeds

Tomato seeds are more suitable objects to study the localization

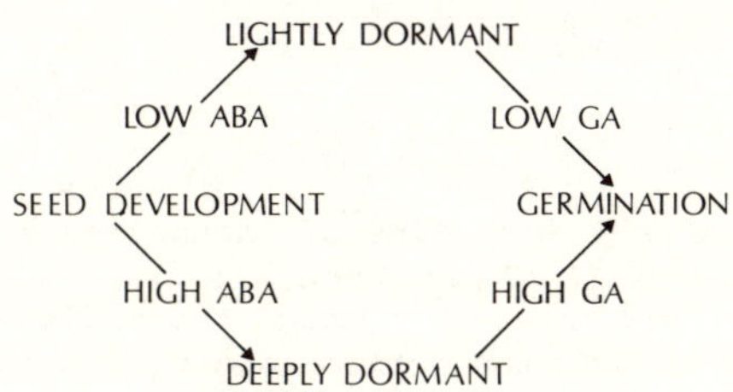

Figure 4. Schematic presentation of the control of seed dormancy and germination by ABA and GA (from Karssen and Laçka, 1986).

of the hormone action in the different seed parts than the very tiny Arabidopsis seeds. Experiments with wild-type and ga-1 seeds of tomato show that the primary action of endogenous GAs is localized in the endosperm cell layers around the tip of the embryo radicle. In wild-type seeds the mechanical resistance of the endosperm decreases prior to the start of visible germination (Figure 5). In ga-1 seeds weakening of the endosperm only occurred when the seeds were soaked in GA_{4+7}. It has been shown that in wild-type seeds the weakening of the endosperm resistance is controlled by a factor, probably GA, that is produced by the embryo (Groot, unpublished results). When the testa and endosperm layers that surround the radicle tip are removed the embryo starts to grow in the absence of GAs.

ABA-deficient sit^{w} seeds have a mechanical resistance similar to wild type and ga-1 (Groot, unpublished results). Nevertheless, sit^{w} seeds germinate in a much more negative water potential than wild-type seeds (Koornneef et al., 1985b). Thus, the ABA deficiency must have caused a change in the growth potential of the embryo. In tomato, ABA and GA not only seem to be active in different developmental periods, but also in different seed parts.

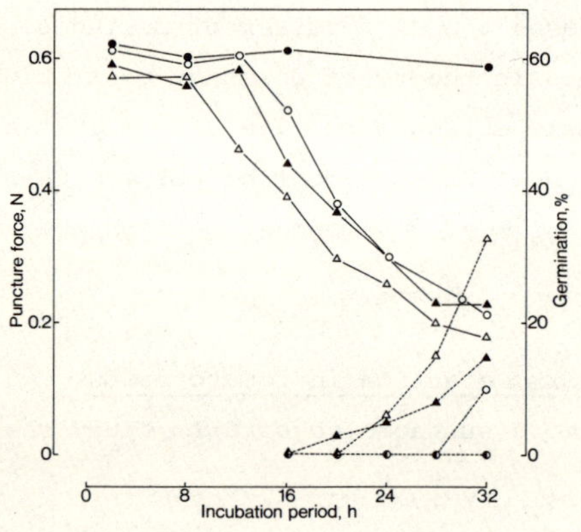

Figure 5. Change with time of incubation in water (closed symbols) or 10 µM GA_{4+7} (open symbols) at 26°C of the puncture force required to break through the endosperm and testa layers (solid lines) and the germination of intact seeds (broken lines) of wild-type (triangles) and ga-1 mutants (circles) seeds of tomato cv. Moneymaker. The puncture force was measured by an Instron Universal testing instrument in seed halfs from which the radicle was removed and replaced by a needle of similar diameter and form (Groot and Karssen in preparation).

CONCLUSIONS

In conclusion, our studies with the ABA- and GA-mutants of Arabidopsis thaliana and tomato leave no doubt that ABA and GAs play an active role in the regulation of dormancy and germination, respectively. They also indicate that the environmental factors light and low temperature primarily act on the sensitivity of the seeds to GA. It has also been shown that hormone mutants are very useful to determine when and where hormones are active.

The use of hormone mutants has not been extended yet to the study of the molecular mechanism of hormone action. The excellent properties of Arabidopsis thaliana for experiments in molecular genetics (Meyerowitz and Pruitt, 1985) raise great expectations that in the future the hormone mutants of that species will serve as important tools in the molecular study of hormone action in plants.

REFERENCES

Barendse, G.W. & Koornneef, M. (1982) Biosynthesis of the gibberellin precursor ent-kaurene in cell-free enzyme preparations from gibberellin-sensitive Arabidopsis mutants. Arabidopsis Information Service, 19, 25-8.

Barendse, G.W.M., Kepçzyński, J., Karssen, C.M. & Koornneef, M. (1986) The role of endogenous gibberellins during fruit and seed development: Studies on gibberellin-deficient genotypes of Arabidopsis thaliana (L.) Heynh. Physiol. Plant., 67, 315-19.

Bewley, J.D. & Black, M. (1982) Physiology and Biochemistry of Seeds in Relation to Germination, vol. 2. Berlin, Heidelberg, New York: Springer.

Cornish, K. & Zeevaart, J.A.D. (1985) Accumulation and catabolism of abscisic acid in relation to water stress in wilting mutants of tomato. In: Twelfth International Conference on Plant Growth Substances, Abstracts, ed. M. Bopp, B. Knoop & W. Rademacher, p.27, August 1985. Heidelberg, F.R.G.

Hedden, D. & Phinney, B.O. (1979) Comparison of ent-kaurene and ent-isokaurene synthesis in cell-free systems from etiolated shoots of normal and dwarf-5 maize seedlings. Phytochem., 18, 1475-9.

Hopp, H.E., Favret, G.C. & Favret, E.A. (1981) Control of barley development using dwarf mutants. In: Induced mutations - A Tool in Plant Research, ed. P. Howard Kitto, pp. 1-7. Vienna: IAEA-SM 251.

Ingram, T.J., Reid, J.B., Murfet, I.L., Gaskin, P., Willis, C.L. & MacMillan, J. (1984). Internode length in Pisum. The Le gene controls the 3 -hydroxylation of gibberellin A_{20} to gibberellin A_1. Planta. 160, 455-63.

Karssen, C.M. (1982) Role of endogenous hormones during seed development and the onset of primary dormancy. In: Plant Growth Substances 1982, ed. P.F. Wareing, pp. 623-32. London: Academic Press.

Karssen, C.M., Brinkhorst-van der Swan, D.L.C., Breekland, A.E. & Koornneef, M. (1983) Induction of dormancy during seed development by endogenous abscisic acid: studies on abscisic acid deficient genotypes of Arabidopsis thaliana (L.) Heynh. Planta, 157, 158-65.

Karssen, C.M. & Laçka, E. (1986) A revision of the hormone-balance theory of seed dormancy: Studies on gibberellin and/or abscisic acid deficient mutants of Arabidopsis thaliana. In: Plant Growth Substances 1985, ed. M. Bopp, pp.315-323, Heidelberg: Springer.

Koornneef, M. (1986) Genetic aspects of abscisic acid. In: Genetic approach to Plant Biochemistry, Plant Gene Research, vol. 3, pp. 35-54. New York, Vienna: Springer.

Koornneef, M., Cone, J.W., Karssen, C.M., Kendrick, R.E., van der Veen, J.H. & Zeevaart, J.A.D. (1985a) Plant hormone and photoreceptor mutants in Arabidopsis and tomato. In: Plant Genetics UCLA Symposia on Molecular and Cellular Biology, New Series, vol. 35, ed. M. Freeling, pp. 103-14. New York, NY: Alan R. Liss. Inc.

Koornneef, M., Elgersma, A., Hanhart, C.J., van Loenen-Martinet, E.P., van Rijn, L. & Zeevaart, J.A.D. (1985b) A gibberellin-insensitive mutant of Arabidopsis thaliana. Physiol. Plant. 65, 33-9.

Koornneef, M., Jorna, M.L., Brinkhorst-van der Swan, D.L.C. & Karssen, C.M. (1982) The isolation of abscisic acid (ABA)-deficient mutants by selection of induced revertants in non-germinating gibberellin-sensitive lines of Arabidopsis thaliana (L.) Heynh. Theor. Appl. Genet., 61, 385-93.

Koornneef, M., Reuling, G. & Karssen, C.M. (1984) The isolation and characterization of abscisic acid insensitive mutants of Arabidopsis thaliana. Physiol. Plant., 61, 377-83.

Koornneef, M. & van der Veen, J.H. (1980). Induction and analysis of gibberellin-sensitive mutants in Arabidopsis thaliana (L.) Heynh. Theor. Appl. Genet., 58, 257-263.

Koornneef, M. van der Veen, J.H., Spruit, C.J.P. & Karssen, C.M. (1981) The isolation and use of mutants with an altered germination behaviour in Arabidopsis thaliana and tomato. In: Induced Mutation - a Tool in Plant Research, ed. P. Howard Kitto, pp. 227-32. Vienna, IAEA-SM 251.

Meyerowitz, E.M. & Pruitt, R.E. (1985) Arabidopsis thaliana and plant molecular genetics. Science, 229, 1214-8.

Neill, S.J. & Horgan, R.J. (1985) Abscisic acid production and water relations in wilty tomato mutants subjected to water deficiency. J. Exp. Bot., 36, 1222-1231.

Phinney, B.O. & Spray, C. (1982) Chemical genetics and gibberellin pathway in Zea mays L. In: Plant Growth Substances 1982, ed. P.F. Wareing, pp. 101-10.

Potts, W.C. & Reid, J.B. (1983) Internode length in Pisum III. The effect and interaction of the Na/na and Le/le gene differences on endogenous gibberellin-like substances. Physiol. Plant., 57, 448-54.

Sponsel, V.M. (1985) Gibberellins in Pisum sativum - their nature, distribution and involvement in growth and development of the plant. Physiol. Plant., 65, 533-8.

Taylor, I.B. & Linforth, R.S.T. (1985) Genetic analysis of ABA accumulation in tomato mutants. In: Twelfth International Conference on Plant Growth Substances, Abstracts, ed. M. Bopp, B. Knoop & W. Rademacher, p. 27, August 1985, Heidelberg, F.R.G.

Zeevaart, J.A.D. (1983) Environmental control of plant development and its relation to hormones. In: 18th Annual Report MSU-DOE Plant

Research Laboratory, pp. 155-9.
Zeevaart, J.A.D. (1984) Environmental control of plant development and its relation to hormones. In: 19th Annual Report MSU-DOE Plant Research Laboratory, pp. 138-46.

MUTANTS AND GRAVITROPISM

Jeremy A. Roberts,
Department of Physiology & Environmental Science,
University of Nottingham, Faculty of Agricultural Science,
Sutton Bonington, Nr. Loughborough, Leics. LE12 5RD.

INTRODUCTION

The gravitropic bending of a plant organ is the culmination of a highly co-ordinated sequence of events. These can be broadly separated into gravity perception, transduction of the stimulus, and finally the growth response. It is the relationship between these events which dictates whether an organ will be positively or negatively gravitropic, or grow at an angle to the gravity vector. If the site of action of a mutation affects any step in the programme, then the chances of the gravitropic response being impaired in some manner are high. Such a modification in the sequence could ultimately lead to a change in gravitropic orientation or sensitivity, or even the inability to respond to a gravitropic stimulus. As an increasing number of plant mutants are characterised, examples of each of these categories of response are being documented. The aim of this review is to draw together all the information available on mutants and consider how the results of these studies assist our understanding of the mechanism of gravitropism in plants.

AGRAVITROPIC PLANTS

By far the largest group of gravitropic mutants documented is composed of plants whose ability to respond to gravity has been partially lost. The size of this class may be a reflection of the ease with which any malfunction in the gravitropic sequence can render an organ agravitropic or alternatively that the prostrate root or shoot growth induced by such mutations makes them readily apparent.

Pea mutants

One of the earliest gravitropic mutants discovered was described by Blixt *et al*. in 1958. This pea mutant originated from an X-ray irradiated population of seeds and was termed *Pisum sativum ageotropum* (*agt*). A more detailed study of agt was undertaken by Scholdeen and Burstrom (1960) who concluded that whilst roots of the mutant were agravitropic, the shoots exhibited some degree of negative gravitropism albeit to a limited extent. Further work by Ekelund and Hemberg (1966) demonstrated that decapitated roots of agt responded positively to gravity if 1.5 mm long tips from normal roots were placed in contact with the mutant stumps. If the converse experiment was carried out and agt tips placed on decapitated normal roots, then no bending was induced by gravistimulation. Although no firm conclusions were drawn from these observations, it is clear from the results that roots from agt plants do not lack the capacity for differential growth in response to a gravistimulus.

In a more recent study the agravitropic nature of roots of the mutant has been confirmed (Figure 1) and an anatomical study of the root carried out (Olsen and Iversen, 1980a,b). These workers examined the ultrastructure of the columella cells (which are thought to house the gravitropic perception machinery) from normal and mutant roots. Although the pattern of movement of amyloplasts was identical in both, they discovered that there was some difference in the distribution of rough endoplasmic reticulum between agt and non-mutant tissues. These observations may provide some support for the hypothesis that a crucial event in graviperception is an interaction between sedimenting amyloplasts and the endoplasmic reticulum (Sievers and Volkmann, 1972).

Arabidopsis mutants

A species that is particularly suitable for the rapid isolation of mutants is Arabidopsis thaliana. From a group classified on the basis of their resistance to the auxin herbicide 2,4-dichlorophenoxyacetic acid (2,4-D), has originated aux-1, a mutant with agravitropic roots (Maher and Martindale, 1980). Like the pea mutant, shoots of aux-1 respond to gravity (Figure 2) indicating that to a certain extent the gravitropic responses of shoots and roots are under separate genetic control. With this in mind however, preliminary studies of another 2,4D resistant mutant Dwf, have shown that at an early seedling stage, neither roots nor shoots of this mutant respond to gravity (Mirza et al., 1984). Some common mechanism in the gravitropic response of these two tissues could therefore exist.

In a comparison of the anatomy of normal roots with those of aux-1, Olsen et al., (1984) found that the rate of amyloplast sedimentation in non-mutant root caps was more than twice that of the mutant (6.0 µm h^{-1}

compared to 2.4 μm h^{-1}). After 40 minutes inversion, amyloplasts in the columella cells of aux-1 roots had only moved 40% of the distance observed in normal plants. These data provide us with a possible clue as to the nature of the mutants agravitropic response since sedimenting amyloplasts are strongly implicated in graviperception (Wilkins,1984). Further work is necessary to determine the reason for the reduced rate of amyloplast sedimentation in the columella cells of aux-1 plants, and whether enforced sedimentation by centrifugation might render the roots graviresponsive.

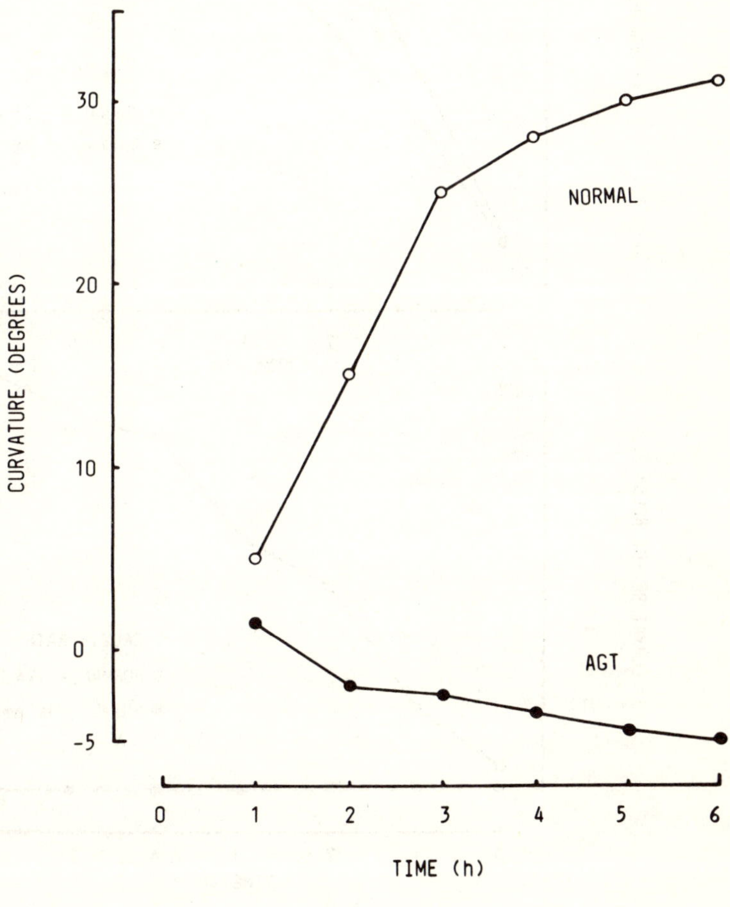

Figure 1. Gravitropic bending of pea roots from normal or ageotropum (agt) seedlings. (Redrawn from Olsen and Iversen, 1980).

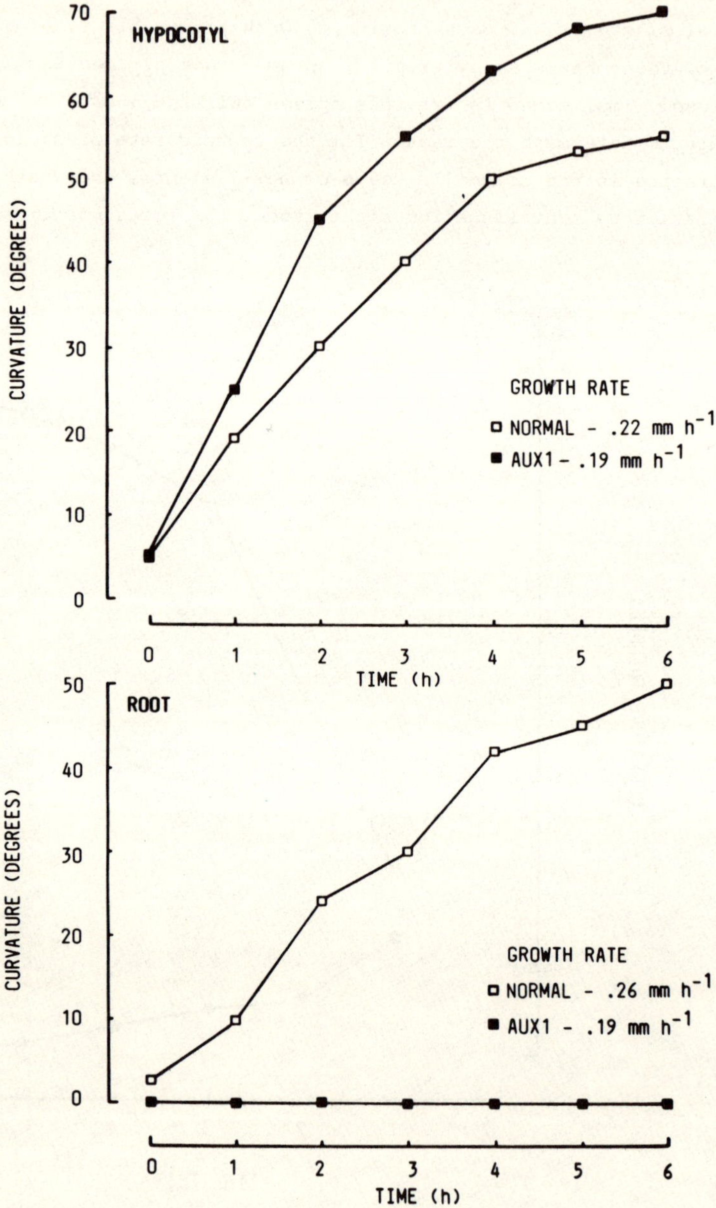

Figure 2. Gravitropic bending in normal or aux-1 *Arabidopsis thaliana* seedlings (Redrawn from Mirza *et al.*, 1984).

Maize mutants

Roots of the maize mutant *agt* are not strictly agravitropic since they display a weak positive response to gravity. Adopting the experimental approach utilised by Shaw and Wilkins (1973), Pilet excised half of the root tip of vertically maintained *agt* roots and compared the bending initiated by this procedure with roots from non-mutant plants (Pilet, 1983a). He found that after 12 hours the angle of curvature towards the intact tip of normal roots was 2-3 times that of the mutant. Coupling these results with the observation that the rate of elongation of intact *agt* roots was significantly greater than that of non-mutant plants, Pilet concluded that a consequence of the *agt* mutation was a reduced level of growth inhibiting compounds in the root extension zone. If this hypothesis is correct, freeze-decapping should be less effective in stimulating elongation growth in *agt* roots than normal roots (Pilet *et al.*, 1982). The identity of the hypothetical inhibitor is unknown. However, either abscisic acid (ABA) (Pilet, 1983b) or the neutral inhibitor described by Feldman (1981b) would be appropriate candidates.

Roots of some cultivars of maize fail to respond to gravity when grown in darkness. Following a brief period of illumination however, the roots become positively gravitropic (Scott and Wilkins, 1969). Investigating the site of perception of light, Feldman (1981a) found that if isolated caps were illuminated and then replaced on decapped roots maintained in darkness, bending could occur. This response was dependent on caps being treated with indole acetic acid (IAA) prior to illumination and was inhibited by cycloheximide (Feldman, 1983). This latter observation implies that protein synthesis is necessary to confer graviresponsiveness and the level of at least one low molecular weight protein has been found to be rapidly enhanced by illumination of isolated caps (Feldman and Gildow, 1984). Accompanying this rise in protein synthesis is an increase in the inhibitor content of cap tissue. Optimal inhibitor production requires the presence of IAA within the culture medium (Feldman, 1981a). Application of ABA has been reported to induce gravitropic curvature in maize roots in the absence of a light stimulus (Wilkins and Wain, 1975), and recent evidence indicates that this plant growth regulator is redistributed from the root cap to the subjacent apex on illumination (Feldman *et al.*, 1985). These studies implicate both IAA and ABA in root gravitropism but whether their roles are primary or secondary is presently unknown (Feldman, 1985).

The agravitropic behaviour of light sensitive maize cultivars has been further exploited to examine the role of calcium in root gravitropism. Studying the distribution of applied $^{45}Ca^{2+}$ to roots, Lee and Evans (1985) found that no polar movement of this ion occurred across the elongation zone unless the roots had been previously illuminated. Linking a rapid redistribution of calcium with the development of gravitropic competence provides strong support for the hypothesis that it plays an important role in root gravitropism (Lee et al., 1983).

Tomato mutants

Although Zobel (1974) describes roots of the tomato mutant diageotropica (dgt) as plageotropic, a recent report has indicated that they are agravitropic (Woods et al. 1984). The dgt phenotype is also associated with aberrant shoot gravitropism (see later) which can be reverted to normal by the exposure of plants to low levels of ethylene (Zobel, 1973). Treatment of roots of this mutant with ethylene has no effect on their agravitropic behaviour nor does it stimulate the production of laterals from the tissue (Roberts, unpublished). An anatomical study of dgt roots indicates that sedimenting amyloplasts are present within the columella cells and the rate of elongation of the mutant root tissue is indistinguishable from normal plants (Roberts, unpublished).

A tomato mutant with agravitropic aerial growth has been described in a paper by Roberts (1984). Etiolated hypocotyls of lazy-1 grow more slowly than those of normal plants and do not reorientate after gravistimulation. Dark grown seedlings of the mutant have a conventional hypocotyl hook which has the ability to respond to gravity (I.R. Gilbert, pers. comm.), and this demonstration supports the assertion that gravity plays an integral role in hook formation (Macdonald et al., 1983). Hypocotyls of lazy-1 are positively phototropic and in accord with their agravitropic nature achieve a significantly greater angle of bending compared to non-mutant seedlings when irradiated unilaterally (Roberts, 1984).

A study of the anatomy of lazy-1 has revealed that in contrast to normal seedlings sedimenting amyloplasts are absent from the endodermal cells in etiolated hypocotyl tissue (Roberts, 1984). In contrast, the pattern of sedimenting starch grains within the hook region of lazy-1 seedlings is similar to that found in normal plants and may account for the graviresponsiveness of the tissue. This sequence of observations strongly supports the hypothesis that sedimenting starch grains act as statoliths in the gravitropic response.

Seedlings of lazy-1 offer a system of considerable potential for the study of tropic curvatures. By comparing the biochemical and physiological changes associated with bending of non-mutant hypocotyls with those of lazy-1 the critical events in gravitropism may be elucidated. Furthermore the mutant offers an opportunity to investigate phototropism in the absence of gravitropism, and other physiological phenomena where an interaction with gravitropism has been established.

Barley mutants

A mutant of barley has been recently identified which has agravitropic roots. In the original report of the discovery Wendt et al. (1984) stated that root caps of the mutant treated with potassium iodide stained more slowly, less intensely and to a different pattern when compared with normal plants. A detailed study of the root anatomy by Moore (1985) however, has failed to distinguish any differences between the structure of the columella cells or the rate of amyloplast sedimentation in root caps from normal or mutant plants.

PLANTS WITH MODIFIED GRAVITROPIC SENSITIVITY

Only one mutant so far has been characterised as having a modified gravitropic sensitivity and this is the mutant amylomaize (am). Coleoptiles of this maize mutant contain amyloplasts reduced significantly in size compared to non-mutant corn. Examining the premise that sedimentation of starch grains played an integral role in graviperception, Hertel et al. (1969) investigated the gravitropic responsiveness of the mutant. In accord with Stokes law these workers found that after reorientation, the smaller amyloplasts from am were displaced more slowly than those from normal coleoptiles and the proportion that sedimented was reduced (Figure 3). Although no determination of the gravitropic presentation time was recorded, the angle of curvature of the mutant was significantly smaller than that of normal coleoptiles after 90 minutes gravistimulation (Figure 3). Unfortunately no data is presented as to whether am coleoptiles have the capacity to eventually achieve as much bending as that exhibited by non-mutant plants.

The physiology of this mutant was further studied in order to examine the role of auxin transport in tropic bending. Agar blocks containing ^{14}C IAA were asymmetrically applied to coleoptile sections and the distribution of radiolabel determined after 4 hours vertical or horizontal orientation of the tissue. The results demonstrate that the gravitropic mutant has a reduced capacity to redistribute ^{14}C label to the lower side of a coleoptile section after horizontal orientation. In this experiment no time course of redistribution was carried out and the identity of the redistributed label was not confirmed.

PLANTS WITH A MODIFIED RESPONSE TO GRAVITY

One of the most extensively documented gravitropic mutants is dgt. This mutant of tomato is primarily characterised by horizontal growth of the stem but has other associated features such as dark green hyponastic leaves, rigid stems and prominent root apical dominance (Zobel, 1973). The

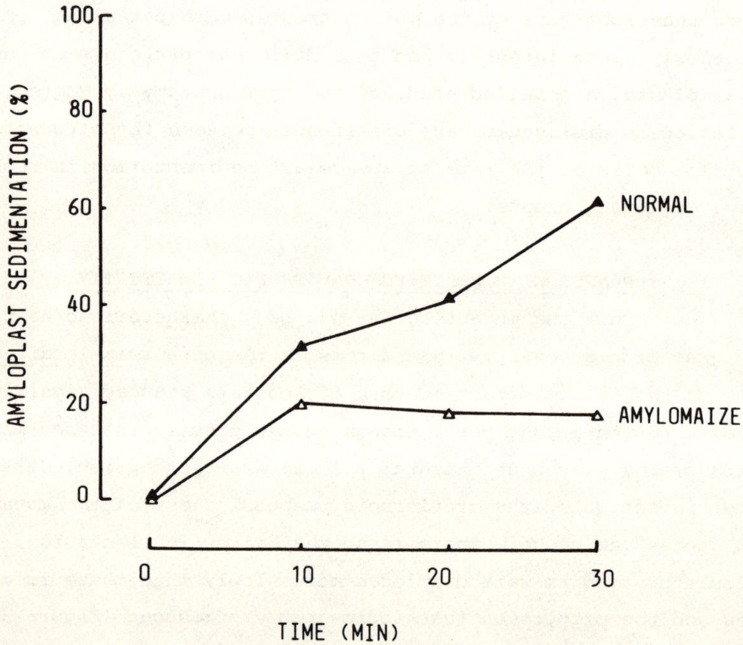

	AMYLOPLAST DIAMETER	GRAVITROPIC CURVATURE
NORMAL	5.2 μm	28°
AMYLOMAIZE	4.0 μm	18°

Figure 3. Sedimentation of amyloplasts in normal or amylomaize coleoptiles after gravistimulation. (Redrawn from Hertel et al., 1969).

mutant has proved of particular interest because its horizontal growth habit can be 'corrected' by low levels of ethylene (Zobel, 1973). Since dgt tissues have a reduced capacity to produce ethylene in response to IAA treatment, Zobel hypothesized that the mutant was ethylene deficient. This view has been challenged by Jackson (1979) on the basis that the basal level of ethylene production by the mutant is indistinguishable from normal plants. Jackson also argued that if the dgt phenotype was the result of an ethylene deficiency then the application of inhibitors of ethylene action to normal plants should induce the mutant phenotype. No evidence for this has been found (Harrison and Pickard, 1986). Although Jackson's results offer no support for the hypothesis that ethylene plays a key role in shoot gravitropism, Harrison and Pickard (1984) have reported that one of the earliest responses to gravistimulation of tomato hypocotyl tissue is a transitory rise in production of the gas. Other workers have documented a rise in production of ethylene after gravitropic bending is under way (Wheeler and Salisbury, 1981). The role for ethylene in "correcting" the gravitropic response of dgt shoots may be indirect and lie in its ability to modify auxin transport or metabolism. Certainly Severi and Laudi (1977) have demonstrated that the pattern of distribution of IAA applied to dgt stems is different from that found in non-mutant plants.

Another tomato mutant which exhibits a modified response to gravity is lazy-2. Plants of this mutant grown under greenhouse conditions have short internodes and are positively gravitropic (Figure 4). The older leaves of lazy-2 plants also display pronounced epinastic curvatures. In order to unravel the gravitropic properties of the mutant, dark and light grown seedlings have been reorientated and the rate of curvature monitored. A time course of the bending of hypocotyls from etiolated plants reveals that the response of the mutant is indistinguishable from normal seedlings. Bending begins within 20 minutes and is complete after 90 minutes stimulation (Figure 5). Gravitropic responsiveness of the roots is also unaffected by the mutation. However, if the seedlings are grown under continuous light the hypocotyl displays pronounced positive gravitropism. On gravistimulation of light grown hypocotyls of the mutant there is an initial negative response to gravity which is discernible within 30 minutes of stimulation, and then this is counteracted by a positive response (Figure 6). A final angle of bending is achieved within approximately 12 hours. The graviresponse of lazy-2 roots is unaffected by light.

If light grown lazy-2 seedlings are placed in the dark then the hypocotyls develop the capacity to respond negatively to gravity within about 48 hours (I. Gilbert, pers. comm.) indicating that although light can dictate the graviresponsiveness of lazy-2 seedlings, it does not irreversibly switch their gravitropic expression.

ABSCISIC ACID-DEFICIENT MUTANTS

The role of plant growth regulators in gravitropism has been extensively discussed in a recent review by Pickard (1985) and therefore will not be repeated here. Since it has been strenuously argued by some workers that ABA plays an important role in root gravitropism (Pilet, 1983b), a further approach to probe this developmental response may be to study mutants deficient in this plant growth regulator.

Although Moore and Smith (1985) could detect no ABA in the roots of the carotenoid deficient maize mutants W-3, vp-5 and vp-7, the gravitropic responses of roots of these seedlings were the same as non-mutant plants. Similarly, Weyers (1985) found that after 48 hours gravistimulation the angle of bending of the ABA deficient tomato mutant flacca was insignificantly different from normal roots. Whilst these results offer no support for a key role for ABA in root gravitropism, it would be unwise to conclude from them that ABA plays no role at all. As Weyers points out, the level of ABA in specific cellular compartments within the root may be of crucial importance. Also measurements of ABA were made on whole roots and not restricted to the tissues of the cap.

Figure 4. Normal, Lazy-1 and Lazy-2 plants grown for 6 weeks under greenhouse conditions with an 18h photoperiod and $15/22^{\circ}C$ minimum night/day temperatures.

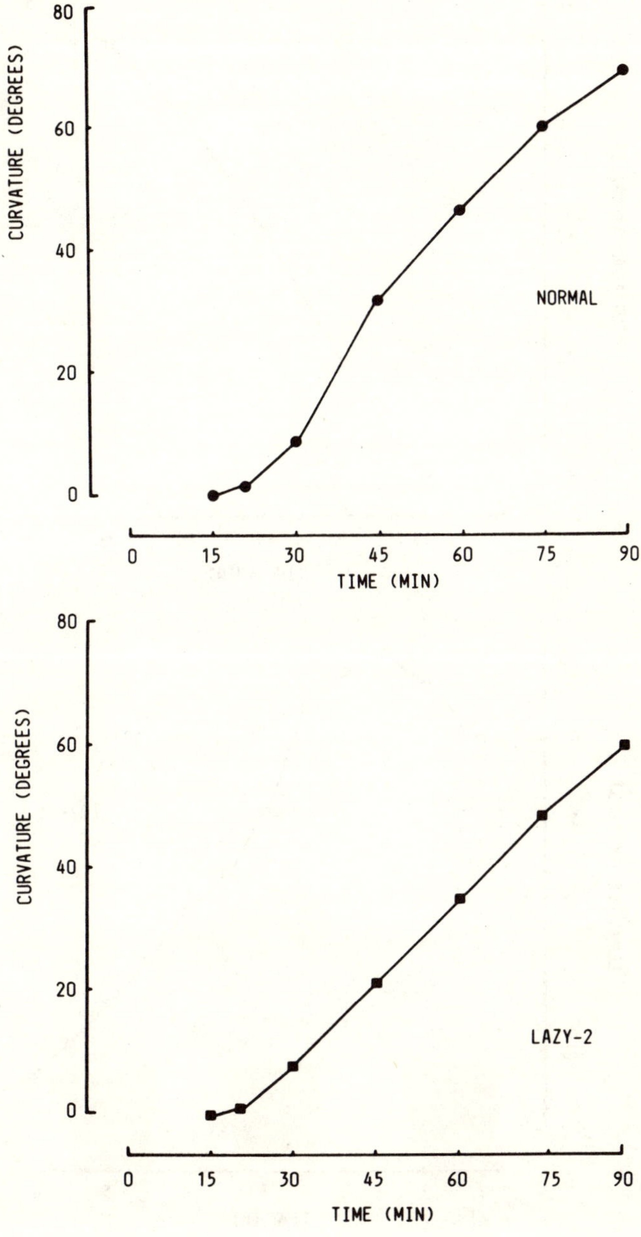

Figure 5. Gravitropic bending of hypocotyls from etiolated seedlings of normal and Lazy-2 tomato plants.

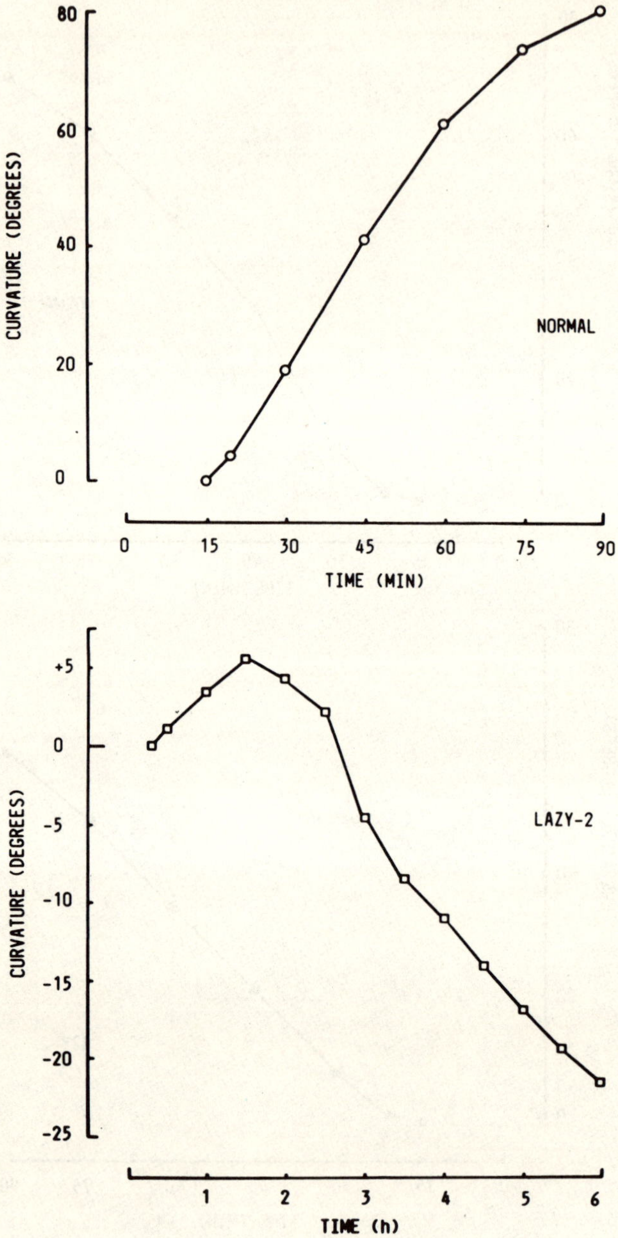

Figure 6. Gravitropic bending of hypocotyls from light grown seedlings of normal and Lazy-2 tomato plants.

CONCLUSIONS FROM STUDIES ON MUTANTS

Although the use of mutants to investigate the mechanism of gravitropism is in its comparative infancy, certain conclusions may be drawn from the studies undertaken so far.

The hypothesis that gravity is perceived by sedimenting amyloplasts is strongly supported by the anatomical studies of lazy-1 hypocotyls, roots of the Arabidopsis mutant aux-1 and amylomaize coleoptiles. In each of these mutants, the most plausible explanation for the gravitropic responses of the tissues concerned lies at the level of graviperception. That is, statoliths are either absent, reduced in diameter, or fail to sediment for some reason. A role for amyloplasts in graviperception is also supported by a comparison of the gravitropic bending of etiolated and light grown tomato seedlings. Dark grown hypocotyls bend at a fixed distance from the apical hook, while light grown hypocotyls reorientate from the base (see Figure 7).

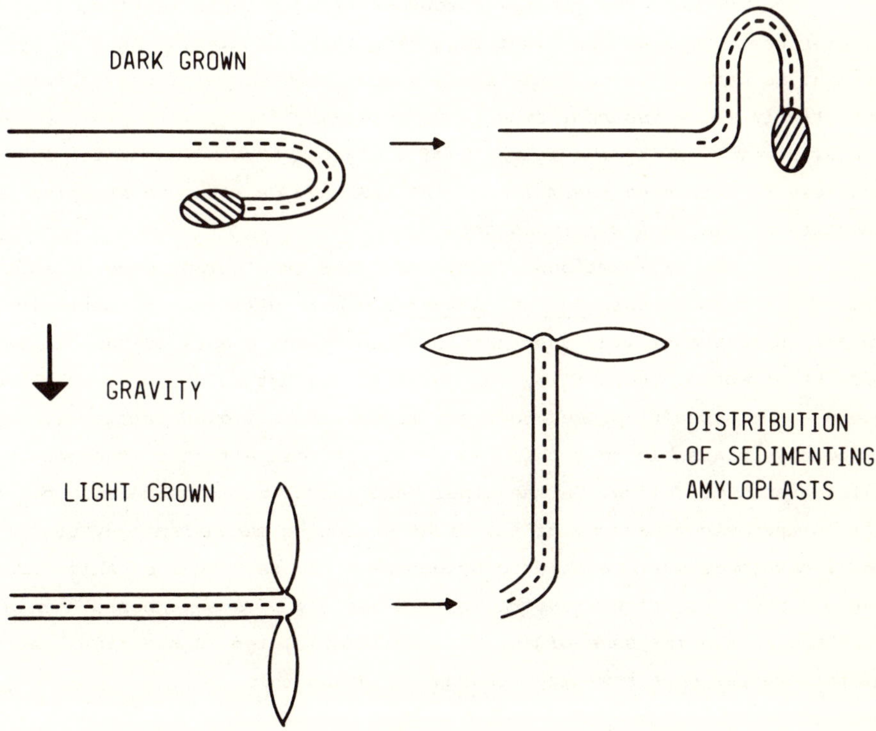

Figure 7.- Diagrammatic representation of the distribution of sedimenting starch grains and site of gravitropic bending in dark or light grown tomato hypocotyls.

Sedimenting amyloplasts in light or dark grown seedlings are only present within the endodermal cells as far as the site of bending. This observation implies that after graviperception, the stimulus inducing differential bending is transmitted laterally and not longitudinally.

Clearly the effect of the mutation in the plants described is not solely restricted to graviperception. For instance, evidence has been presented that the root cap of the maize mutant agt lacks a growth inhibitor prerequisite for gravitropism. Such a deficiency could also account for the agravitropic behaviour of the pea mutant ageotropeum since elongation of horizontally placed roots is significantly greater than those from non-mutant plants. Certainly the effects of this mutation are most likely to originate at the level of stimulus transduction since agt roots can execute a bend in response to a gravistimulus from normal root caps (Ekelund and Hemberg, 1966).

Perhaps the greatest progress using mutants to study root gravitropism has come from the light requiring maize cultivars. In this system we must be careful to separate those events necessary for gravitropic sensitivity to be induced, from those necessary for gravitropism to take place. However, using the facility of a light switching mechanism it may be possible to determine the roles of IAA, ABA and Ca^{2+} in gravitropism and the identity of the root cap inhibitor.

The explanations for the modified growth responses of shoots from the tomato mutants dgt and lazy-2 are more difficult to determine. Both mutants perceive gravity and therefore subsequent events in the gravisequence must be affected. The ability to 'correct' the dgt mutation by ethylene may prove to be a useful probe. However, it should be viewed cautiously since it is equally plausible that this may be an indirect effect of the gas and a blind avenue to follow. On the other hand lazy-2 has considerable potential for further study because of its light switching mechanism. Any gravitropic model developed must be able to accommodate the responses of this mutant and explain how a positive gravitropic response may be superimposed upon a negative response within the same organ. The model formulated in the final section of this review goes some way towards achieving this.

A GRAVITROPIC MECHANISM FOR HYPOCOTYL BENDING

The endodermal cells within the root play a crucial role in regulating the import of molecules into the vascular tissue. This is achieved by the localization of the casparian strip within the root endodermis which restricts apoplastic movement and ensures symplastic transport into the vascular tissue. Continuous with the endodermis of the root is the starch sheath which contains large numbers of sedimenting amyloplasts. The evidence that has been presented in the foregoing part of this chapter indicates that these amyloplasts are responsible for the perception of gravity. In view of the proposal that hypocotyl growth is primarily regulated by the rate of elongation of the peripheral cell layers (Firm and Digby, 1977) it seems paradoxical to separate the site of graviperception from the site of response. Indeed some workers have attempted to bring these two processes within the same cells (Firm and Digby, 1977). One plausible explanation for this apparent paradox might be that just as the root endodermal cells act to regulate import into the vascular tissue, the shoot endodermal cells act to regulate export out of the vascular tissue. This hypothesis may be developed into a model to account for gravitropism in hypocotyl tissue (see Figure 8).

In this model sedimenting amyloplasts act as a non-specific barrier to the outflow of solutes including plant growth regulating compounds. It is envisaged that these are primarily moving in a basipetal direction since decapitation reduces or abolishes gravitropic bending (Macdonald and Hart, 1985). When a hypocotyl is maintained in a vertical orientation (Figure 8a) then the lateral distribution of growth regulating compounds would be equally balanced and even elongation would proceed. On reorientation by 90° (Figure 8b) the amyloplasts sediment and form a barrier restricting symplastic flow into the upper cortical tissue. A role for amyloplasts in regulating movement through plasmodesmata has been previously proposed by Juniper (1976). The distribution of sedimenting starch grains on the lower endodermal cells would present no barrier to flow from the vascular tissue. Once the growth regulating compounds have gained access into the endodermal cells further lateral flow may be mediated by symplastic and/or apoplastic movement. This simple model would account for the common observations that growth of the upper side of a gravistimulated hypocotyl terminates rapidly after reorientation and that differential growth which brings about gravicurvature begins simultaneously along the length of the organ (Digby and Firn, 1979). It would also explain why there is a limited longitudinal transmission of graviperception as described earlier.

If the proposed model does operate then a gravistimulated hypocotyl may rapidly redistribute a number of compounds after reorientation. Such a hypothesis would explain why there are reports of a lateral translocation of both IAA and gibberellic acid and of such compounds as fluorescein which has been shown to be distributed to the lower side of mung bean hypocotyls within 2.5 minutes of gravistimulation (Tanada, 1978). The model may also be used to account for the gravitropic responses of the tomato mutant lazy-2. Etiolated hypocotyls of this mutant would respond negatively to

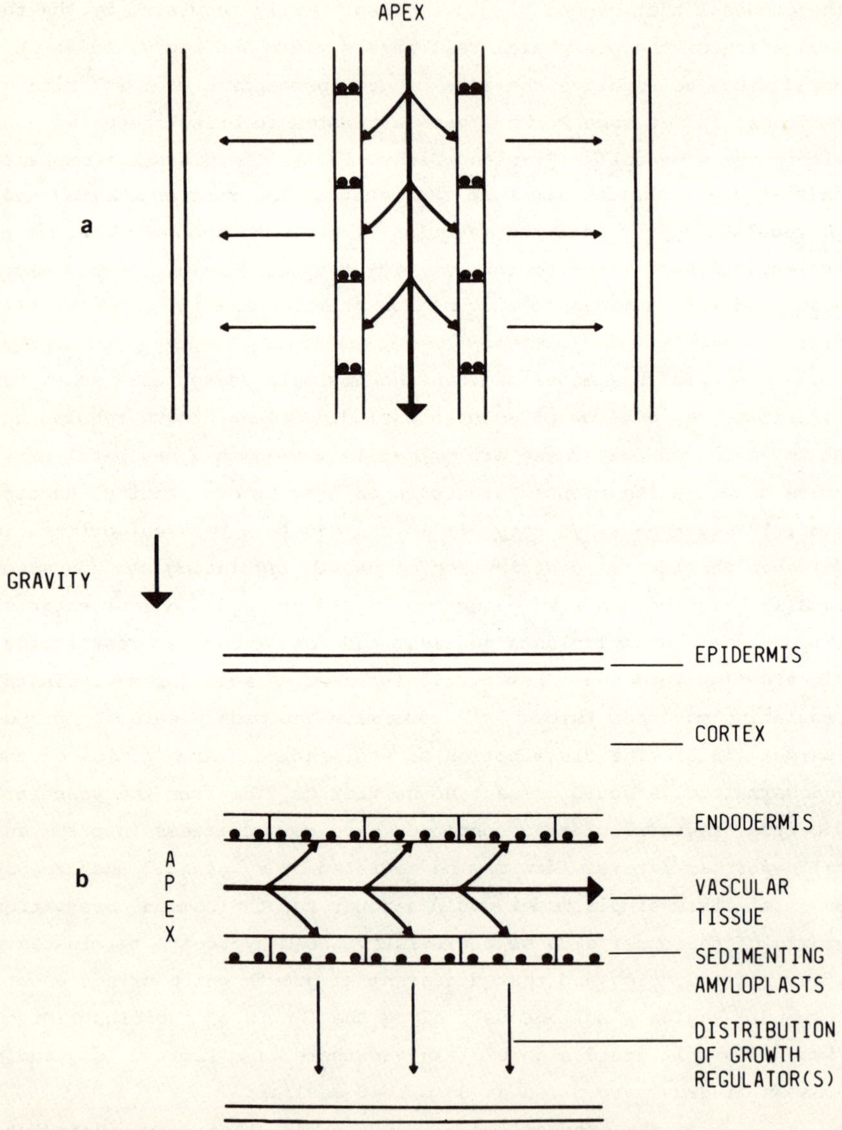

Figure 8. Hypothetical mechanism of gravitropism in (a) vertical and (b) horizontally-placed hypocotyl tissue.

gravity as a result of a downward movement of a growth promoting compound(s). If the apical tissues are induced to synthesize elevated levels of a growth inhibitor by light, then on gravistimulation this could also be redistributed. The net effect would be for growth on the lower surface to be gradually inhibited and a positive gravitropic response would overwhelm the initial negative one. In support of an elevated growth inhibitor level in light grown lazy-2 plants is the observation that internodal elongation of this mutant is significantly reduced.

In conclusion the use of mutants is having a considerable impact on our knowledge of gravitropism. We now have a number of systems which can be studied to elucidate not only the mechanism of gravitropic bending but also determine the role of gravity in other developmental processes such as circumnutation and epinasty. A simple model has been presented to account for gravitropic bending in hypocotyls and this warrants further examination.

ACKNOWLEDGEMENT

Some of the work described has been supported by the University of Nottingham Research Fund.

REFERENCES

Blixt, S., Ehrenberg, L. & Gelin, D. (1958) Quantitative studies of induced mutations in peas. I. Methodological investigations. Agri Hortique Genetica, 16, 238-50.

Digby, J. & Firn, R.D. (1979) An analysis of the changes in growth rate occurring during the initial stages of geocurvature in shoots. Plant Cell Environ., 2, 145-8.

Ekelund, R. & Hemberg, T (1966) A comparison between geotropism and geo-electric effect in Pisum sativum and its mutant ageotropum. Physiol. Plant. 19, 1120-4.

Feldman, L.J. (1981a) Root cap inhibitor formation in isolated root caps of Zea mays. J. Exp. Bot., 32, 779-88.

Feldman, L.J. (1981b) Light-induced inhibitors from intact and cultured caps of Zea roots. Planta, 153, 471-5.

Feldman, L.J. (1983) Light-enhanced protein synthesis in gravitropically stimulated root caps of corn. Plant Physiol., 72, 833-6.

Feldman, L.J. (1985) Root gravitropism. Physiol. Plant., 65, 341-4.

Feldman, L.J., Arroyave, W.J. & Sun, P.S. (1985) Abscisic acid, xanthoxin and violaxanthin in the caps of gravistimulated maize roots. Planta, 166, 483-9.

Feldman, L.J. & Gildow, V. (1984) Effects of light on protein patterns in gravitropically stimulated root caps of corn. Plant Physiol., 74, 208-12.

Firn, R. & Digby, J. (1977) The role of the peripheral cell layers in the geotropic curvature of sunflower hypocotyls: A new model of shoot geotropism. Aust. J. Plant Physiol., 4, 337-347.

Harrison, M.A. & Pickard, B.G. (1984) Burst of ethylene upon horizontal placement of tomato seedlings. Plant Physiol., 75, 1167-9.

Harrison, M.A. & Pickard, B.G. (1986) Evaluation of ethylene as a mediator of gravitropism by tomato hypocotyls. Plant Physiol., 80, 592-5.

Hertel, R., De la Fuente, R.K. & Leopold, A.C. (1969) Geotropism and the lateral transport of auxin in the corn mutant amylomaize. Planta, 88, 204-14.

Jackson, M.B. (1979) Is the diageotropic tomato ethylene deficient? Physiol. Plant., 46, 347-51.

Juniper, B.E. (1976) Geotropism. Ann. Rev. Plant Physiol., 27, 385-406.

Lee, J.S. & Evans, M.L. (1985) Polar transport of $^{45}Ca^{2+}$ across the elongation zone of gravistimulated roots. Plant Cell Physiol., 26, 1587-95.

Lee, J.S., Mulkey, T.J. & Evans, M.L. (1983) Reversible loss of gravitropic sensitivity in maize roots after tip application of calcium chelators. Science, 220, 1375-6.

Macdonald, I.R., Gordon, D.C., Hart, J.W. & Maher, E.P. (1983) The positive hook: the role of gravity in the formation and opening of the apical hook. Planta, 158, 76-81.

Macdonald, I.R. & Hart, J.W. (1985) The role of the apex in normal and tropic growth of sunflower hypocotyls. Planta, 163, 549-53.

Maher, E.P. & Martindale, S.J.B. (1980) Mutants of A. thaliana with altered responses to auxin and gravity. Biochem. Genet., 18, 1041-53.

Mirza, J.I., Olsen, G.M., Iversen, T-H & Maher, E.P. (1984) The growth and gravitropic responses of wild-type and auxin resistant mutants of Arabidopsis thaliana. Physiol. Plant., 60, 516-22.

Moore, R. (1985) A morphometric analysis of the redistribution of organelles in columella cells in primary roots of normal seedlings and agravitropic mutants of Hordeum vulgare. J. Exp. Bot., 36, 1275-86.

Moore, R. & Smith, J.D. (1985) Graviresponsiveness and ABA-content of roots of carotenoid deficient mutants of Zea mays. Planta, 164, 126-8.

Olsen, G.M. & Iversen, T-H. (1980a) Growth and curvature in seedlings of Pisum sativum and an ageotropic mutant. Physiol. Plant., 50, 269-74.

Olsen, G.M. & Iversen, T-H. (1980b) Ultrastructure and movements of cell structures in normal pea and an ageotropic mutant. Physiol. Plant., 50, 275-84.

Olsen, G.M., Mirza, J.I., Maher, P. & Iversen, T-H. (1984) Ultrastructure and movements of cell organelles in the root cap of agravitropic mutants and normal seedlings of Arabidopsis thaliana. Physiol. Plant., 60, 523-31.

Pickard, B.G. (1985) Role of hormones, protons and calcium in geotropism. In: Hormonal regulation of development II, eds. R.P. Pharis & D.M. Reid. Berlin: Springer-Verlag.

Pilet, P.E. (1983a) Elongation and gravireactivity of roots from an agravitropic maize mutant: Implications of growth inhibitors. Plant Cell Physiol., 24, 333-6.

Pilet, P.E. (1983b) Control of root growth by endogenous IAA and ABA. In: Growth regulators in root development, B.P.G.R.G. monograph 10, eds. M.B. Jackson & A.D. Stead, pp. 15-24.

Pilet, P.E., Mayor, G., Jaunin, F. & Nicollier, F. (1982) Freeze-decapping: A new technique for analysing root growth and gravireaction. Plant Sci. Lett., 28, 121-8.

Roberts, J.A. (1984) Tropic responses of hypocotyls from normal tomato plants and the gravitropic mutant Lazy-1. Plant Cell Environ., 7, 515-20.

Scholdeen, C. & Burstrom, H. (1960) Physiological studies of an ageotropic pea mutant. Physiol. Plant., 13, 831-8.

Scott, T.K. & Wilkins, M.B. (1969) Auxin transport in roots. IV Effects of light on IAA movement and geotropic responsiveness in Zea roots. Planta, 87, 249-58.

Severi, A. & Laudi, G. (1977) Transport of labelled IAA in normal or Lazy stems of tomato. Z. Pflanzenphysiol., 84, 313-20.

Shaw, S. & Wilkins, M.B. (1973) The source and lateral transport of growth inhibitors in geotropically stimulated roots of Zea mays and Pisum sativum. Planta, 109, 11-26.

Sievers, A. & Volkmann, D. (1972) Verursacht differentieller Druck der Amyloplasten auf ein kumplexes Endomembransystem die Geoperzeption in Wurzeln. Planta, 102, 160-72.

Tanada, T. (1978) A rapid gravitational effect on the translocation of fluorescein in mung bean hypocotyl sections. Planta, 142, 221-3.

Wendt, L., Tagliani, L., Eslick, R. & Foley, M. (1984) Selection of an ageotropic barley root mutant. Plant Physiol., 75, S180.

Weyers, J.D.B. (1985) Germination and root gravitropism of flacca, the tomato mutant deficient in abscisic acid. J. Plant Physiol., 121, 475-80.

Wheeler, R.M. & Salisbury, F.B. (1981) Gravitropism in higher plant shoots. I. A role for ethylene. Plant Physiol., 67, 686-90.

Wilkins, H. & Wain, R.L. (1975) Abscisic acid and the response of the roots of Zea mays seedlings to gravity. Planta, 126, 19-23.

Wilkins, M.B. (1984) Gravitropism. In: Advanced Plant Physiology, ed. M.B. Wilkins, pp. 163-85. London: Pitman.

Woods, S.L., Roberts, J.A. & Taylor, I.B. (1984) Ethylene-induced root coiling in tomato seedlings. Plant Growth Reg. 2, 217-25.

Zobel, R.W. (1973) Some physiological characteristics of the ethylene-requiring tomato mutant diageotropica. Plant Physiol., 52, 385-9.

Zobel, R.W. (1974) Control of morphogenesis in the ethylene-requiring tomato mutant, diageotropica. Can. J. Bot., 52, 735-43.

GENETIC AND HORMONAL REGULATION OF STATURE

John L. Stoddart
Welsh Plant Breeding Station, Aberystwyth, Dyfed. SY23 3EB

INTRODUCTION

Plant stature is a primary factor in survival because of advantages conferred in terms of preferential light interception, ease of pollination and more effective seed dispersal. Dominant species in climax stands developed in most habitats usually have the greatest height. At the extreme, the ability to produce and extend an indeterminate series of stem internodes is crucial to the development of a climbing habit.

In economic terms, stature is a primary component of yield. Maximum production in crop plants frequently has a close relationship to stature, or to the success with which the plant breeder can manipulate that parameter. Increases in height can enhance yield directly where biomass is the harvested element but the relationship can be more subtle with criteria such as grain yield or vegetative quality. In ornamentals, the ability to manipulate stature is often crucial for marketability.

Control of plant height is achieved by varying the number and/or the extension rate of sub-apical internodes, and the process is amenable to intervention in either the division or elongation phases of the cell cycle. Changes in the stature of a given species can be achieved by two recognised routes, the first being genetic and the second via the use of growth regulating chemicals, or the regulated exposure to specific environmental factors (e.g. low temperature or photomorphogenic light).

Traditionally, manipulation of height in the major crop species has been achieved by the use of selection and hybridisation procedures (i.e the genetic route) and in wheat, for example, current crop statures are around 40% of those prevailing in agricultural practice at the end of the second World War. This trend has been driven by the need to employ high fertiliser inputs to achieve satisfactory economic returns and by the enhanced grain producing potential emanating from breeding strategies aimed at shifting carbon-partioning processes in favour of the grain. A further factor has been the pressure to use entirely mechanised harvesting procedures in order to reduce the labour cost element. Thus, crop stature can become a central theme in a complex economic exercise.

Chemical strategies for the modification of extension growth and plant form have developed rapidly over a similar period, and have spawned a vast literature. Advances were consequent upon the production of synthetic auxins, the development of large-scale fermentative procedures for GAs (see Phinney, 1983) and, latterly, the discovery of growth retardants and ethylene-generating compounds (see Sembdner _et al._, 1980). A daunting array of effects and interactions can be extracted from the literature by those with the necessary courage!

This review is restricted to a consideration of genetic effects on stature, together with discussion of underlying hormonal, or other, mechanisms where these are understood. The examples used in the various sections are drawn from a restricted taxonomic range and reflect the heavy emphasis placed on stature control in crop species during recent years.

GENETIC CONTROL

Stature reduction

Pisum sativum, _Zea mays_, _Triticum aestivum_, _Oryza sativa_, _Phaseolus vulgaris_ and _Pharbitis nil_ are species in which genetic dwarfing systems have been characterised extensively. However, at the outset, it should be appreciated that control varies from simple single-site events to complex allelic interactions at multiple loci and that a temporal control of gene expression (itself regulated by genetic factors) can be involved.

Pisum sativum

Genetic control of height in _Pisum_ is fundamentally related to the _Le_ locus (White, 1917) but at least four other loci (_La_, _Cry_, _Na_ and _Lm_) are known to contribute to the determination of internode length. The effect of _le_ is, however, confined to reductions in the lengths of the upper stem internodes (Reid _et al._, 1983). A number of workers (e.g. Rasmusson, 1927; Lamm, 1937) have shown interdependent effects at these loci, exemplified by the finding that in a _le/le_ background either _La_ or _Cry_ will produce a dwarf phenotype, whereas the opposite stature is produced in the presence of homozygous dominant alleles. The mutant _na_ (_nana_) allele produces a sharp reduction in internode length in the presence of both dominant and recessive combinations at the _Le_ locus (Wellensiek, 1969). Two recessive alleles of _Cry_

(cry^{c} and cry^{s}) are expressed, respectively, as cryptodwarf and slender phenotypes when in combination with le and la.

A further recessive mutant (lm) results in a micro phenotype (Lindquist, 1951) with shorter internodes than those produced in Lm genotypes. A full range of micro stature forms (micro-tall, micro-dwarf, micro-cryptodwarf and micro-slender) are produced in the presence of various alleles at the Le, La and Cry loci.

Recently, three further dwarfing mutants in peas have been described, designated lh, ls and lk. They are not allelic with the genes at the five established loci, already described (Reid and Potts, 1985).

Le, La, Cry and Na exert their effects on internode length in a similar manner. All modulate a large change in the number of cells per internode for the outer cortical tissue of the stem and only influence cell length to a lesser extent (Reid et al., 1983). However, in the epidermis, the reverse situation applies. Extensive grafting experiments have been conducted to examine the nature of the genetic controls on stature in peas and a number of workers have suggested that the effects of the Le locus are not graft-transmissible (e.g. McComb and McComb, 1970; Lockhard and Grunwald, 1970). Reid et al. (1983) extended this finding to cover the La and Cry loci and they suggest that the three loci are affecting processes localised within the meristematic zones of the apical regions. In contrast, a considerable transmissible effect from Na scions, capable of promoting growth in na stocks, was observed. This suggests that na blocks some step in the biosynthetic pathway of a translocatable growth regulating compound and, contrary to earlier findings, there are also suggestions that Le may interact in similar processes (Reid, et al., 1983).

The correspondence between phenotype and various allelic combinations at height determining loci has been described by Reid et al., (1983) and this information is reproduced in Table 1.

There is now considerable evidence to support the hypothesis that a number of the pea dwarfing genes operate by modifying aspects of the gibberellin (GA) biosynthesis pathway in the subapical region. For example, Potts et al. (1982) examined the effect of the Le/le locus on endogenous GA-like substances using non-homologous bioassays. The approach was based upon the observation that quaternary ammonium growth retardants, known to inhibit GA biosynthesis (Kende et al., 1963), had dwarfing effects on tall pea cultivars and that these effects could be reversed by exogenous GA.

Consistently large differences were found in the GA contents of shoots of light-grown tall and dwarf lines. The major fraction of the biological activity in tall (Le) material was present in a polar form and this component was essentially undetectable in the dwarf (le) lines. A further zone of biological activity, cochromatographing with GA_{20}, was invariate between lines. The polar component was shown to be present predominantly in the actively elongating sub-apical internode tissues and was virtually absent in mature stems and leaves. Further studies (Ingram et al., 1983) using combined gas chromatography-mass spectrometry (GCMS) have

Table 1.

The relationship between stature phenotype and allelic combinations at height-determining loci in Pisum (from Reid et al., 1983)

Phenotype	Genotype	
TALL	Le La Cry Na Lm.	Le La cry^{c} Na Lm.
	Le La cry^{s} Na Lm	Le la Cry Na Lm.
DWARF	le La Cry Na Lm	le La cry^{c} Na Lm.
	le La cry^{s} Na Lm	le la Cry Na Lm.
CRYPTODWARF	le la cry^{c} Na Lm.	
CRYPTOTALL	Le la cry^{c} Na Lm.	
SLENDER	Le la cry^{s} Na Lm	le la cry^{s} Na Lm.
MICROCRYPTODWARF	le la cry^{c} Na lm.	
MICROTALL	As for TALL except lm.	
MICRODWARF	As for DWARF except lm.	
MICROSLENDER	As for SLENDER except lm.	
NANA (COMPACTUM)	As for TALL and DWARF, except na.	

demonstrated that the polar compound is GA_1 (Figure 1) and have confirmed the virtual absence of this GA in le plants. Exogenous supplies of GA_1 were effective in overcoming dwarfism due to the Le locus. Both GA_1 and its immediate biosynthetic precursor GA_{20} promoted increased elongation when applied to Le plants but, in contrast, only GA_1 stimulated growth in le genotypes. This suggested that the Le/le locus should be concerned with the enzymology of conversion of GA_{20} to GA_1. This possibility was directly addressed by Ingram et al., (1984) using GA_{20} doubly labelled with ^{13}C and ^{3}H. The metabolic fate of (^{13}C, ^{3}H) GA_{20} was followed by GCMS after feeding to Le/le and Na/na lines. Apical tissues of Le containing plants produced three metabolites (^{13}C, ^{3}H) GA_1, (^{13}C, ^{3}H) GA_8 and (^{13}C, ^{3}H) GA_{29}, whereas only (^{13}C, ^{3}H) GA_8 and (^{13}C, ^{3}H) GA_{29} and a catabolite of that GA, were detected in le homozygotes. The Na and na lines showed similar patterns of

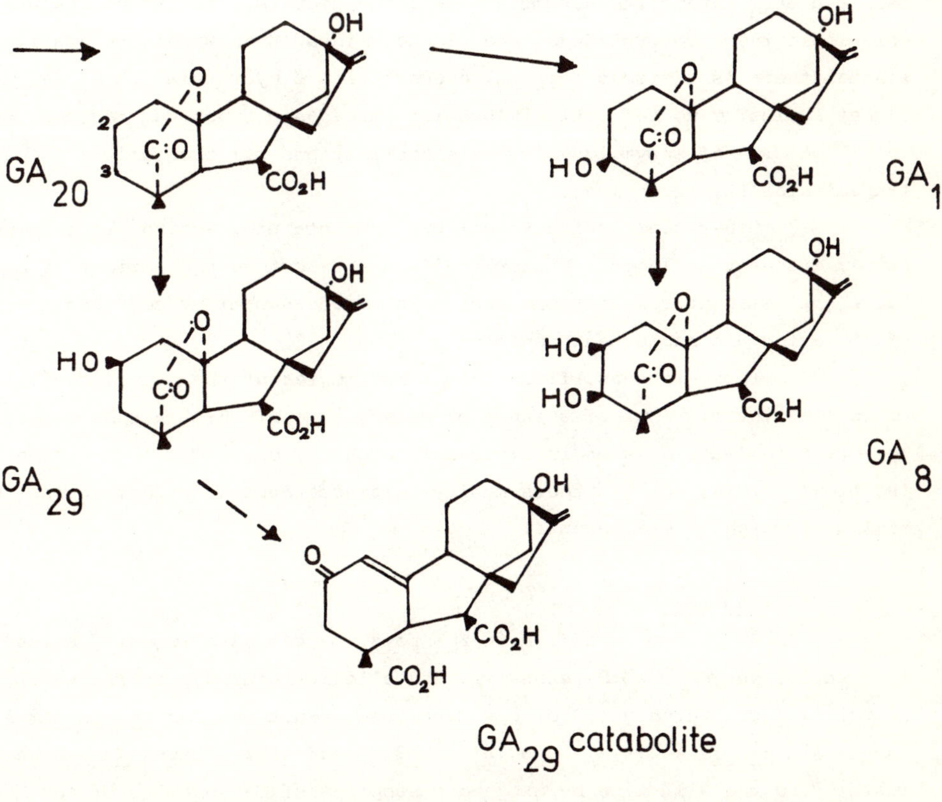

Figure 1. Structures and metabolic interrelationships of some gibberellins occurring in vegetative tissues of *Pisum sativum* L.

metabolism but there was a higher dilution of the GA samples with endogenous ^{12}C forms in the extracts of material containing the dominant allele, confirming that the na allele is affecting biosynthetic steps prior to GA_{20} (Potts and Reid, 1983).

These data constitute strong evidence for the Le/le locus being expressed as a 3 β hydroxylase controlling the ultimate activation step in the GA biosynthetic pathway in Pisum, and for the Na/na locus having a similar influence at an earlier, uncharacterised, point in the pathway.

The new non-allelic dwarfs lh and ls are responsive to applications of GA_1 and are thought, therefore, also to result from interferences with the GA pathway. Their points of action are, as yet, unexplored (Reid and Potts, 1985).

For those with an interest in the temporal control of gene action, it is noteworthy that the Le locus is not expressed in seed tissues, where GA_{29} and GA_{29}-catabolite accumulate in large amounts. The catabolite was only found in microcryptodwarf and slender lines. Even within vegetative tissues there is a strong temporal element with 3 hydroxylation of GA_{20} being almost exclusive to the actively growing apical and subapical regions. Such tight location of enzymology suggests explanations for the failure of graft transmissibility experiments.

Other gene loci have effects upon stature, acting via internode number rather than length. Murfet (1971) has described the actions of Sn and Hr, where dominant combinations modify internode number by delaying the onset of flowering and subsequent senescence.

Thus, in peas, final height may represent the sum of the actions or complex interactions of a range of genes, some of which act by modifying the extension rate of stem internodes throughout, or in defined portions of the growth cycle, whilst others modify internode number by influencing the timing of morphogenetic events.

Zea mays

Studies on dwarfism in Zea mays (maize) have revealed a less interactive pattern, with recessive, non-allelic mutations operative at a number of loci. Phinney (1956) has described phenotypes for anther-ear 1 (an-1) and dwarfs 1,2,3,5 and 8 (d1,d2,d3,d5 and d8). All but d8 can be restored to the wild type by exogenous supplies of GA_1 or GA_3. Of the known mutant genes in maize, Phinney and Spray (1982) claim that thirty have effects upon plant stature.

In contrast to the pea mutants, *d1*,*d2*,*d3* and *d5* remain dwarfed when grown in darkness and express degrees of etiolation which are proportional to their dwarfness in the light. The tightly defined responses to applied GA suggested that the mutations could act by modifying supplies of GA to the elongating internode tissues and a series of subsequent elegant studies have refined this suggestion to the point of certainty. The operative points for each of the mutant alleles are shown on the biosynthetic pathway for GAs in maize depicted in Figure 2.

The *d1* mutation disables the final β-hydroxylation step in the biosynthesis of GA_1 and is, therefore, analogous to the action of *le* in *Pisum*. Only GA_1 shows significant biological activity when applied to *d1* seedlings. Consecutive steps between GA_{12} aldehyde and GA_{53} are blocked by the *d3* and *d2* mutations, the former being a C-13-hydroxylation and the latter a C-7-oxidation. All intermediates following GA_{12} aldehyde therefore have some effectiveness in stimulating growth of *d3* seedlings whilst GA_{53} is the earliest effective compound for *d2*. *d5* responds to all cyclic intermediates but effectiveness decreases for earlier points on the pathway. Detailed studies have shown that *d5* blocks the cyclisation process yielding *ent*-kaurene from copalyl pyrophosphate (CPP) (Hedden and Phinney, 1979). The enzyme exists in A and B forms, acting as shown in Figure 3. Duncan and West (1981) have shown that the two sites reside on separate polypeptides but interact closely with preferential transport of CPP between the A and B sites. The *d5* mutation affects the kaurene synthetase B step leading,

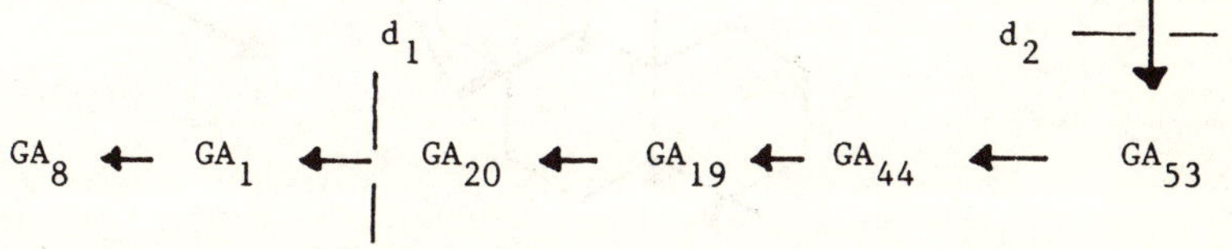

Figure 2. Postulated biosynthetic pathway for gibberellin A_1 production in *Zea mays* L., with action sites for four mutant dwarfing genes (after Phinney and Spray, 1982).

unusually, to an aberrant product rather than a complete functional blockage. Mutant seedlings produce ent-isokaurene which is either incapable of being utilised by the next enzyme in the biosynthetic sequence (ent-kaurenol synthesis) or proceeds to yield a final GA with the 'wrong' configuration for biological activity. This point is uncertain. Hedden and Phinney (1979) indicate that d5 mutants synthesise less than 25% of the ent-kaurene found in wild-type seedlings and infer that this is the basis for the reduced growth rate in the mutant. However, this statement also indicates that d5 is a 'leaky' mutation.

The classical work by Phinney and co-workers on the maize mutants has been used as principal evidence to support the important assertion that GA_1 is the only active endogenous GA, and that all other GA structures extractable from maize tissues are biologically active only to the extent to which they can be metabolically converted to GA_1 (Phinney and Spray, 1982).

The regulatory steps in the GA pathway are, however, considered to precede the synthesis of GA_{20} and are not thought to be resident in the final hydroxylation to yield GA_1 (Spray et al., 1984).

A series of genetically-directed mutations in maize, depending

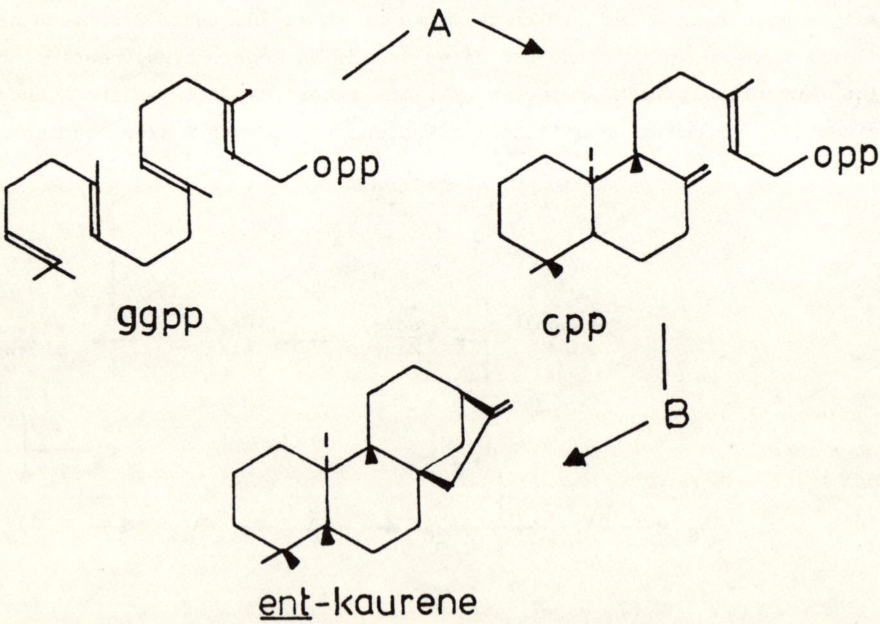

Figure 3. Reaction scheme for ent-kaurene synthetases A and B.

upon the random insertion of a transposable genetic element, has been described recently (Barker et al., 1984).

Robertson's (1978) Mutator system in maize is characterised by a high induction frequency for mutations, many of which are somatically unstable or are incapable of transmission through a seed generation. The transposable element Mu1 has been characterised by Barker et al. (1984) using a mutant allele of alcohol dehydrogenase (Adh1-S) which is expressed as low enzyme and mRNA levels (ca 40%). An insertion element has been isolated from a clone of this gene and shown to contain long terminal inverted repeats, with duplications of the target DNA sequence adjacent to the transposable element. Maize lines displaying Mutator activity can contain up to 60 copies of the element per diploid genome.

Structurally, Mu1 consists of about 1400 base pairs, divided into symmetrical halves each containing long inverted terminal and direct internal repeats. There are two open reading frames in each half and a region of high GC content, which is very different from that characterising the maize genome.

The substantially random manner in which such elements can be inserted permits the generation of a wide spectrum of mutations, many of which may concern nuances of cellular biochemistry and architecture, and may not be expressed in visible external changes. However, Phinney (personal communication) has indicated that dwarf phenotypes can arise from insertions at appropriate points in the genome, some of which will certainly differ from the sites affected by the characterised undirected mutations (d1, d5, etc.). Even in cases where Mu1 is inserted into known genes, it will provide a means of mapping these regions by utilising clones of Mu1 labelled by nick translation.

Temperate cereals

Stature determination in temperate cereals has been of primary concern to crop geneticists. Dwarf plants in *Triticum aestivum* L. (wheat) have been categorised into 'grass dwarfs' and short-statured or semi-dwarf types (Fick and Qualset, 1973). Grass dwarfs have an extremely short compact habit, are high tillering and have short dark-green leaves. Extreme internode shortening also occurs near the base of the plant, leading to an overall growth habit more reminiscent of a perennial grass than a cereal. It is thought that the grass dwarf phenotype results from the interactions of three dominant genes (D1,D2,D3). Homozygous dominance of heterozygosity at all three loci resulting in dwarfism, which also occurs in the absence of D3 if the genotype is homozygous for D2 and at least heterozygous for D1. A fourth gene (D4) is suspected to be present in some lines and the expression of all the genes is variable in certain crosses. There are indications that grass dwarfing genes may also contribute to stature determination in some semi-dwarf genotypes.

Height in semi-dwarf lines is also thought to be regulated by a limited number of factors and the interactions with grass dwarfing genes have received little attention, other than in the form of suggestions that grass dwarf types may be the ancestors of semi-dwarf lines (Reitz and Salmon, 1968). However, inheritance studies on semi-dwarfing genes themselves are abundant (e.g. Powell and Schleuhuber, 1967; Johnson et al., 1966) and the Norin-10 lines, emanating from Japan have formed the basis of the range of modern short-statured cultivars leading to the 'Green Revolution'. Norin-10 has two genes for dwarfness (Fick and Qualset, 1973), currently designated as Rht1 and Rht2 which result in statures around 70% of those typifying 'tall' cultivars. A third dwarfing factor, the so called 'Tom Thumb' gene (Rht3) confers full dwarfness, with statures of 40% or less of those in tall controls. These plants are dark green and small-leaved but tillering patterns are not abnormal, as in grass-dwarfs. In combination, the Rht1, Rht2 and Rht3 loci have been used to titrate crop stature in a sensitive manner and their utility is increased by the fact that, unlike the dwarfing factors discussed for pea and maize, they exhibit dominance.

The Norin-10 genes resemble Rht3 in their mode of action. Seedlings containing any of the three genes are insensitive to exogenous GA, showing none of the characteristic acceleration of growth rate exhibited by non-dwarf cultivars after such treatment. They are, therefore, classified as 'gibberellin insensitive dwarfs' and this term was the basis for the Gai nomenclature used in earlier publications on the topic (e.g. Gale and Marshall, 1975). However Rht1 and Rht2 differ significantly from Rht3 in terms of the GA responsiveness of α-amylase synthesis in the aleurone layer of the germinating grain. Norin-10 and similar semi-dwarf lines show a total α-amylase release over 72 h, and a dose/response pattern towards GA_3, which closely parallels that of tall cultivars, whereas Tom Thumb grains release much smaller quantities of enzyme (ca 25%) and have a shallower response curve (Gale and Marshall, 1975). Seedling growth rates are also characteristically slower in Rht3 lines. A detailed consideration of the biochemical aspects of the Tall/Rht3 comparison for GA response in aleurone layers has been provided by Ho et al., (1981), using D6899 as the Rht3 source. They found no differences in uptake and metabolism of GA_3 between talls and dwarfs and suggested that the blockage could be concerned with the primary cellular receptor for GAs. Ho et al. (1981) also drew attention to the interesting corollary that the inability to produce high amounts of α-amylase in the cereal grain did not appear to prejudice the survival of the seedling. This is a valid observation for the monoculture state prevailing in cereal crops but one wonders, for example, what the ultimate survival patterns would be like in mixtures of tall and dwarf cultivars.

The vegetative tissues of all dwarf lines reveal an interesting, and biologically very significant, phenomenon. Radley (1970) showed that, contrary to expectations at the time, seedlings of Norin-10 and related dwarf lines contained about ten times as much extractable free GAs as did tall lines at a comparable age. These differences were maintained over a range of growth stages and were common to all parts of the seedlings, although shoot apices of dwarfs tended to exceed talls by a factor of four, rather than the ten to twenty-fold excess measured in leaves. Growth of seedlings in the presence of the quaternary ammonium growth retardant CCC reduced seedling endogenous GA levels to near zero and abolished the tall/dwarf difference. Measurements of GA_1 pool sizes in tall and Rht3 seedlings, using a radio-immunoassay technique, indicated that dwarfs had approximately thirteen times as much endogenous GA_1 (Stoddart, 1984). The basis for the buildup of such high GA levels is a matter for speculation. If, as has been suggested, GA insensitive dwarfs exist because of disablement of the GA response mechanism this does not, of necessity, predicate supra-normal endogenous GA levels, unless the act of association with the receptor moeity consumes or inactivates GA - and there is no evidence to suggest this, beyond speculations that 2-β-hydroxylation and receptor function may be properties of the same protein. Furthermore, there are a number of studies which show that the rates and nature of GA metabolism do not differ between talls and Rht3 seedlings (Ho et al., 1981; Stoddart, 1984). For example, when labelled GA_1 is supplied to Rht3 seedlings both GA_8 and a polar conjugate of GA_8 become labelled. The rate of production of both components was higher in tall segregates and thus would be compatible with a buildup of GAs due to reduced metabolic depletion of the GA_1 pool in dwarfs. However, when the dilution effect of the initially larger GA pool size in Rht3 seedlings is taken into account, there is no significant difference in GA_1 metabolism between talls and dwarfs. The alternative explanation of enhanced GA biosynthesis in dwarfs is possible but seems improbable. One of the most notable features of the Radley (1970) investigations was that they constituted the first definitive demonstration that no obligate link existed between extractable GA content and growth rate, and effectively terminated an era which sought to establish a causal relationship between rates of GA synthesis and growth. The *Rht* genes provide the plant physiologist with an important complementary tool to that furnished by the genetic blockages in the biosynthetic pathway and are likely to be of significance in establishing the mode of action of GAs in particular and endogenous growth regulatory processes in general.

In the latter context, there have been suggestions that the effects of *Rht3* in suppressing the GA response of aleurone layers can be overcome by a period of pretreatment at 5°C and that the sensitivity of aleurones containing *Rht1* and *Rht2* could be enhanced by similar treatment (Singh and Paleg, 1984; 1985a,b). Changes in phospholipid composition were detected during this period and were temporally correlated with the attainment of GA sensitivity, possibly supporting the interpretation that the *Rht* blockage of α-amylase synthesis/secretion are mediated *via* changes in membrane composition and consequent physical properties. These findings are surprising in so far as they are not mirrored by reversal of the lesion by low temperature treatment in vegetative tissues (Author, unpublished observations).

Hopp *et al.* (1981) have described a barley (*Hordeum vulgare* L.) dwarf which is insensitive to GA_3 applications in both vegetative and aleurone tissues. They designate the mutation as *Ga-ins* but it is also known as MC90. This material shows no increase in growth rate in darkness and is also less environmentally sensitive. Dwarfs showed lower levels of cell wall invertase, whilst both free and bound peroxidase failed to show the expected depression following addition of GA_3. Phenylalanine-ammonia lyase (PAL) was unaffected by the mutation. The authors consider *Ga-ins* to be analogous to Norin-10.

There is a temptation to assume that GA-insensitive mutants relate directly to the properties of the primary subcellular receptor for GAs but this need not necessarily be the case. For example, if we envisage a chain of events between this primary interaction and the ultimate biological response (in this case a change in growth rate or an increase in α-amylase production) a lesion at either end, or at any point between perception and action, would effectively confer GA insensitivity. A series of insensitive mutations can, therefore, be envisaged, allowing a stepwise examination of the response pathway in a manner analogous to that described for GA biosynthesis. Some indications along these lines have already been gained from studies on cell wall properties. It is considered that the ultimate site of GA modification of growth rate resides within the cell wall, with growth regulator action modifying wall extensibility, probably through the plastic component (e.g. Keegstra et al., 1973; Darvill et al., 1977). Sakurai et al. (1984) have studied the growth and stress-relaxation parameters for 10 dwarf barley strains and have compared these with normals. Methanol-preserved coleoptile sections of the 10 isogenic dwarf lines were compared by the Instron technique which allowed stress-relaxation times and apparent Young's moduli to be calculated. The minimum stress-relaxation times and relaxation rates were found, generally, to be correlated with growth rates and the apparent Young's modulus was related to final coleoptile length. Correlations between these parameters and cellulosic or non-cellulosic cell wall components were derived, but these tended to be complex in nature. All the strains tested were single gene dwarfs but Uzu, a wild-type dwarf barley grown in western Japan, did not fit any of the correlations, indicating that modifications of cell wall properties cannot be a universal explanation for dwarfism and supporting the expressed contention that action at a range of sites could have similar phenotypic expression.

The examples used thus far have emphasised the control of stature by the interference with promotive principles (GAs being used as the prime example). An equally valid route to height determination can be envisaged via the regulation of endogenous inhibitory processes. Abscisic acid (ABA) is regarded as the major endogenous inhibitory growth regulator and has been extensively implicated in dormancy and drought-reaction mechanisms. Quarrie (1982) has presented data showing ABA responses in a range of drought stressed spring wheat cultivars, mostly of normal stature. In response to drought stress all lines accumulate high endogenous contents of ABA but it is possible to select 'high' and 'low' accumulators and this character is correlated with similar changes in stomatal conductance. In these studies, selection was also practised for endogenous ABA levels under non-stress conditions. Plant height at maturity in high ABA (688 ng g^{-1}) plants was 84% of that measured in low ABA (357 ng g^{-1}) material, although flowering, and anthesis were earlier at the higher ABA level. Treatment of spring wheat seedlings with exogenous ABA over a 3 week period also reduced plant height to 83% of that prevailing in controls.

Other species

Dwarf phenotypes have been reported in a number of other genera and have been characterised to varying degrees.

A range of dwarf rice (Oryza sativa L.) lines, including Norin 25, Waito C, Tan-ginbozu, Sankei-10 and Kotake-tamanishiki have been shown to be single gene recessives (Morinaga and Fukushima, 1943; Suge and Murakami, 1968). Crosses between Tan-ginbozu and Kotaki-tamanishiki yield double recessive dwarfs. The stature ratio is approximately 3:2:1 for normals, dwarfs and double dwarfs.

In all cases, the single gene recessive dwarfs can be restored to the normal phenotype by treatment with GA_3 and similar responses have been reported after applications of (-)-kaurene, (-)-kauren-19-ol, (-)-kauren-19-oic acid and steviol (Murakami, 1968a,b), suggesting a pre-cyclisation block in the GA biosynthetic pathway in each case. There are no reports of the GA responsiveness of double dwarf lines but bioassays have shown a sharp reduction in endogenous levels of free GA-like activity (Murakami, 1972). The spectrum of response towards a range of GA structures was found to vary in some dwarfs, with GA_5, GA_9 and GA_{20} being highly active on Tan-ginbozu seedlings and of low activity in Waito-C. This is effective specificity for 3 β-hydroxylation.

The relationship between the endogenous levels of free GA-like substances and stature has also been examined in Japanese morning glory (Pharbitis nil. Choisy) using tall ('Violet') and dwarf ('Kidachi') cultivars. Levels of free diffusible or extractable GA-like activity in apical buds did not differ significantly between varieties (Barendse and Lang, 1972) but 'bound' GAs (i.e. biological activity released by acid hydrolysis of the aqueous phase remaining after solvent partitioning) were less in the dwarfs. This difference was greatest in light grown plants and indicates that dwarfism could be related to endogenous GA synthesis. The authors postulate that lowered rates of synthesis in dwarfs may make the system more sensitive to light-stimulated GA binding processes.

Liu and Loy (1972) have described two single-gene dwarfs in watermelon (Citrullus lanatus) Thunb. Matsu) designated dw-1 and dw-2. Both developed normal phenotypes when treated with GA_3 solutions. Later studies on the monorecessive dwarf line WB-2 (Loy and Liu, 1974) indicated that GA_1, GA_3 and a GA_{4+7} mixture were capable of reversing dwarfism, with optimal concentrations around 10^{-5} M. Roots and hypocotyls showed differential responses. Increasing GA concentrations promoted hypocotyl elongation in

WB-2 but "Sugar-baby" controls were essentially insensitive. Control root growth was unaffected but dwarf mutants showed a strong inhibition of elongation with increasing GA concentration. However, as untreated dwarf root elongation was up to 100% greater than in controls, the effect of GA treatment was to mediate a return to lengths characteristic of controls. Comparable effects of dwarfing genes in other species on root morphology have not been reported but the implied suggestion that expression may produce opposite effects in above and below-ground tissues is interesting. Alternatively, and equally validly, it could be argued that, in watermelon, the WB-2 locus is not expressed in roots, leading to enhanced growth on the basis of source/sink effects related to reduced shoot demand for assimilates.

A radiation-induced single gene mutant (Dwarf 1) of *Phaseolus vulgaris* L. cv. Mexico 80-R has been described (Proano and Greene, 1968). It develops successively more compacted internodes above an elongated hypocotyl and, therefore, does not develop the prominent vertical growth characteristic of normal genotypes. The mutant condition was reversed by applications of GA_3 and extraction studies indicated lower levels of GA-like activity in dwarfs.

A genetic dwarf of *Lolium perenne* L. was described by Jenkin (1954) and further characterised by Cooper (1958). The dd genotype was returned to an almost fully normal stature by a single foliar spraying with 2×10^{-5}M GA_3. Treatment with indole-3-acetic acid was ineffective. Gibberellin-enhanced growth was expressed as large differences in leaf sheath and lamina growth. Interstomatal distances were trebled but guard cell length was almost unaffected. Cell numbers per epidermal file were not collected, but current knowledge of relative effects of GAs on cell division and elongation, suggests a differential effect on the elongation of epidermal and guard cells. Crown root and tiller numbers were depressed in dwarfs and returned to the normal frequency by GA_3.

The dd locus was found to be in the same linkage group as Rr for anthocyanin synthesis and Yy, governing chlorophyll formation.

OVERGROWTH ('SLENDER') MUTANTS

Pisum sativum

Slender mutants in pea are characterised by excessive elongation of stem internodes, leading to the development of extremely tall phenotypes. In *Pisum* the trait was first described by de Haan (1927).

Maximum expression is obtained in the presence of the la cry^s

combination but similar, slightly lesser effects are seen with *la cry*^{c}. Slender plants produce extremely thin internodes, smaller leaves, long peduncles and a higher growth rate overall than normals. Effects on seed setting, including a variable incidence of parthenocarpy have been described (e.g. Potts *et al*., 1985). Both *la cry*^{s} and *la cry*^{c} combinations are epistatic to the *Na/na* locus, which confers a dwarf phenotype in their absence, or in the presence of the *La* allele. Thus, the genotypes *La cry*^{c} *na*, *La cry*^{s} *na*, *la cry*^{s} *na* and *la cry*^{s} *Na* have statures in the ratio 1 : 1 : 20 : 20. Mean stem lengths between internodes 1 and 4 are around 8 cm for *La cry*^{s} *na* and 36 cm for *la cry*^{s} *na* (Potts *et al*. 1985). Presence of the *na* allele does not modify expression of the slender phenotype.

Estimates of gibberellin-like activity in shoot tissues showed levels in slender segregates (*le la cry*^{s} *Na Lm*) which were significantly lower than those prevailing in dwarfs (*le la Cry Na Lm*). The extracted biological activity had the same qualitative characteristics in each case, the main peak co-eluting with GA_{20} on reverse-phase HPLC. The quantitative difference was largest in immature (apical) tissues (Potts *et al*. 1985). *Na* slender plants yielded substantially larger amounts of GA-like activity in comparison to *na* slender individuals and a similar difference existed between *Na* and *na* cryptodwarfs (*la cry*^{c}).

Lycopersicon esculentum

The *procera* (*pro*) mutation in the tomato cultivar 'Alisa Craig' is described as having long, thin stem internodes, fewer folioles and less indented major leaflets. Plants average twice control height and the floral parts are frequently over-extended, leading to exposed stigmas and poor fruit setting. Mature fruits have an elongated shape (Darby *et al*., 1977). At present, there are no reports on the GA economy of this material.

Hordeum vulgare

Barley slender mutants have been reported by Smith (1951), Favret *et al*. (1975) and Foster (1977). The two earlier publications described vine and gigas phenotypes, distinguished by elongated leaf sheaths, leading to a tall of slender growth habit. These mutations were recessive and lethal, being further characterised by functional nodal rooting and light-stimulated anthocyanin synthesis in ligules and leaf sheaths. Phenocopies of the gigas mutant could be produced by repeated spraying with 10^{-5}M GA_3.

The slender mutation (Cb 3014) described by Foster (1977) was attributed to recessive alleles (sln) at a locus on chromosome 4, probably close to the centromere. Because of extreme extension of the rachis and floral parts, these mutants are sterile, necessitating propagation by hybridisation of heterozygotes. Generally, the plants are pale, mechanically weak and exhibit extreme overgrowth of leaf sheaths and stem internodes. Typically, sub-apical internodes will also extend, carrying the shoot apex well clear of the soil surface. In common with all the described 'overgrowth' mutations, these plants behave as though continuously saturated with GA. At 15°C emerging leaves of normal seedlings grow at a rate of 1.5 mm h^{-1}, compared with 3.7 mm h^{-1} for slender segregates (Stoddart and Lloyd, 1986).

Determinations of endogenous GA-like activity in 3 leaf seedlings have indicated a small increase (ca 30%) in slender segregates (Author, unpublished observations). When it is borne in mind that normal GA dose/ response relationships require a tenfold increase in concentration for a doubling in growth rate, this difference cannot be regarded as physiologically significant.

One of the most notable effects of the sln mutation in barley concerns the basic growth/temperature behaviour of the plants. Measurements of leaf growth rate under conditions where the leaf meristem temperature can be continuously varied (Stoddart and Lloyd, 1986) indicate that the single-point mutation has a substantial effect on growth capacity at low temperatures. In Figure 4 growth rate versus temperature relationships are shown for programmed cooling and warming of leaf meristem in normal and slender Cb 3014 seedlings. As already indicated growth rates for slender seedlings are considerably higher at all temperatures but both classes of segregate showed a progressive decline in rate with reducing temperature. Growth ceased at +6°C in normal seedlings but continued down to -4°C in slender material. The extent of chilling damage is indicated by the warming (dotted) curves. Strong positive hysteresis on rewarming, as shown by the normals, indicates lack of damage, whereas the depressed growth rates exhibited by slender seedlings reflect substantial frost injury to meristematic tissues. When cooling ceases around freezing point slender seedlings exhibit a more normal recovery pattern (Stoddart and Lloyd, 1986).

Cell size measurements (Table 2) show that increased leaf length in slenders is achieved by virtue of longer, thinner cells but there also tend to be more cells per cross-sectional area. Interstomatal distance is kept almost constant by elongation of both guard and intervening cells in the stomatal ranks.

When tension is applied to the tip of the emerging leaf in Cb 3014 seedlings, normals can tolerate loadings in excess of 20 g without apparent mechanical damage to the basal meristem. Slender meristems rupture with tensions of the order of 5 to 10 g, indicating a drastically reduced mechanical strength in the meristematic cell walls (Author, unpublished observations).

Effects of growth retardants on slender mutants

Effects of a range of GA biosynthesis inhibitors on growth in slender mutants have been assessed. Slender pea segregates were almost unaffected by seed treatment with 100 μg of AMO 1618 or 20 ug of PP333. Both compounds produced substantial stature reductions when applied to tall and dwarf genotypes at similar dose rates (Potts et al., 1985).

Growth in the gigas mutant of barley is reported to be reduced

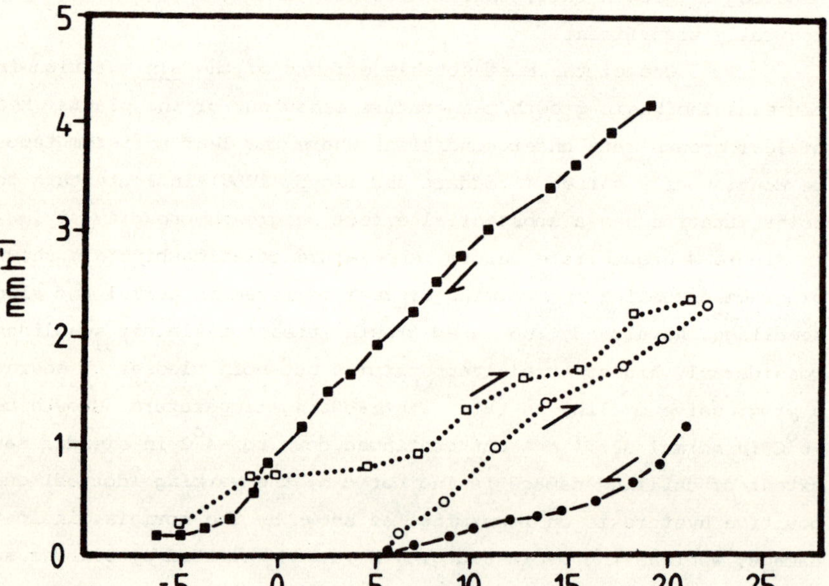

Figure 4. Growth rate versus meristem temperature curves for normal (●——●, o---o) and slender Cb 3014 (■——■, ---) seedlings of Hordeum vulgare. (Data from Stoddart and Lloyd, 1986).

Cooling and warming trends are indicated by solid and dotted lines respectively. Rate of temperature change: ca 1.0°C/min in all cases.

with CCC (2-chloroethyl trimethylammonium chloride) or 'stopped' with abscisic acid. Flowering occurs in the presence of CCC (concentration unspecified) but the rachis internodes are elongated, reaching 30 to 70 mm, compared with about 6 mm in normals. The flowers are male sterile but ovule fertile. More detailed studies on Cb 3014 barley have shown a clear response to CCC treatment, applied as a soil drench, in both slender and normal segregates. Increasing concentrations of the retardant, up to a maximum of 5×10^{-2}M, resulted in a height reduction of 70% in normals. Response was linearly progressive with concentration. Slender seedlings were unresponsive to CCC up to a concentration of 5×10^{-4}M but thereafter showed a linear response to give a stature reduction of 38% at the highest concentration. From 5×10^{-4} to 5×10^{-2}M the CCC response relationships for slender and normal segregates were parallel (Stoddart, 1979). Concentrations higher than 5×10^{-2}M were universally lethal.

Untreated slenders were approximately 100% taller than normal segregates at the two leaf stage but although absolute statures were greatly reduced, this difference had risen to almost 200% at the highest CCC

Table 2.

Cell size data derived from fully-expanded second leaves of normal and slender segregates of Cb 3014 barley

	Normal	Slender	Ratio
Cells/mm^2*	92.4	136.9	1 : 1.48
Cell length (mm)**	0.74	1.44	1 : 1.94
Cross sectional area (um^2)	739	466	1 : 0.63
Interstomatal distance (mm)	0.14	0.18	1 : 1.28
Leaf length (cm)	17.8	29.8	1 :1.67

* In transverse section

** Interstomatal rows

concentration.

It seems evident that pea and barley slender segregates have different behaviours towards GA biosynthesis inhibitors. However, the overall nature of the barley responses suggests that the expression of the slender locus is not being mediated via direct effects on GA supply through modulations of biosynthesis. Even at the highest concentrations the height differences are maintained. Indeed, the insensitivity of Cb 3014 to low CCC concentrations suggests that growth reductions in the higher treatment range may result from the known generalised effects of retardants on protein synthesis and may not represent specific effects on GA biosynthesis. The lack of retardant effects in peas is consistent with the observation that slender segregates are essentially insensitive to applied GAs.

Mechanistic aspects of slender mutations

The spectacular and 'uncontrolled' nature of the slender mutants leads to a strong conclusion that they have great potential importance for the elucidation of stature control mechanisms. The very low endogenous GA-like activity levels and substantial insensitivity to both exogenous GA and retardants indicate that the lesion is not directly related to GA supply or metabolism. The most evident possibility is interference with some aspect of the chain of response following the primary subcellular interaction with GA. In this sense they could be regarded as being akin to the Rht3 dwarf lines in wheat. Brian (1957) postulated that dominant alleles at the La and Cry loci governed the production of an endogenous inhibitor in peas and that this substance was absent in la cry^{s} (slender) plants. Potts et al. (1985) generally agree with Brian in considering that the most direct explanation lies in an effector/repressor model where the product of La or Cry functions directly as a repressor and GA as an effector at the receptor site. There is no evidence to support differential contents of inhibitory or 'repressor' compounds in slender and normal segregates and, indeed, the biological assay data presented by Potts et al. (1985) give no indication of strong inhibitory zones on HPLC profiles. Undoubtedly, this area will repay further study.

The immediate implication of the GA receptor site seems poorly supported. Ultimate GA action in extension growth resides in a modification of cell wall rheology (e.g. Stuart and Jones, 1977) and the changed mechanical properties of Cb 3014 slender seedlings, described earlier, indicate that the mutation involves changes in cell wall tensile strength. Expression of the slender locus could be exerted at any point in the chain of action between primary receptor site and the ultimate modification of cell wall physical structure. Presumably, a number of genetically distinct slender mutations are theoretically possible. For example, using sodium azide mutagenesis, Foster (personal communication) has produced two additional slender barley phenotypes (sln 2 and sln 3) which are distinguishable from Cb 3014 slenders. These remain to be characterised physiologically.

ENVIRONMENTAL EFFECTS ON STATURE MUTATIONS

Many of the stature mutations described in this chapter show strong interactions with environmental variables.

Tall and dwarf peas are of similar height when grown in total darkness and, although light inhibits growth in both cases, the effect is considerably greater in the dwarf genotypes (Lockhart, 1959). Dwarfing genes are expressed during growth in red light, but not in far-red illumination (Lockhart, 1956). Light inhibition can be overcome by GAs but this effect is greatest in dwarf lines and there is no effect when GA_1 or GA_3 are applied in darkness. At the time of writing there are no published reports of sensitivity to GA_{20} under these conditions. Thus, it is possible that the Le/le locus, governing 3 β-hydroxylation of GA_{20} may be light (phytochrome) regulated. Alternatively, the basal sensitivity of the pea stem internode tissue towards GAs may be modified in the presence of light, either at the level of the receptor site or as a consequence of the synthesis of a competing inhibitor. Dark-grown WB2 dwarfs in watermelon were found to respond to lower concentrations of GA_{4+7} than those grown in the light (Loy and Liu, 1974). Light reduces stem growth in tall and dwarf genotypes of *Pharbitis nil* and, as in the other cases, this inhibition can be overcome by GA_3 (Ogawa, 1962).

Gigas barley mutants are phenotypically identical to etiolated normal seedlings suggesting that the mutation is concerned with those processes imposing light inhibition on cell elongation (Favret *et al*., 1975). The effects on growth may, of course, be remote from the initial action of the allele, possibly, for example, being concerned with light-driven processes which may remotely govern wall extensibility such as phenylpropanoid synthesis or cell wall enzyme activity (Hopp *et al*., 1981).

Interactions with temperature have also been reported in Cb 3014 slender barley (Stoddart and Lloyd, 1986) and these are detailed in Figure 4.

At the other extreme, the Rht3 GA-insensitive dwarfing mutation raises the temperature below which growth is constrained in wheat (Stoddart and Lloyd, 1986) and a similar effect is seen with the d5 mutation in maize.

CONCLUDING REMARKS

In this chapter emphasis has been placed upon genetically-dependent aspects of stature control, with development into consequent endogenous hormonal processes where information exists. Modification of growth patterns by external agents, even where these augment or interfere with normal metabolic processes, has been excluded. The reasoning behind this approach is twofold.

Firstly, durable stature control with the lowest cost and the highest environmental acceptability will, inevitably, need to be based upon stable genetic components, possibly modified by recourse to modern DNA-manipulative techniques.

Secondly, progress towards a better understanding of growth control in plants will be greatly aided by the availability of a range of well-characterised mutations for component processes of the cell cycle and the subsequent cell enlargement processes. These will complement the various chemical probes currently available (e.g. the GAs and Fusicoccin). The substantial benefits accruing from the genetic approach to biosynthetic problems is well, and appropriately, exemplified by the indispensibility of the range of single gene mutations used to elucidate the GA biosynthetic pathway in *Gibberella fujikuroi* (e.g. Bearder, 1983).

Existing mutant material with evident relationships to the response site of growth regulator action already provides exciting possibilities for progress in our understanding of the chain of cellular events succeeding the primary hormone/receptor association. It is hoped that the realisation of the potential value of such material will stimulate effort in the labour-intensive search for further mutations. The importance of communicating their existence, in the form of preliminary physiological descriptions, cannot be over-emphasised.

REFERENCES

Barendse, G.W.M. & Lang, A. (1972) Comparison of endogenous gibberellins and the fate of applied radioactive gibberellin A_1 in a normal and a dwarf strain of Japanese morning glory. Plant Physiol. 49, 836-841.

Barker, R.F., Thompson, D.V., Talbot, D.R., Swanson, J. & Bennetzen, J.L. (1984) Nucleotide sequence of the maize transposable element Mul. Nucleic Acids Res., 12, 5955-5967.

Bearder, J.R. (1983) In vivo diterpenoid biosynthesis in Gibberella fujikuroi. The pathway after ent-kaurene. In: Biochemistry and Physiology of Gibberellins. ed. A. Crozier. Vol. 1, 251-387. New York: Praeger.

Brian, P.W. (1957) The effects of some microbial metabolic products on plant growth. Symp. Soc. Exp. Bot. II 166-182. Cambridge Univ. Press.

Cooper, J.P. (1958) The effect of gibberellic acid on a genetic dwarf in Lolium perenne. New Phytol., 57, 235-238.

Darby, L.A., Ritchie, D.B. & Taylor, I.B. (1977) Isogenic lines of the tomato 'Ailsa Craig'. Rep. Glasshouse Crops Res. Inst. 1977, 168-184.

Darvill, A.G., Smith, C.J. & Hall, M.A. (1977) Auxin induced proton release, cell wall structure and elongation growth: A hypothesis. In: Regulation of cell membrane activities in plants. ed. E. Marre & O. Ciferri. pp. 275-282. Elsevier/North Holland.

de Haan, H. (1927) Length factors in Pisum. Genetica 12, 481-497.

Duncan, J.D. & West, C.A. (1981) Properties of kaurene synthetase from Marah macrocarpus endosperm: evidence for the participation of separate but interacting enzymes. Plant Physiol. 68, 1128-1134.

Favret, E.A., Favret, G.C. & Malvarez, E.M. (1975) Genetic regulatory mechanisms for seedling growth in barley. In: Barley Genetics III. ed. H. Gaul, pp. 37-42. Munchen: Verlag Karl Thiemig.

Fick, G.N. & Qualset, C.O. (1973) Inheritance and distribution of grass-dwarfing genes in short-statured wheats. Crop Sci. 13, 31-33.

Foster, C.A. (1977) Slender: an accelerated extension growth mutant of barley. Barley Genet. Newslett. 7, 24-27.

Gale, M.D. & Marshall, G.A. (1973) Insensitivity to gibberellin in dwarf wheats. Ann. Bot. 37, 729-735.

Hedden, P. & Phinney, B.O. (1979) Comparison of ent-kaurene and ent-isokaurene synthesis in cell-free systems from etiolated shoots of normal and dwarf-5 maize seedlings. Phytochem., 18, 1475-1479.

Ho, T.D., Nolan, R.C. & Shute, D.E. (1981) Characterisation of a gibberellin-insensitive dwarf wheat, D6899. Plant Physiol. 67, 1026-1031.

Hopp, H.E., Favret, G.C. & Favret, E.A. (1981) Pleiotropic effects of a dwarfing gene causing insensitivity to gibberellin on some key enzymes. Barley Genetics IV pp. 544-549. Edinburgh: Univ. Press.

Ingram, T.J., Reid, J.B., Murfet, I.C., Gaskin, P., Willis, C.L. & MacMillan, J. (1984) Internode length in Pisum. The Le gene controls the 3 β-hydroxylation of gibberellin A_{20} to gibberellin A_1. Planta 160, 455-463.

Ingram, T.J., Reid, J.B., Potts, W.C. & Murfet, I.C. (1983) Internode length in Pisum IV. The effect of the gene Le on gibberellin metabolism. Physiol. Plant 59, 607-616.

Jenkin, T.J. (1954) Interspecific and intergeneric hybrids in herbage grasses. VII Lolium perenne L. with other Lolium species. J. Genet. 52, 300-317.

Johnson, V.A., Biever, K.J., Haunold, A. & Schmidt, J.W. (1966) Inheritance of plant height, yield of grain and other plant and seed characteristics in a cross of hard red winter wheat. Triticum aestivum L. Crop Sci. 6, 336-338.

Keegstra, K., Talmadge, K.W., Bauer, W.D. & Albersheim, P. (1973) The structure of plant cell walls III. A model of suspension cultured sycamore cell walls based on the interconnections of macromolecular components. Plant Physiol. 51, 188-196.

Kende, H., Ninnemann, H. & Lang, A. (1963) Inhibition of gibberellic acid biosynthesis in Fusarium moniliforme by AMO 1618 and CCC. Naturwiss. 50, 599-600.

Lamm, R. (1937) Length factors in dwarf peas. Hereditas 23, 38-48.

Lindquist, K. (1951) The mutant 'micro' in Pisum. Hereditas 37, 389-420.

Liu, P.B.W. & Loy, J.B. (1972) Inheritance and morphology of two dwarf mutants in watermelon. J. Am. Soc. Hort. Sci. 97, 745-748.

Lockhard, R.G. & Grunwald, C. (1970) Grafting and gibberellin effects on the growth of tall and dwarf peas. Plant Physiol. 45, 160-162.

Lockhart, J.A. (1956) Reversal of the light inhibition of pea stem growth by gibberellin. Proc. Natl. Acad. Sci., USA 42, 841-848.

Lockhart, J.A. (1959) Studies on the mechanism of stem growth inhibition of visible radiation. Plant Physiol. 34, 457-460.

Loy, J.B. & Liu, P.B.W. (1974) Response of seedlings of a dwarf and normal strain of watermelon to gibberellins. Plant Physiol. 53, 325-330.

McComb, A.J. & McComb, J.A. (1970) Growth substances and the relationship between genotype and phenotype in Pisum sativum. Planta 91, 235-245.

Morinaga, T. & Fukushima, E. (1943) Heritable characters in rice. I. Abnormal mutant characters and their mode of inheritance. Gakaugei Zasshi 10, 301-339.

Murakami, Y. (1968a) Gibberellin-like activity of (-)-kaurene, (-)-kauren-19-ol and (-)-kauren-19-oic acid in leaf sheath elongation of 'Tanginbozu' dwarf of Oryza sativa. Bot. Mag. Tokyo 81, 100-102.

Murakami, Y. (1968b) Gibberellin-like activity of steviol in leaf sheath elongation of the 'Tan-ginbozu' dwarf of Oryza sativa. Bot. Mag. Tokyo 81, 464-466.

Murakami, Y. (1972) Dwarfing genes in rice and their relation to gibberellin biosynthesis. In: Proc. 7th Internat. Conference on Plant Growth Substances. ed. D.J. Carr, pp. 166-174. Berlin: Springer-Verlag.

Murfet, I.C. (1971) Flowering in Pisum. Three distinct phenotypic classes determined by the interaction of a dominant early and a dominant late gene. Heredity 26, 243-257.

Ogawa, Y. (1962) Quantitative differences of gibberellin-like substances in normal and dwarf varieties of Pharbitis nil Chois. Bot. Mag. Tokyo 78, 474-480.

Phinney, B.O. (1956) Growth response of single-gene dwarf mutants in maize to gibberellic acid. Proc. Natl. Acad. Sci. USA 42, 185-189.

Phinney, B.O. (1983) The history of gibberellin. In: The Biochemistry and Physiology of Gibberellins ed. A. Crozier. Vol.1, pp. 19-52. New York: Praeger.

Phinney, B.O. & Spray, C. (1982) Chemical genetics and the gibberellin pathway in Zea mays L. In: Plant Growth Substances 1982. ed. P.F. Wareing pp. 101-110. London: Academic Press.

Potts, W.C. & Reid, J.B. (1983) Internode length in Pisum III. The effect and interaction of the Na/na and Le/le gene differences on endogenous gibberellin-like substances. Physiol. Plant 59, 448-454.

Potts, W.C., Reid, J.B. & Murfet, I.C. (1982) Internode length in Pisum I. The effect of the Le/le gene difference on endogenous gibberellin-like substances. Physiol. Plant 55, 323-328.

Potts, W.C., Reid, J.B. & Murfet, I.C. (1985) Internode length in Pisum. Gibberellins and the slender phenotype. Physiol Plant 63, 357-364.

Powell, J.B. & Schleuhuber, A.M. (1967) Components of height inheritance of the semi-dwarf straw character in wheat Triticum aestivum L. Crop Sci. 7, 511-516.

Proano, V.A. & Greene, G.L. (1968) Endogenous gibberellins of a radiation induced single gene dwarf mutant of bean. Plant Physiol. 43, 613-618.

Quarrie, S.A. (1982) The role of abscisic acid in the control of spring wheat growth and development In: Plant Growth Substances 1982. ed. P.F. Wareing, pp. 609-619. London: Academic Press.

Radley, M. (1970) Comparison of endogenous gibberellins and response to applied gibberellin of some dwarf and tall wheat cultivars. Planta 92, 292-300.

Rasmusson, J. (1927) Genetically changed linkage values in Pisum. Hereditas 10, 100-150.

Reid, J.B., Murfet, I.C. & Potts, W.C. (1983) Internode length in Pisum II. Additional information of the relationships and action of loci Le, La, Cry, Na and Lm. J. Exp. Bot. 34, 349-364.

Reid, J.B. & Potts, W.C. (1985) Internode length in Pisum. Further mutants controlling GA synthesis and sensitivity. In: 12th Internat. Conf. on Plant Growth Substances, Heidelberg. Abstract No. 0266.

Reitz, L.P. & Salmon, S.C. (1968) Origin, history and use of Norin 10 wheat. Crop Sci. 8, 686-689.

Robertson, D.S. (1978) A transposable element governing grain characteristics in Zea mays L. Mutat. Res. 51, 21-28.

Sakurai, N., Kuraishi, S., Inouhe, M. & Masuda, Y. (1984) Growth and stress relaxation parameters for the cell wall of normal and 10 dwarf barley strains. Plant Cell Physiol. 25, 721-729.

Sembdner, G., Gross, D., Liebisch, H.W. & Schneider, G. (1980) Biosynthesis and metabolism of plant hormones. In: Hormonal regulation of development: Molecular aspects of plant hormones. Encyc. Plant Physiol. ed. J. MacMillan, Vol. 9 pp. 289-444. Berlin, Heide Heidelberg, New York: Springer-Verlag.

Singh, S.P. & Paleg, L.G. (1984) Low temperature-induced GA_3 sensitivity in wheat. I Characterisation of the low temperature effect on isolated aleurone of kite. Plant Physiol. 76, 139-142.

Singh, S.P. & Paleg, L.G. (1985a) Low temperature induced GA_3 sensitivity of wheat. III Comparison of low temperature effects on α-amylase production by aleurone tissue of dwarf and tall wheat. Aust. J. Plant Physiol. 12, 269-275.

Singh, S.P. & Paleg, L.G. (1985b) Low temperature induced GA_3 sensitivity of wheat. IV Comparison of low temperature effects on the phospholipids of aleurone tissue of dwarf and tall wheat. Aust. J. Plant Physiol 12, 277-289.

Smith, L. (1951) Cytology and genetics of barley. Bot. Rev. 17, 1-51.

Spray, C., Phinney, B.O., Gaskin, P., Gilmour, S.J. & MacMillan, J. (1984) Internode length in Zea mays L.: The dwarf-1 mutation controls the 3 β hydroxylation of gibberellin A_{20} to gibberellin A_1. Planta 160, 464-468.

Stoddart, J.L. (1979) Endogenous gibberellin relationships in slender and normal barley seedlings. In: 10th Internat. Conf. on Plant Growth Substances, Madison, Wisc. 1979. Poster abstract G/22.

Stoddart, J.L. (1984) Growth and gibberellin-A_1 metabolism in normal and gibberellin-insensitive (Rht3) wheat (Triticum aestivum L.)

seedlings. Planta 161, 432-438.

Stoddart, J.L. & Lloyd, E.J. (1986) Modification by gibberellin of the growth/temperature relationship in mutant and normal genotypes of several cereals. Planta 167, 364-368.

Stuart, D.A. & Jones, R.L. (1977) The roles of extensibility and turgor in gibberellin and dark-stimulated growth. Plant Physiol. 59, 61-68.

Suge, H. & Murakami, Y. (1968) Occurrence of a rice mutant deficient in gibberellin-like substances. Plant Cell Physiol. 9, 411-414.

Wellensiek, S.J. (1969) The physiological effects of flower forming genes in pea. Z. Pflanzenphysiol. 5, 388-402.

White, O.E. (1917) Studies of inheritance in *Pisum*. II The present state of knowledge of heredity and variation in peas. Proc. Ann. Phil. Soc. 56, 487-588.

GRAVITROPISM AND PHOTOTROPISM IN THE MOSS, PHYSCOMITRELLA PATENS

D.J. Cove and C.D. Knight,
Department of Genetics, University of Leeds,
Leeds LS2 9JT.

INTRODUCTION

The inclusion of work on a bryophyte in a symposium devoted to developmental mutants of higher plants needs some explanation. Although *Physcomitrella patens* can be handled by essentially microbiological techniques including sterile growth in Petri-dishes or in a fermenter, it is suitable as a model system for higher plant development since it shares with higher plants many developmental features including, for example, the involvement of auxins and cytokinins in developmental regulation and response to light stimuli mediated by phytochrome. In addition it is amenable to both conventional and molecular genetic analysis aided by the haploidy of its gametophyte generation.

In this contribution, we will concentrate on tropic responses. These are examples of morphogenetic processes which generate the shape and form of a plant. Plant morphogenesis involves the control of the timing and plane of cell division as well as the regulation of cell size and shape. Very little is known about the molecular mechanisms which determine these processes which are so crucial to development. The tropic responses shown by apical cells of protonemal filaments of *P. patens* appear to be excellent material for the study of how external factors such as light and gravity are detected by cells and transduced to regulate cell shape.

The development of the haploid gametophyte in *P. patens* starts with a filamentous stage, the protonema. Spore germination (or the regeneration of almost any tissue) gives rise to primary chloronemal filaments. These grow by division of tip cells and branch by division of sub-apical cells. Chloronemal tip cell division can also give rise to a second cell type, caulonemata. Caulonemal filaments also grow by division of the tip cell and branch by sub-apical cell division but caulonemal side branch initials show three possible developmental fates: secondary chloronemal filaments, caulonemal filaments or gametophores, the leafy shoots which give rise to the gametes.

Following fertilisation, zygotes develop into diploid sporophytes, which produce haploid spores by meiosis. Figure 1 shows the morphology of part of a gametophyte and illustrates the relationship between caulonemal filaments and their various types of side branch. Much information has now been

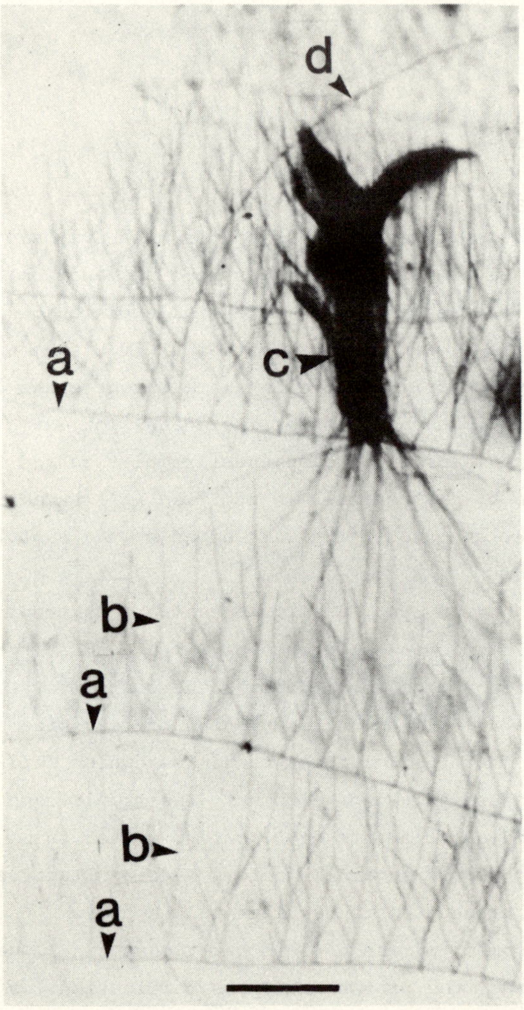

Figure 1. Physcomitrella patens gametophyte morphology. Lateral view of a part of a wild-type culture which has been grown for 28 d at 25°C in white light at ca. 10 W m^{-2} (Ashton and Cove, 1977). Key: a, caulonemal filaments growing lateral to the direction of incident light; b, compound secondary chloronemal filaments showing positive phototropism; c, young gametophore showing positive phototropism; d, side branch showing transition from secondary chloronemal to caulonemal filament and concomitant change in phototropic resonse.

accumulated about the regulation of the course of development in P. patens and the analysis of mutant phenotypes has played a central role particularly in establishing the roles of auxins and cytokinins (Ashton et al., 1979; Cove and Ashton, 1984). Figure 2 gives a summary of our present state of understanding of the factors influencing the development of P. patens with more emphasis given to protonemal development than to that of the gametophore and sporophyte.

Light and gravity influence the direction of growth of apical cells of protonema and of gametophores. It is therefore possible to study tropic responses in both single cells and in a multicellular organ although we have so far concentrated on the protonemal responses.

PHOTOTROPISM AND POLAROTROPISM

The results of studies of phototropism and polarotropism in P. patens have already been published (Cove et al., 1978; Jenkins and Cove, 1983b, c) and will therefore only be summarised here. Different phototropic (and polarotropic) responses are shown by various protonemal cell types and by gametophores and these are listed in Table 1. The responses of the apical cells of primary chloronemata originating from germinating spores, have been studied in most detail (Jenkins and Cove, 1983b, c). In red light, alternative phototropic and polarotropic responses occur depending on light intensity. At low light intensities, filaments show positive phototropism or orient perpendicular to the electrical vector of polarised light (the E vector). At high light intensities, filaments grow perpendicular to the direction of incident light and orient parallel to the E vector. The transition from low to high light responses is also seen for the blue/green region of the spectrum where it occurs at similar light intensities but at other wavelengths only the "low" light responses are observed even at the highest light intensities available. A detailed difference in the phototropic response of primary chloronemata in blue light from that in red light is that at higher intensities of blue light (e.g. above ca. 10 µmol quanta m^{-2} s^{-1} at 442 nm), a majority of filaments show negative phototropism whereas at the highest intensities of red light available (110 µmol quanta m^{-2} s^{-1} at 665 nm) the majority of filaments still align perpendicular to the light direction (Henderson, Jenkins and Cove, unpublished data). There is evidence that phytochrome is involved in the detection of light intensity for the switch between the low and high intensity phototropic and polarotropic responses of primary chloronemata in red light (Jenkins and Cove, 1983b).

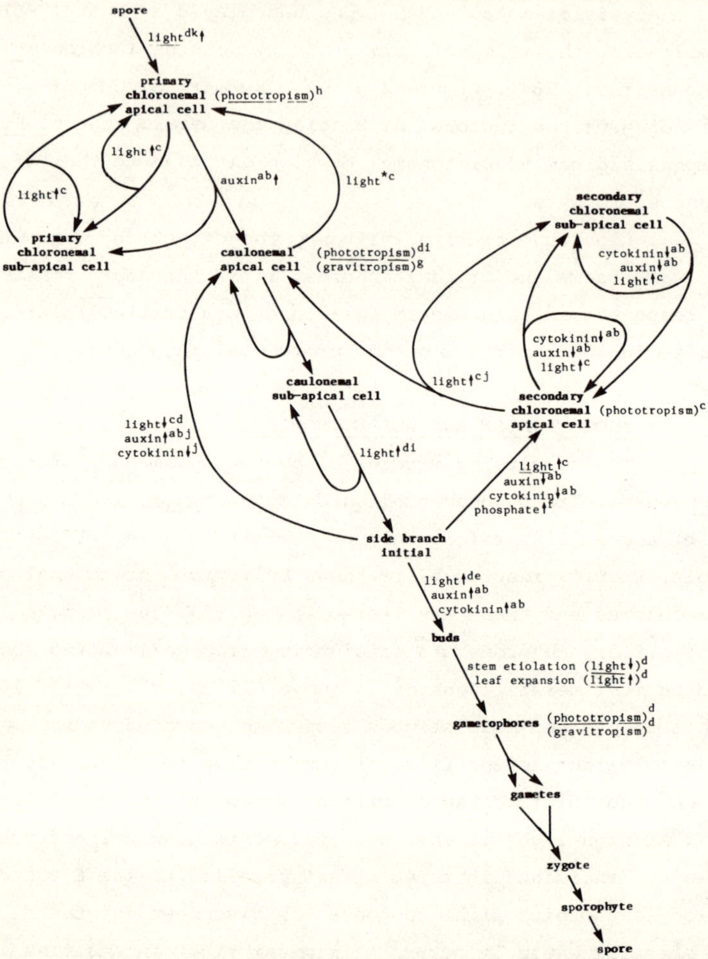

Figure 2. Diagram of development in *Physcomitrella patens*. An arrow pointing upwards indicates that the step is enhanced by the stimulus, an arrow pointing downward that it is depressed by the stimulus, underlining indicates phytochrome involvement.
*this transition occurs when caulonemal apical cells which have been growing in darkness are transferred to high levels of light.
References: a, Ashton *et al.*, 1979; b, Cove, 1984; c, Cove, unpublished data; d, Cove *et al.*, 1978; e, Futers, Wang and Cove, unpublished data; f, Grimsley and Ashton, unpublished data; g, Jenkins *et al.*, 1986; h, Jenkins and Cove, 1983b; i, Jenkins and Cove, unpublished data; j, McClelland and Cove, unpublished data; k, Schild and Cove, unpublished data.

It may be however that the switch to negative phototropism which occurs in blue light is mediated by a different photoreceptor.

A series of mutants which show abnormal phototropic responses have been isolated. Most of these mutants were identified originally because their gametophores showed no phototropic response (see Figure 3) but it was also found that their caulonemal apical cells were aphototropic and apolarotropic (Cove et al., 1978). Genetic analysis using both conventional methods and complementation tests on somatic hybrids derived following protoplast fusion have shown that at least two genes can mutate to give this phenotype (Grimsley et al., 1980). The phototropic and polarotropic responses of primary chloronemal apical cells of three of these mutants, ptrA1, ptrB2 and ptrB3 have also been studied in detail (Jenkins and Cove, 1983c). Surprisingly, none had lost its ability to show either a "low intensity" or "high intensity" phototropic or polarotropic response but all were similarly

Table 1. Summary of tropic responses shown by gametophytes of *Physcomitrella patens*.

	PHOTOTROPISM (at 660/5 nm)	POLAROTROPISM (at 660/5 nm)	GRAVITROPISM
PRIMARY CHLORONEMA	+VE at low intensities LAT. at high int.[f]	PERP. to E at low int. PAR. to E at high int.[f]	No growth in darkness No response in light.[a,b]
CAULONEMA	+VE at low intensities LAT to -VE at high int.[b]	Either E + 16° or E - 16° (? at high int. only)[a]	-VE but over-ridden by phototropism above ca. 200 nmol quanta m^{-2} s^{-1} (660nm)[b,d]
SECONDARY CHLORONEMA	+VE at all intensities[b]	?	No growth in darkness No response in light.[a,b]
CHLORONEMA FROM REGENERATING PROTOPLASTS	?	PERP. to E at all int.[c]	No growth in darkness. No response in light.[a,e]
GAMETOPHORES	+VE at all intensities[b]	?	-VE but over-ridden by phototropism above ca. 100 nmol quanta m^{-2} s^{-1} (660nm)[b,d]

References: a: Cove, unpublished data; b: Cove et al., 1978; c: Henderson, Jenkins and Cove, unpublished data; d: Jenkins et al., 1986; e: Jenkins and Cove, 1983a; f: Jenkins and Cove, 1983b.

abnormal in that the light intensity required to switch from one response to another was lower for the mutants than the wild-type. For example, in the wild-type the switch from perpendicular to parallel polarotropism occurs in red light (665 nm) at an intensity of about 1 μmol quanta m^{-2} s^{-1}, whereas the three mutants all show a parallel response even at the lowest intensities of red light at which growth occurs (330 nmol quanta m^{-2} s^{-1}). At 542 nm, the mutants switch from a perpendicular to parallel orientation at about 3 μmol quanta m^{-2} s^{-1} but the wild-type shows a perpendicular response even at the highest intensities available (100 μmol quanta m^{-2} s^{-1}). Only at 498 nm has the switch been observed in both wild-type and the mutants, occurring at about 1 μmol quanta m^{-2} s^{-1} in the mutants and at 5 μmol quanta m^{-2} s^{-1} in the wild-type. Although the studies of phototropism in these mutants are less extensive, being confined so far largely to studies at 665 nm. essentially similar results have been obtained. The mutants show the "high intensity" (lateral phototropic) response at the lowest light intensities at which growth occurs.

We can now ask what we have so far learned about phototropism as a result of the study of these mutants. The initial conclusion that the _ptrA_ and _ptrB_ gene products are components required for the phototropic response of both caulonemal apical cells and gametophore stems (Cove _et al_.,

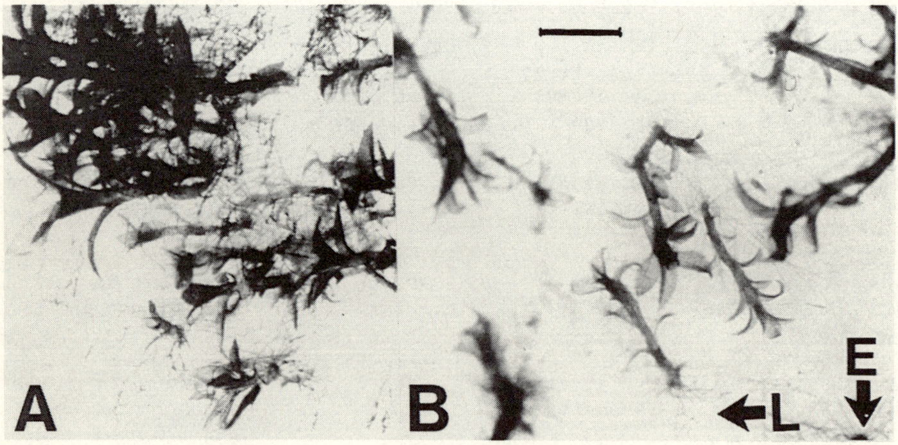

Figure 3. Details of wild-type (A) and _ptrA1_ (B) cultures which have been grown for 31 days at 25°C in lateral white light (5 W m^{-2}). The arrow labelled E points to the direction of the earth, the arrow labelled L points towards the light source. Scale bar = 1 mm.

1978) still stands but the situation must be more complex since the mutants do not show the same simple abnormality for primary chloronemal phototropism. It is difficult to see how the two aspects of the mutant phototropic phenotype, loss of response from caulonemal tip cells and gametophores and lowering of the light intensity required to switch the primary chloronemal tip cells between their alternative responses, relate to one another but study of the mutants has already alerted us to complexity which was not previously expected.

A number of other conclusions may also be drawn from the studies of the ptr mutants. The parallel between the phototropic and polarotropic phenotype of the mutants supports the idea that phototropism and polarotropism are two ways of observing the same phenomenon. It is improbable that the ability to orient to polarised light is of biological significance. Polarotropism is instead likely to be a consequence of the anisotropic arrangement of the photoreceptors involved in the phototropic response. Since these mutants appear to be normal for other light-dependent developmental responses such as bud induction (Cove, unpublished data), the ptr genes cannot be required for some general light detection system. A further interesting conclusion that may be drawn from studies of ptr mutants relates to the relationship between phototropism and gravitropism but discussion of this will be left until later in this paper.

GRAVITROPISM

Negative gravitropic responses are shown by caulonemal apical cells and by gametophore stems. Both responses are routinely studied in darkness since phototropism overrides gravitropism at all but the lowest of light intensities (see below). The response of caulonemata to reorientation with respect to the gravitational vector can be observed in two ways:

Method I. A number of cultures can be inoculated, cultured and reoriented in parallel and the response of a sample of caulonemata from cultures removed from darkness at different times after reorientation can be measured (Jenkins *et al.*, 1986).

Method II. The response of an individual caulonemal filament can be observed continuously using time-lapse video microscopy. Growing cultures were observed with an infra-red sensitive television camera (TC2014 fitted with Ultricon 4833U tube, R.C.A., Lancaster, Pennsylvania, U.S.A.), attached to a microscope using light from a tungsten halogen bulb (12 v, 50 W, M52, Thorn E.M.I. Lighting, U.K.) which had been passed through an infra-red transmitting filter (RG 850, Schott Glaswerke, Mainz, F.R.G.). Growth was recorded using a time-lapse video recorder (Panasonic/National NV-8050, Matsushita Electric Trading Co., Osaka, Japan). Cultures for time-lapse observation were started by inoculating small protonemal fragments (Ashton and Cove, 1977) onto a 1 mm thick layer of agar medium on a piece of glass, 24 mm x 50 mm 100 µm in thickness. The medium used was as described by Futers, Wang and Cove, 1986 except that it was gelled with 0.6% (w/v) Special Agar-Noble (Difco Laboratories, Detroit, Michigan, U.S.A.) and was supplemented with nicotinic acid, p-amino benzoic acid and thiamine (Ashton and Cove, 1977) and 0.5% (w/v) sucrose. The medium was covered with uncoated fine grade cellophane (Walter E. Canning, Bristol, U.K.) and inverted over a well, cut in agar contained in a 90 mm diameter Petri dish. Petri dishes were incubated for 6 days at 25°C in white light (Ashton and Cove, 1977) and then incubated on edge in darkness for a further 10 days. Care was taken that all procedures to this point were carried out sterilely. The culture together with the glass support was then transferred to a specially constructed chamber (see Figure 4) and placed on a rotatable stage on a microscope which was laid on its side, so that the caulonemal filaments retained their orientation with respect to gravity. The screen of the television monitor was covered with 1 layer of no. 46 Cinemoid (Rank Strand Electrics, London, U.K.) and 1 sheet of 3 mm thick blue Plexiglass (GS Blau 602, Roehm Ltd., London, U.K.) to reduce transmission below wavelengths of 710 nm. Although wavelengths above 710 nm have some morphogenetic effects on P. patens development, brief exposure to this light has no detectable effect on gravitropism and the screened monitor was used as a safelight during the transfer of the culture to the microscope stage. During the observation period, the microscope was contained within a light-proof box. After observing a filament apex for about two hours to establish its growth rate, the chamber was reoriented by 90° and observations continued. No change in growth rate is observed upon reorientation and cultures can continue to grow at a constant rate for up to 50 h although it is not uncommon for growth to cease earlier.

Figure 5A compares the kinetics of a gravitropic response to a 90° reorientation using the two methods of observation. Although the response of an individual filament shows more variation, the kinetics are essentially similar at least for 30 h following reorientation. Less detailed observation of other filaments indicates that the slowing down of the response after 30 h of the single filament shown in Figure 5A is likely to be artefactual. The negative gravitropic response of gametophores has not been studied in detail but their response to a 90° reorientation appears to be broadly similar to that of caulonemal filaments (Jenkins et al., 1986).

A number of gravitropically-abnormal mutant strains have been isolated (Jenkins et al., 1986) and analysis of these both genetically and physiologically is at an early stage. A total of 19 mutants showing altered caulonemal gravitropism has now been obtained. All mutants so far examined show normal gametophore gravitropism indicating that the gravitropic responses of caulonemal tip cells and of gametophores may be more independent than their phototropic responses. No attempt has yet been made to isolate directly mutants with abnormal gametophore gravitropism.

The majority of gravitropically-abnormal mutants show some degree of impairment of their caulonemal response. Some mutants, e.g. gtrB1,

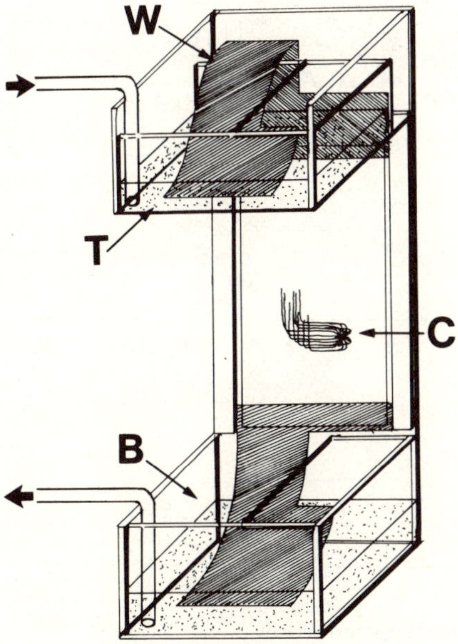

Figure 4. Growth chamber for time-lapse microscopy of P. patens. The chamber is constructed from 1 mm thick glass and is 100 x 35 x 25 mm. The orientation shown is after 90° rotation of the culture (C). The top reservoir (T) contains about 5 ml of medium which is fed to the culture growing on agar by a wick (W) of filter paper. Medium is pumped into the top reservoir and removed from the bottom reservoir (B) at 2.5 ml h^{-1}.

are almost completely agravitropic, while most show some degree of negative response, having an essentially "leaky" phenotype e.g. gtrA12. Complementation analysis is at present being carried out to determine how many genes can mutate to give this impaired response but it would be surprising if this group of leaky mutants was not heterogeneous.

Analysis of the gravitropic response of a number of mutants has begun. The response of an individual filament of gtrA12 to 90° reorientation, analysed by means of time-lapse video recording (method II), is shown in Figure 5B. The immediate response of this mutant to reorientation is similar to the wild-type but, although growth continues at the same rate, bending stops once a 10 to 15° bend has been achieved. No response is obtained if a filament is reorientated by 60° or by 120°. It is possible that this phenotype results from a greatly reduced sensitivity to gravity such that only when the gravity vector is near to perpendicular to the axis of the apical cell can a response be elicited. We are investigating this possibility by

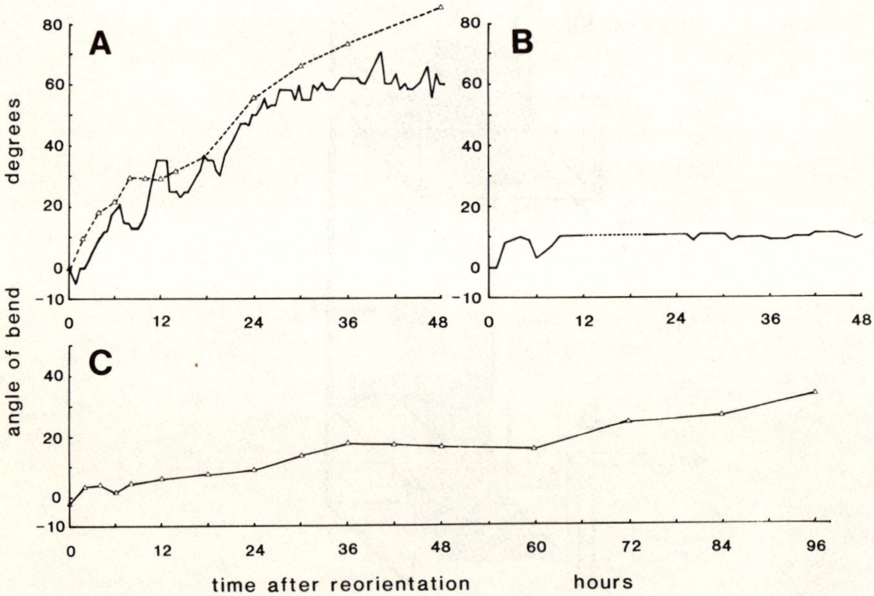

Figure 5. Gravitropic response of caulonemal apical cells to 90° reorientation of wild-type and gtr mutant strains. A: wild-type; dashed line, measured using method I; continuous line, measured using method II, the bend was measured at 30 min intervals. B: gtrA12; measured using method II, no recording was available for the dashed portion of the line. C: gtr6; measured by method I.

growing cultures on a centrifuge to see, for example, if the mutant responds to 60° reorientation in an enhanced gravity field. The similarity in the rate of the initial response of the wild-type and the response of gtrA12, led us to investigate an alternative possibility, that the mutant was slow to replace a resource which was used up during a gravitational response. However when a filament of gtrA12 which has been reoriented by 90° and has responded by bending ca. 15°, is reoriented once again so that the apical cell is again at right angles to the gravity vector, a further 10 to 15° response occurs. This possibility is therefore unlikely.

There is some evidence that the wild-type response is biphasic (Jenkins et al., 1986) comprising an initial rapid response and a subsequent slower phase. If this is so, another explanation of the phenotype of gtrA12 is that it might be impaired in the second stage but normal for the initial rapid response. It is therefore of interest to analyse in detail the phenotype of further "leaky" mutants since some might be abnormal only in the first phase of response. We have so far analysed only one other mutant, gtr6. Figure 5C shows the response of this mutant to 90° reorientation, measuring the mean bend of populations of caulonemal filaments sampled at different times after reorientation (method I). This mutant shows no initial rapid response but its subsequent response is also much slower than the wild-type. It does however continue to bend until it eventually achieves a full 90° response. This is not the phenotype which would be predicted of a mutant which was abnormal only in an initial rapid response and so no support is provided by it for the two-phase response model. However there are many more mutants yet to be analysed and this work is now in hand.

A further class of gravitropically-abnormal mutant comprising so far four independently-isolated strains, has caulonemal apical cells which show a positive gravitropic response. We have not yet made a detailed analysis of the kinetics of the response to reorientation of any of this class of mutant but it appears from their gross phenotype that they may respond to gravity in a similar way to the wild-type except the direction of bending is reversed. This class of mutant may therefore have normal gravity perception but be impaired in the interpretation perhaps of a gravity-induced intracellular gradient. Although there appears to be no support from other systems for the involvement of an intracellular gradient in gravitropism, Hartmann (1984) reports that in apical cells of the protonema of Ceratodon purpureus, an asymmetrical distribution of Ca^{++} ions is associated with phototropic bending. There is also evidence for a cellular gradient of Ca^{++} ions across the elongation zone of gravistimulated maize roots (Lee et al., 1983; Lee and Evans, 1985b). We have so far been unable to carry out studies with calcium chelators on P. patens because of its sensitivity even to low levels of EDTA and no radioactivity experiments have so far been attempted.

Lee and Evans (1985a) link the Ca^{++} gradient with the polar transport of auxin. Studies of mutants of P. patens have provided no evidence to support an involvement of auxin in gravitropism in this species. Attempts to repair the abnormal gravitropic phenotype of gtrA3, gtr 4 and gtr 5 by treatment with auxin have been unsuccessful (Jenkins and Cove, unpublished data) although it may be argued that if the gravitropic response requires the establishment of an auxin gradient, the exogenous supply of auxin cannot achieve this. However, further evidence against the involvement of auxin in gravitropic responses in P. patens is provided by the study of mutants the phenotype of which is consistent with their having impaired auxin synthesis (Ashton et al., 1979) since such mutants exhibit normal gravitropism (Courtice, unpublished data). Our findings contrast with those in Arabidopsis thaliana where mutants isolated as resistant to auxin analogues show altered root gravitropism (Mirza et al., 1984).

INTERACTION BETWEEN GRAVITROPISM AND PHOTOTROPISM

The interaction between gravitropism and phototropism has been studied in a number of ways. Wild-type cultures which have been grown in undirectional monochromatic red light (660 nm) at right-angles to the gravity vector for prolonged periods (up to 28 d) show that at intensities down to 200 nmol quanta m^{-2} s^{-1}, caulonemal tip cell growth is positively phototropic (Cove et al., 1978; Jenkins et al., 1986). Under the same conditions but at light intensities of ca. 90 nmol quanta m^{-2} s^{-1}, caulonemal apical cells appear to make a choice between positively phototropic growth and negatively gravitropic growth (Cove, unpublished data) - see Figure 6. However if caulonemal filaments which have been growing in darkness are then irradiated with red light so that the light direction and the direction of gravity are at right-angles the filaments grow at an angle intermediate between a negatively gravitropic and a positively phototropic response, a light intensity of ca. 60 nmol quanta m^{-2} s^{-1} at 660 nm leading to a 45° response (Jenkins and Cove, unpublished data). Study of the reaction of aphototropic mutants to low light intensities indicates that light turns off the gravitropic response. In darkness, aphototropic mutants show a negative gravitropic response similar to the wild-type (see Figure 7A

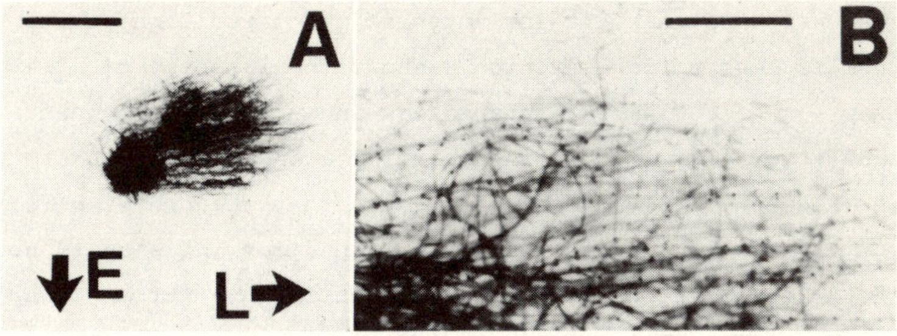

Figure 6. Wild-type culture grown at 25°C in white light at 15 W m^{-2} for 7 d and then grown for a further 26 d in unidirectional monochromatic red light (660 nm) at 90 nmol quanta m^{-2} s^{-1}. The arrow labelled E points to the direction of the earth, the arrow labelled L points towards the light source. Scale bar, A = 5 mm; B = 1 mm.

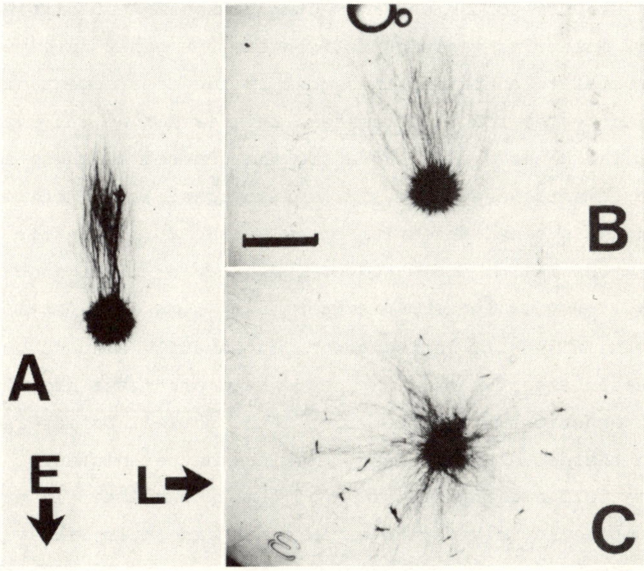

Figure 7. Cultures of ptrA1 grown under the same conditions as described in the legend to figure 6 except that culture A was grown in darkness, B at 90 nmol quanta m^{-2} s^{-1} (c.f. wild-type culture in figure 6) and C at 1.4 μmol quanta m^{-2} s^{-1}. The arrow labelled E points to the direction of the earth, the arrow labelled L points towards the light source. Scale bar = 5 mm.

for response of ptrA1). At low intensities of red light the gravitropic response is less strong. Figure 7B shows the response of ptrA1 at 90 nmol quanta m^{-2} s^{-1}, 660 nm. At intensities above ca. 1 μmol quanta m^{-2} s^{-1}, no response to gravity is observed in ptrA1 even though no phototropic response occurs (Figure 7C). It therefore appears that the preferential orientation towards light which occurs as light intensities increase is not only because the phototropic response is stronger than the gravitropic response but also the result of the active inhibition by light of the gravitational response.

THE WAY AHEAD FOR STUDIES OF PLANT MORPHOGENESIS

Although developmentally-abnormal mutants of *P. patens* have already been invaluable in allowing the complexity of the processes involved in development to be appreciated, they have so far been of little use in understanding the molecular mechanisms which bring about morphogenesis. An obvious next step is to use developmentally-abnormal mutants and the techniques of molecular genetics to identify the genes involved and their products. Crucial to achieving this goal is the development of an efficient transformation system for *P. patens* and this is not so far available. The isolation of the genes would also allow the control of their transcription in space and time to be studied. It is also possible to determine directly whether changes in gene expression occur during morphogenetic responses and work of this type is in hand. However the study of the transcriptional control of genes involved in tropic responses or even the identification of their products, will not, in isolation, necessarily advance our understanding of morphogenesis significantly. Morphogenetic processes are likely to require molecular interactions at the cellular level as well as interactions between cells. It is therefore necessary to combine the techniques of molecular genetics with cell biology. The role of the cytoskeleton in determining the shape of plant cells is becoming established and it is likely that our understanding of morphogenetic processes will be advanced considerably by the study of the cytoskeleton. Techniques have already been devised which allow the visualisation of the microtubular cytoskeleton in the cells of both protonemata and gametophores of *P. patens* (Doonan *et al.*, 1985 & unpublished data) and we are extending these studies both to determine the earliest changes to the cytoskeleton which can be associated with a particular morphogenetic event and to see if any of our mutants are abnormal in either the organisation or components of their cytoskeleton. Morphogenesis even in a simple model system such as *P. patens* is likely to be an extremely complex process involving a variety of different mechanisms.

The study of developmentally-abnormal mutants provides a useful tool in our efforts to understand these mechanisms but it is likely to require the combination of a variety of analytical approaches, some old and some new, if significant progress is to be made.

REFERENCES

Ashton, N.W. & Cove, D.J. (1977) The isolation and preliminary characterisation of auxotrophic and analogue resistant mutants of the moss, Physcomitrella patens. Mol. Gen. Genet., 154, 87-95.

Ashton, N.W., Grimsley, N.H. & Cove, D.J. (1979) Analysis of gametophytic development in the moss, Physcomitrella patens, using auxin and cytokinin resistant mutants. Planta, 144, 427-435.

Cove, D.J. (1984) The role of cytokinin and auxin in protonemal development in Physcomitrella patens and Physcomitrium sphaericum. J. Hattori Bot. Lab., 55, 79-86.

Cove, D.J. & Ashton, N.W. (1984) The hormonal regulation of gametophyte development in bryophytes. In: The Experimental Biology of Bryophytes, ed. A.F. Dyer & J.G. Duckett, pp. 177-201. London: Academic Press.

Cove, D.J., Schild, A., Ashton, N.W. & Hartmann, E. (1978) Genetic and physiological studies of the effect of light on the development of the moss, Physcomitrella patens. Photochem. Photobiol., 27, 249-254.

Doonan, J.H., Cove, D.J. & Lloyd, C.W. (1985) Immunofluorescence microscopy of microtubules in intact cell lineages of the moss, Physcomitrella patens. I. Normal and CIPC-treated tip cells. J. Cell Sci., 75, 131-147.

Grimsley, N.H., Featherstone, D.R., Courtice, G.R.M., Ashton, N.W. & Cove, D.J. (1980) Somatic hybridization following protoplast fusion as a tool for the genetic analysis of development in the moss Physcomitrella patens. In: Advances in Protoplast Research. ed. L. Ferenczy & G.L. Farkas, pp. 363-376. Budapest: Akademiai Kiado.

Hartmann, E. (1984) Influence of light on phototropic bending of moss protonemata of Ceratadon purpureus (Hedw.) Brid. J. Hattori Bot. Lab., 55, 87-98.

Jenkins, G.I., Courtice, G.R.M. & Cove, D.J. (1986) Gravitropic responses of wild-type and mutant strains of the moss, Physcomitrella patens. Plant Cell Envir., (in press).

Jenkins, G.I. & Cove, D.J. (1983a) Light requirements for regeneration of protoplasts of the moss, Physcomitrella patens. Planta, 157, 39-45.

Jenkins, G.I. & Cove, D.J. (1983b) Phototropism and polarotropism of primary chloronemata of the moss, Physcomitrella patens: responses of the wild-type. Planta, 158, 357-364.

Jenkins, G.I. & Cove, D.J. (1983c) Phototropism and polarotropism of primary chloronemata of the moss, Physcomitrella patens: responses of mutant strains. Planta, 159, 432-438.

Lee, J.S. & Evans, M.L. (1985a) Polar transport of auxin across gravistimulated roots of maize and its enhancement by calcium. Plant Physiol., 77, 824-827.

Lee, J.S. & Evans, M.L. (1985b) Polar transport of $^{45}Ca^{2+}$ across the elongation zone of gravistimulated roots. Plant Cell Physiol., 26, 1587-1595.

Lee, J.S., Mulkey, T.J. & Evans, M.L. (1983) Reversible loss of gravitropic sensitivity in maize roots after tip application of calcium chelators. Science, 220, 1375-1376.

Mirza, J.I., Olsen, G.M., Iverson, T.-H. & Maher, E.P. (1984) The growth and gravitropic responses of wild-type and auxin-resistant mutants of *Arabidopsis thaliana*. Physiol. Plant., 60, 516-522.

ABA-DEFICIENT TOMATO MUTANTS

I.B. Taylor,
Department of Physiology and Environmental Science,
University of Nottingham, School of Agriculture,
Sutton Bonington, Loughborough, Leics. LE12 5RD, U.K.

INTRODUCTION

The tomato has been subjected to an intensive programme of cytogenetic and classical genetic analysis. It can be ranked along with maize, barley and the pea as one of the four genetically best characterised crop species. Large numbers of mutants have been generated in the tomato. Many of these have been mapped and assigned to one of the twelve cytologically identifiable chromosomes (tomato genome formula, 2n=2x=24). A linkage map was last published by the Tomato Genetics Cooperative in 1980 (Rick, 1980). In view of these solid foundations, it is to be hoped that the tomato will also become one of the more favoured plant species for molecular genetic analysis.

A number of the tomato mutants are associated with interesting abnormalities of plant growth. Consequently, they have frequently attracted the attention of plant physiologists. Understanding the precise biochemical basis of some of these genetic lesions should significantly advance our knowledge of the control of plant development. Amongst the best characterised 'developmental mutants' of the tomato are those which are deficient in the phytohormone, abscisic acid (ABA).

THE ABA-DEFICIENT GENOTYPES

Three single gene mutants of the tomato can reasonably be regarded as ABA-deficient. These are:-

i) notabilis (not not) - Chromosome 7; 40 map units

ii) flacca (flc flc) - Chromosome 7; 59 map units

iii) sitiens (sit sit) - Chromosome 1 ; 32 map units

All three mutant homozygotes have a 'wilty' phenotype. However they should not be confused with two other tomato mutants, wilty (wt wt) and wilty dwarf (wd wd), which have not been shown to be ABA-deficient. References to 'the wilty tomato mutants' (often meaning flc homozygotes) can be misleading.

The mutants *notabilis*, *flacca* and *sitiens* may be legitimately termed ABA-deficient because:-

i) The 'basal' (i.e. not deliberately stressed) endogenous ABA concentrations are lower than in the non-mutant isogenic control (Tal and Nevo, 1973; Taylor and Rossall, 1982; Neill and Horgan, 1985; Linforth, 1986). The percentage reductions are undoubtedly dependent on the conditions under which the plants are grown. The results shown in Table 1 are the best available estimates of the true 'basal' levels, since the plants were grown in a controlled environment cabinet at high humidity (95% RH).

ii) The mutant genotypes do not respond to sudden losses of turgor with the usual massive increase in ABA accumulation. Taylor *et al* (1984) state that 'ABA concentrations (analysed by combined gas liquid chromatography - mass spectrometry) increased by less than 30% in *flacca* following a severe water stress treatment which induced a 7-fold increase in the isogenic control'. The dramatically reduced response of all three mutants to water stress has been confirmed by Neill and Horgan (1985).

iii) Correcting the ABA deficiency with a supply of exogenous ABA, results in a dramatic reversion to a normal phenotype (Imber and Tal, 1970; Tal and Nevo, 1973). Overcoming the ABA deficiency also removes the secondary morphological symptoms characteristic of each mutant.

The last of these three criteria should be rigorously applied in defining ABA-deficient mutants. Genetic lesions affecting carotenoid biosynthesis (e.g. members of the *vp* series of maize mutants; Moore and Smith, 1985) which also result in low endogenous ABA levels, should not strictly be regarded as ABA-deficient mutants. Phenotypic studies of such mutants should be interpreted with special care.

Of the three ABA-deficient tomato mutants, *flacca* has received by far the most attention. The main reason for the comparative neglect of *notabilis* is that its phenotype is less dramatic than that of *flacca*. Experiments therefore need to be carried out more carefully to establish small differences between *notabilis* and the isogenic control. The *sitiens* mutant has been neglected for opposite reasons. Its phenotype is too severe. Plants homozygous for *sit* are difficult to raise to maturity in a normal glasshouse environment. The flowers usually shed pollen poorly; seed yield is consequently low and therefore seed stocks are more difficult to maintain.

In view of this tendency to study flacca in isolation, it is worth emphasising that the three mutants form a quantitative series. Tal and Nevo (1973) were the first to point out that the mutants can be ranked with respect to declining endogenous ABA concentration in the order notabilis>flacca>sitiens. Those aspects of development which are truly ABA dependent should be progressively affected as the ABA levels decline.

This quantitative series of ABA deficient tomato genotypes has recently been extended. Taylor and Tarr (1984) produced three new double mutant homozygotes i.e. not not flc flc; not not sit sit and flc flc sit sit. Study of the severity of phenotype and the degree of ABA deficiency in these double mutants (see Table 1) has resulted in some interesting observations:-

1) It is possible to produce tomato plants with even lower endogenous ABA levels than those characteristically found in sitiens.
2) Extreme ABA deficiency results in grossly retarded slow growing plants. These symptoms can be partially (but not completely) alleviated by growing the plants at controlled high humidity (95% RH) (Linforth, 1986).

Table 1. Characteristic levels of ABA and the C_{10} dicarboxylic acids (OTA, ODA and OEA) in ABA-deficient tomato genotypes. Plants grown in a controlled environment cabinet at 95% RH for 6 weeks.

Genotype	ABA ($ng.g^{-1}$ F.wt)	OTA ($ng.g^{-1}$ F.wt)	ODA ($ng.g^{-1}$ F.wt)	OEA ($ng.g^{-1}$ F.wt)
+	110	N.D.	74	231
not	52	N.D.	25	157
flc	23	N.D.	115	166
sit	9	N.D.	130	159
flc/sit	10	N.D.	135	177
flc/not	4	N.D.	30	128
sit/not	3	N.D.	47	175

All values represent the mean of duplicate samples; except flc/not where one of the samples was discarded due to contamination.

N.D. = Not detected.

3) The double mutant homozygous for both flc and sit is only marginally worse than sitiens alone. In contrast, the double mutants involving the notabilis gene are much more severely affected (Taylor and Tarr, 1984). This result suggests some important qualitative difference in the biochemical basis of the notabilis gene on the one hand and both flacca and sitiens on the other.

A treble mutant genotype, homozygous for all three mutant alleles, has also been produced (Taylor - unpublished data). Consequently there are now seven different ABA-deficient genotypes available in the tomato. By including the isogenic non-mutant control it is now possible to produce a series of eight point dosage curves, allowing different aspects of phenotype to be correlated with a progressively declining endogenous ABA concentration.

THE VALUE OF THE ABA-DEFICIENT GENOTYPES

The ABA-deficient genotypes are of value to plant physiologists in two major areas of their work. The first involves the function of ABA in regulating different aspects of plant growth and development. Considerable confusion persists as to whether ABA has a major role in controlling such diverse processes as abscission; bud dormancy; seed germination; apical dominance; root geotropism and water stress responses (Walton, 1980). The general preconception that endogenous ABA performs a growth inhibitory role is also questionable. The ABA-deficient tomato genotypes provide a means of progressively lowering endogenous ABA content and identifying those processes which are ABA dependent. The systematic use of this approach should help to clarify the role of ABA *in vivo*.

The second area in which the ABA-deficient genotypes may play a major role, is in understanding the pathway of ABA biosynthesis and the control of ABA accumulation. The remainder of this article will be devoted to an assessment of recent developments and future prospects in this area.

POSSIBLE PATHWAYS OF ABA BIOSYNTHESIS

It is probable that one or more of the three ABA-deficient tomato mutants may be associated with a defective enzyme in the pathway of biosynthesis. It would be useful to be able to ascribe a specific function to the wild-type alleles at each of the three mutant gene loci. It may eventually be possible to assign each genetic lesion to a particular step in

ABA biosynthesis. This task would be relatively easily accomplished if precise details of the ABA biosynthetic pathway in plants had been worked out.

Unfortunately, even the overall framework of ABA biosynthesis has been the subject of much dispute over the last two decades. The details of ABA biosynthesis still remain to be elucidated. Two possible routes for the synthesis of ABA have been proposed:-

a) The direct route (C_{15} pathway)

b) The indirect route (C_{40} pathway)

ABA is a sesquiterpenoid compound and is ultimately formed from three mevalonic acid residues. It is reasonable to suggest that ABA may be synthesised directly via a series of oxidation steps following cyclisation of the C_{15} terpenoid precursor, farnesyl pyrophosphate. It appears that those fungal species which can synthesise ABA (e.g. Cercospera rosicola and C. cruenta), do so using a direct pathway of this type (Neill et al., 1982a; Oritani and Yamashita, 1985). Feeding radiolabelled fungal C_{15} intermediates to higher plants has failed to show that these two groups share a common pathway of ABA biosynthesis (Neill et al., 1982b). In fact Zeevaart (1985) has recently concluded that there is scant evidence to support the idea that higher plants synthesise ABA via the direct (C_{15}) pathway.

As soon as the chemical structure of ABA had been established (Ohkuma et al., 1965; Cornforth et al., 1965), its similarity to the C_{15} end fragments of certain cyclic carotenoids was noted. Since then the suggestion has frequently been made that ABA is an apo-carotenoid, being formed indirectly by oxidative cleavage of a larger (C_{40}) precursor such as violaxanthin. Attempts to refute this suggestion over the last twenty years have never quite succeeded. At one time the C_{40} proposal looked 'dead and buried' when it was shown that the stereochemistry of ABA was incompatible with its formation from violaxanthin (see Milborrow, 1971). However, the indirect route was revived when it was shown that the absolute stereochemistry of ABA was incorrectly assigned and subsequently had to be amended (see Taylor and Burden, 1970a). The feasibility of the C_{40} pathway was established in a number of pioneering papers by Frank Taylor and co-workers (Taylor and Smith, 1967; Taylor and Burden, 1970b; Taylor and Burden, 1970c; Taylor and Burden, 1972; Taylor and Burden, 1973).

In the past it has proved difficult to design experiments which would discriminate unequivocally between the direct and indirect pathways of ABA biosynthesis. Recently two independent lines of investigation have led to the conclusion that higher plants synthesise ABA via the C_{40} route. Following work with the ABA-deficient tomato mutant *flacca*, Taylor *et al.* (1984) stated that the 'results clearly fit in with the indirect C_{40} pathway, and are not in any way compatible with direct biosynthesis'. Creelman and Zeevaart (1984) studied the incorporation of ^{18}O into ABA synthesised following water stress. They also concluded that their results were 'in accord with the indirect pathway, although the existence of two separate, independent pathways, one operating in turgid and one in water-stressed leaves, cannot be ruled out'.

For the remainder of this article it will be assumed that the C_{40} pathway provides the only suitable biosynthetic framework for assigning a precise 'site of action' to the ABA-deficient tomato mutants.

THE PREDICTED EFFECT OF GENETIC LESIONS WITHIN THE C_{40} PATHWAY OF ABA BIOSYNTHESIS

In this section consideration will be given to the likely consequences of mutants whose primary gene products are enzymes of the ABA biosynthetic pathway. The idea is to highlight the fact that there are important differences between mutants acting at the C_{40} level and those affecting the conversion of C_{15} intermediates. In doing this, it is helpful to work to a fairly detailed biosynthetic scheme. One such scheme is shown in Figure 1. A number of variations on the C_{40} theme are clearly possible, given the lack of detailed evidence concerning the individual biosynthetic steps. However, the pathway shown represents a reasonable speculation based on all the information currently available.

Proponents of the indirect pathway have traditionally assumed that all-*trans* violaxanthin is the most likely C_{40} precursor of ABA. Oxidative cleavage of both end groups would yield two molecules of the C_{15} aldehyde, 2-*trans* Xanthoxin. This could conceivably be the first C_{15} intermediate in ABA biosynthesis and has frequently been detected in plant tissues (e.g. Firn *et al.*, 1972). However, the results of feeding plants with ^{14}C radiolabelled isomers of Xanthoxin are inconsistent with this view. The 2-*trans* isomer of Xanthoxin could only be converted to the unnatural 2-*trans* isomer of ABA (Taylor and Burden, 1973). In contrast, the radiolabelled 2-*cis* isomer of Xanthoxin was efficiently converted to the natural isomer of ABA. This appears to indicate that 2-*trans* Xanthoxin is probably not a genuine intermediate in ABA biosynthesis. Why it is present in plants remains an open question.

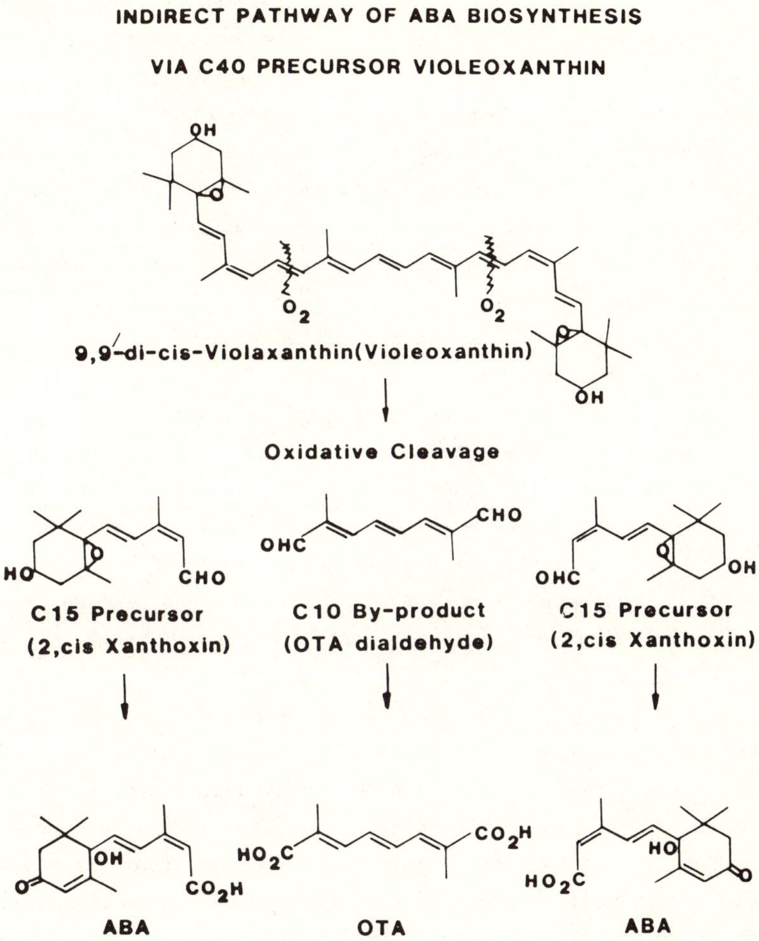

Figure 1. One possible biosynthetic pathway for ABA from a carotenoid precursor. It is likely that this scheme will have to be modified as further details of events at the C_{40}, C_{15} and C_{10} levels become known. The scheme provides a suitable framework for provisionally allocating mutants to different positions.

It is possible that all of the C_{15} intermediates in ABA biosynthesis are 2-cis isomers. This would mean that isomerisation must take place at the C_{40} level. For this reason, 9,9' di-cis-Violaxanthin (or violeoxanthin) is shown as a putative C_{40} precursor of ABA in Figure 1. This compound has recently been shown to be native to at least one plant species (Molnar et al., 1986). Oxidative cleavage would yield two molecules of 2-cis-Xanthoxin, which would consequently be the first C_{15} intermediate of ABA biosynthesis. Only three steps are required to convert this compound to ABA:-

1) Oxidation of the side chain aldehyde to give the corresponding carboxylic acid.
2) Isomerisation to form the γ-hydroxy αβ-unsaturated ring.
3) Oxidation of the 4' alcohol to a ketone.

There are therefore only three conceivable enzyme catalysed steps which could be subject to genetic lesion. If any of the three ABA deficient tomato mutants can be shown to impair the conversion of labelled 2-cis Xanthoxin to ABA, then it should be relatively easy to pinpoint the precise step involved. In contrast, mutants affecting ABA biosynthesis at the C_{40} level may prove more difficult to define.

One feature of the biosynthetic scheme shown in Figure 1 which is often overlooked, is the formation of a by-product. This is of particular relevance to research with which I have been personally involved (Taylor et al., 1984). The way in which the results in the next section are presented owes a great deal to the benefit of 'hindsight' derived from this earlier work. The discussion will be focussed particularly on information which can be gained by studying possible by-products of ABA biosynthesis. It therefore seems appropriate to consider the theoretical effects of different types of genetic lesion on by-product formation, before describing the actual results obtained for the three ABA deficient tomato mutants.

The central event of the indirect pathway of ABA biosynthesis is oxidative cleavage of a C_{40} precursor to form the first C_{15} intermediate. It is not inconceivable that only one of the end groups of the C_{40} precursor is destined to form ABA. A single cleavage event would in this case result in one molecule of the C_{15} intermediate and one molecule of a C_{25} by-product. This would obviously be less efficient than having a dual cleavage event, resulting in two molecules of the C_{15} intermediate and a C_{10} by-product. On theoretical grounds, the most likely initial by-product of ABA biosynthesis would therefore be the C_{10} dialdehyde, 2,7-dimethyl-2,4,6-Octatrienedialdehyde.

It is likely that this compound might eventually be oxidised in vivo to the corresponding dicarboxylic acid (2,7-dimethyl-2,4,6-Octatrienedioic acid, or OTA). The two C_{10} compounds (OTA-dialdehyde and OTA) are shown as by-products of ABA biosynthesis in Figure 1. Such compounds could be produced by oxidative cleavage of a large number of different carotenoids. OTA-dialdehyde has been isolated in 6-8% yield following in vitro oxidation of β carotene (Wendler et al., 1950). Both OTA itself and the dialdehyde are known compounds and have previously been synthesised (Mildner and Weedon, 1953).

Assuming that any by-products of ABA biosynthesis are not instantly metabolised, their concentration would be a reflection of the rate of cleavage of the C_{40} precursor. Measuring by-product levels could therefore provide important information on the rate of operation of this key step in the pathway. Although C_{15} intermediates would simultaneously be produced, these may be more difficult to quantify accurately due to their rapid conversion to ABA and/or various metabolites. The C_{15} intermediates may be inherently more transient than the C_{10} by-products.

It is now possible to consider the consequences of introducing genetic lesions at different points in the putative ABA biosynthetic pathway (Figure 1). Such lesions can be classified into two categories depending on whether they act before or after the oxidative cleavage of the last C_{40} ABA precursor. These categories will now be considered in turn:-

1) Mutant type A - genetic lesions of this type could be termed 'downstream mutants'. The most obvious example would be a mutant gene locus coding for an enzyme responsible for a step in the conversion of the first C_{15} intermediate (possibly 2-cis Xanthoxin) to ABA. This example is illustrated in Figure 2. The primary consequence would be a reduction in endogenous ABA levels. A number of secondary effects would be expected to follow from this. It seems likely that the rate of operation of the biosynthetic pathway would be elevated. This would occur in an 'attempt' to correct the ABA deficiency by increased biosynthesis. This may result directly from the removal of end product inhibition of biosynthesis; or indirectly from the tendency of ABA deficient mutants to wilt, thereby providing a stimulus for additional ABA synthesis.

An increased rate of operation of the biosynthetic pathway as a whole necessarily implies an increased rate of cleavage of the C_{40} precursor. This in turn would lead to an increased rate of formation of any C_{10} by-products. Elevated levels of C_{10} by-products in any of the three ABA deficient tomato mutants would therefore be a clear indication that they are acting 'downstream' of the cleavage step.

2) Mutant type B - genetic lesions of this type could be termed 'upstream mutants'. All mutants in this category can be defined as acting at, or before, the oxidative cleavage step. The most obvious example would be a mutant gene locus coding for the enzyme catalysing the cleavage step itself. This example is illustrated in Figure 3. The effect of such a mutation on the formation of any by-product of C_{40} cleavage would be exactly the

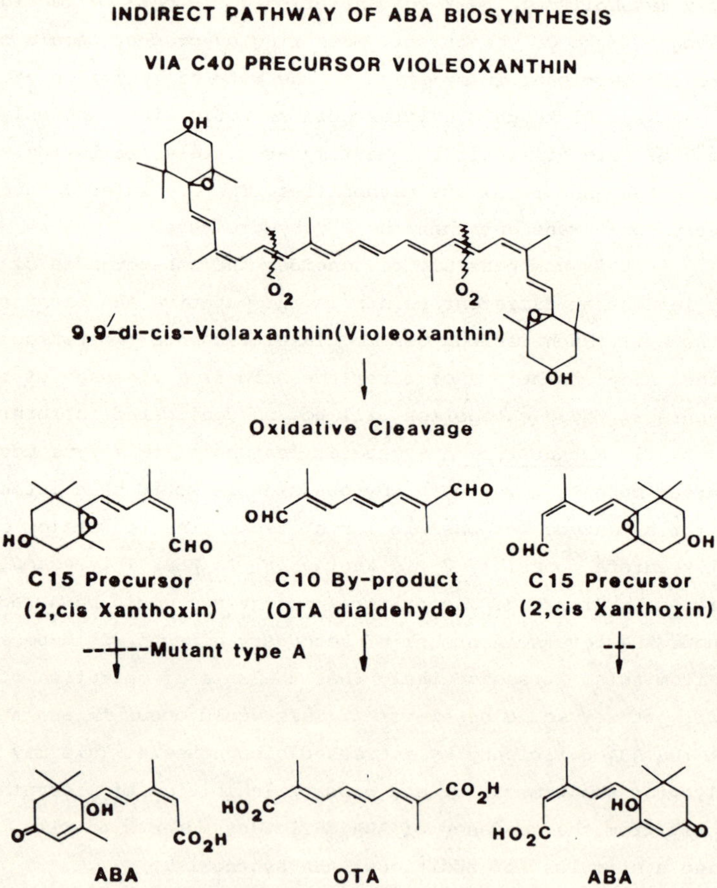

Figure 2. A genetic lesion impairing the conversion of C_{15} intermediates to ABA is illustrated. A mutant at this level (mutant type A) would be likely to increase the overall rate of C_{40} cleavage and to cause excessive accumulation of the C_{10} by-products.

opposite of that of the 'downstream' mutants. Any mutant reducing the rate of C_{40} cleavage would cause a reduction in ABA concentration by limiting the supply of the first C_{15} intermediate. This must simultaneously lead to a reduced rate of formation of any C_{10} by-products of the cleavage step. Any mutant having abnormally low levels of the biosynthetic by-products can therefore be assumed to belong to this 'upstream' category.

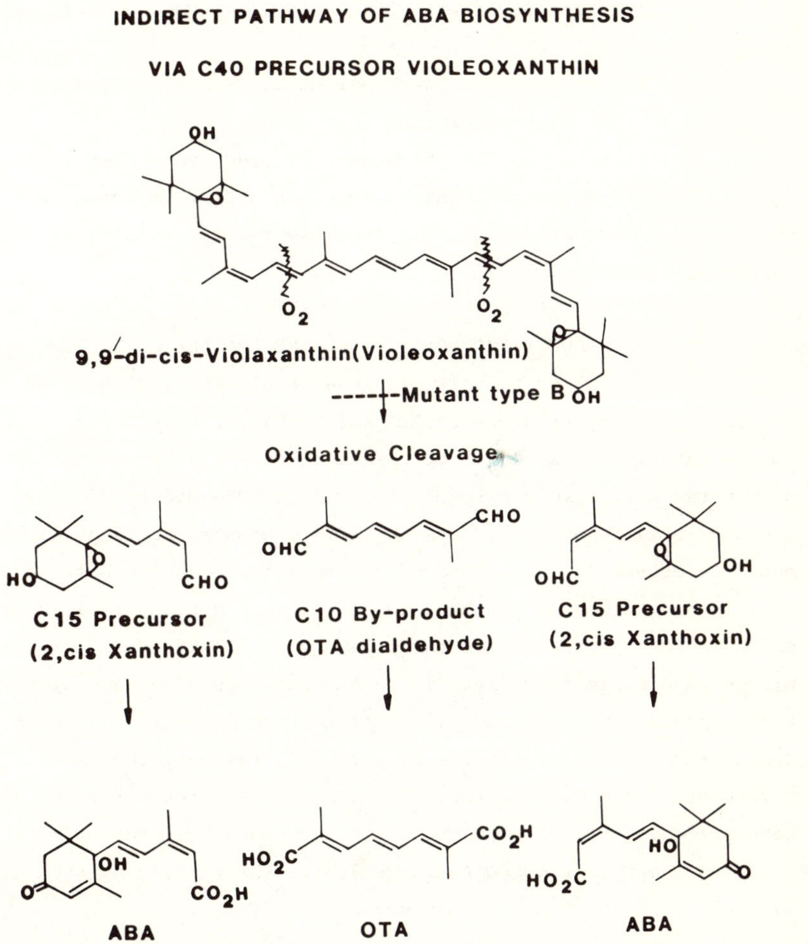

Figure 3. A genetic lesion impairing the oxidative cleavage of a C_{40} precursor is illustrated. A mutant at this level (mutant type B) would be likely to cause a simultaneous fall in the concentration of ABA, the C_{15} intermediates and any C_{10} by-products.

Further confirmation of the relative sites of action of different mutants could be provided by an analysis of by-product levels in double mutant combinations. Any 'upstream' mutant should be dominant (or strictly speaking, epistatic) over the 'downstream' mutants with respect to the level of by-product formation. By virtue of acting earlier in the pathway, the 'upstream' mutants should ensure that relatively low concentrations of the by-products are found in double mutant homozygotes.

As a result of this discussion, two questions immediately follow:-

a) Can any putative by-products of ABA biosynthesis be identified in plant extracts?

b) What are the characteristic endogenous levels of such by-products in the various ABA deficient tomato genotypes?

The evidence currently available on these two questions will be summarised in the next section.

BY-PRODUCT ACCUMULATION IN THE ABA DEFICIENT TOMATO GENOTYPES

As yet it is only possible to give partial answers to questions concerning the by-products of ABA biosynthesis. A fuller picture should emerge shortly, as a result of research currently being carried out. Work on the two most likely by-products, OTA and OTA-dialdehyde, is incomplete. The dicarboxylic acid, 2,7-dimethyl-2,4,6-Octatrienedioic acid (OTA) is shown in Figure 4. This compound has recently been synthesised (Linforth et al., 1986). Authentic standards are therefore available to allow plant extracts to be screened for the presence of OTA by GC-MS. It has so far proved impossible to detect OTA in any extracts of tomato shoot tissue. If the compound exists in the plant, it must be present at extremely low concentration. OTA-dialdehyde is currently being synthesised by 'dibal' treatment of the OTA dimethyl ester. Suitable plant extracts have not yet been screened for the presence of this compound. It therefore remains an open question whether or not OTA-dialdehyde is the initial by-product of ABA biosynthesis via the indirect pathway.

For historical reasons (see Taylor et al., 1984), work with two other C_{10} dicarboxylic acids (ODA and OEA; Figure 4) is further advanced than that on OTA itself. Unlike OTA, both ODA (2,7-dimethyl-2,4-octadienedioic acid) and OEA (2,7-dimethyl-4-octenedioic acid) have been detected in plant tissue (Linforth et al., 1986). Authentic standards of ODA and OEA have been synthesised, along with d_2-OEA (protons at C-2 and C-7 replaced by deuterium) to enable precise quantification by GC-MS. The corresponding dialdehydes of ODA and OEA have not yet been synthesised and so their presence in plant tissue remains to be established.

From the arguments presented so far, there is no reason to suspect that the C_{10} diolefin (ODA) would accumulate as a by-product of ABA biosynthesis rather than the related triolefin (OTA). Nevertheless, the results obtained with the three C_{10} dicarboxylic acids (OTA, ODA and OEA) appear to implicate ODA and not the other two (see Table 1). The pattern shown in Table 1 is entirely typical and is consistent with the results of a whole series of experiments we have performed using widely different methods of analysis (Linforth, 1986). The ODA levels are characteristic of

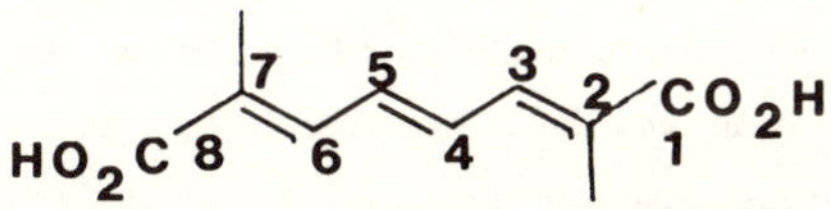

OTA (2,7-dimethyl-2,4,6-octatrienedioic acid)

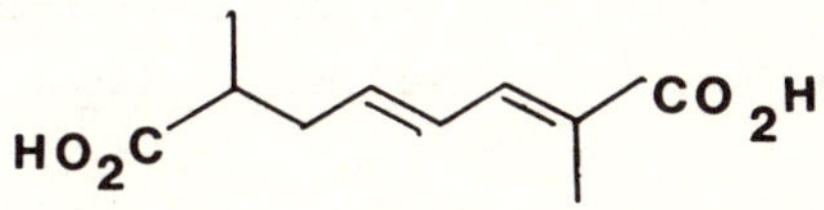

ODA (2,7-dimethyl-2,4-octadienedioic acid)

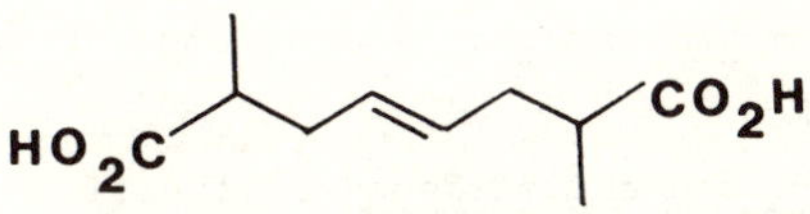

OEA (2,7-dimethyl-4-octenedioic acid)

Figure 4. Three C_{10} dicarboxylic acids are illustrated. One or more of these may be formed as by-products of the biosynthesis of ABA. On theoretical grounds, OTA is the most likely by-product, but there is currently more evidence to implicate ODA.

particular ABA-deficient genotypes and could be used to identify them. These results will be explained in further detail in the remainder of this section.

The first thing to note is that there is a clear inverse correlation between the severity of phenotype of the various wilty mutant genotypes (as reported by Taylor and Tarr, 1984) and their 'basal' ABA concentrations. In contrast, the endogenous levels of the C_{10} dicarboxylic acids do not correlate with either ABA concentration or the severity of phenotype. Any artefact associated with the 'wilty state' might accumulate to abnormally high levels in ABA-deficient mutants compared to the isogenic control. However, it would be expected that the most severe mutant combinations would have the highest levels. None of the C_{10} dicarboxylic acids show such a correlation.

Only one of the three compounds fits the pattern expected for a by-product of ABA biosynthesis. That compound is ODA. It can be seen from Table 1 that unstressed non-mutant plants have an ODA concentration of around 70 $ng.g^{-1}$ fresh weight. The effect of the flacca mutation is to increase the endogenous ODA content to 115 $ng.g^{-1}$ fresh weight. This difference is substantial in spite of being minimised by growing the plants in the virtual absence of water stress. Levels of ODA in flacca plants can be substantially increased by water-stress treatments (Linforth, 1986). Conversely, it has been found that the ODA concentration can be more or less halved, following phenotypic reversion of flacca plants with exogenous ABA (Linforth, 1986).

The sitiens mutation appears to have an almost identical effect to that of flacca. The ODA concentration is possibly even more elevated in sit homozygotes (130 $ng.g^{-1}$ F.wt) than in flacca plants. This may be due to the greater reduction in ABA caused by the sit mutant allele. Double mutants homozygous for both sit and flc also accumulate excessive amounts of ODA (135 $ng.g^{-1}$ F.wt).

Both flacca and sitiens are therefore 'type A' or 'downstream mutants'. Their ABA deficiency leads to an increased rate of cleavage of the C_{40} precursor, resulting in abnormally high levels of the biosynthetic by-product ODA. These mutant gene loci are likely to be involved in the conversion of C_{15} intermediates to ABA (see Figure 5). It can therefore be predicted that flacca and sitiens plants would be less efficient in converting radiolabelled Xanthoxin to ABA than corresponding isogenic controls. It is a pity that after feeding ^{14}C labelled Xanthoxin to tomato plants, Taylor and Burden (1973) did not go on to try the same procedure with flacca and sitiens.*

It can be seen from Table 1 that the third mutant, notabilis, does not increase ODA levels. In fact, the ODA concentration of notabilis was only around one-third of that found in the non-mutant controls. In order to confirm that this was a genuine effect, a further experiment was carried out involving notabilis and control plants grown in a glasshouse

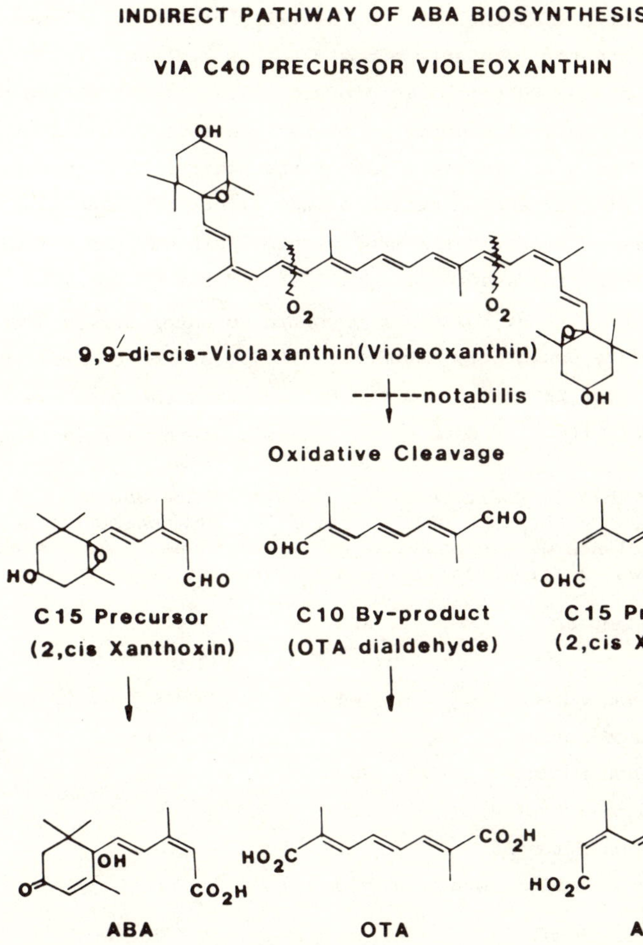

Figure 5. Two of the ABA-deficient tomato mutants (flacca and sitiens) are assigned to the C_{15} level of the biosynthetic pathway. Both mutants are 'downstream' of the oxidative cleavage step.

*See 'Note added in proof' at the end of this chapter

environment. The plants were grown for 6 weeks and then allowed to accumulate ABA following a water-stress treatment involving 10% loss of fresh weight. It was thought that a severe stress treatment would induce a high rate of ABA biosynthesis and therefore cause maximum accumulation of any C_{10} by-products. The levels of ABA, ODA and OEA were quantified by GC-MS and the results are shown in Table 2.

By comparing the ABA concentrations shown in Table 1 and Table 2, it can be seen that the apparent severity of ABA deficiency caused by notabilis is greater in plants exposed to water stress. In 'unstressed' plants grown at 95% RH notabilis produced 47.3% of the ABA found in the non-mutant control. Following exposure to water stress (Table 2) the notabilis mutants could only accumulate 13.2% of the control ABA concentration. It was hoped that a similar amplification of the effect of notabilis on ODA levels would be observed in water stressed plants. This was undoubtedly the case. The ODA concentration was reduced to 33.8% of the control, in notabilis plants grown at 95% RH. Following exposure to water stress (Table 2) the notabilis plants could only accumulate 18.75% of the control ODA concentration. This effect is mainly due to the fact that the ODA levels in the non-mutant control virtually doubled as a result of the stress regime, whereas

Table 2. Characteristic levels of ABA and the C_{10} dicarboxylic acids (ODA and OEA) in notabilis and non-mutant control plants. Plants were glasshouse grown for 6 weeks and subjected to water stress following a 10% loss of fresh weight.

Sample	ABA ($ng.g^{-1}$ F.wt)	ODA ($ng.g^{-1}$ F.wt)	OEA ($ng.g^{-1}$ F.wt)
non-mutant 1	564	100	52
non-mutant 2	400	224	120
non-mutant 3	460	114	98
non-mutant 4	505	136	92
non-mutant 5	571	144	108
non-mutant mean	500	144	94
notabilis 1	51	20	88
notabilis 2	61	30	88
notabilis 3	84	24	96
notabilis 4	78	32	310
notabilis 5	57	30	83
notabilis mean	66	27	133

those of notabilis were more or less unaffected.

The mutant notabilis is therefore a 'type B' or 'upstream' mutant. Its ABA deficiency appears to be associated with a low rate of cleavage of the C_{40} precursor. This would lead to a simultaneous fall in the accumulation of both ABA and any by-products of biosynthesis. It can therefore be predicted that notabilis plants would be just as efficient as the non-mutant controls in converting labelled Xanthoxin to ABA. In this respect it would be expected to differ from both flacca and sitiens. One possible site of action for the notabilis genetic lesion is illustrated in Figure 6; although it is conceivable that notabilis is acting even earlier than this.

In the previous section it was suggested that by acting earlier in the pathway, an 'upstream' mutant should be able to prevent the typical increase in by-product accumulation normally caused by a 'downstream' mutant. It can be seen that the ODA concentrations of the two double mutant genotypes containing notabilis, are lower than that of the non-mutant control (Table 1). These two genotypes have precisely the opposite effect on ODA levels to that of the double mutant homozygous for both flc and sit.

It can be seen that the complex pattern of ODA concentrations in the whole ABA-deficient genotype series can only be satisfactorily explained on the assumption that ODA is a by-product of ABA biosynthesis.* However, this in itself raises further questions.

WHY IS ODA A BY-PRODUCT OF ABA BIOSYNTHESIS?

Work with the ABA-deficient tomato mutants has implicated ODA in ABA biosynthesis. There is no evidence to suggest that either of the other two C_{10} dicarboxylic acids shown in Figure 4 (OTA and OEA) are also involved. Endogenous OEA concentrations seem totally random with respect to genotype and water stress. The reason why plants produce OEA is uncertain, but it seems to have no obvious connection with either ABA or ODA. Failure to detect OTA, means that if it is produced at all as a result of ABA biosynthesis, it must be rapidly metabolised (possibly by reduction to ODA?).

The evidence that ODA is a by-product of ABA biosynthesis is circumstantial, but nevertheless strong. Direct proof may only emerge when full details of the last C_{40} precursor and the oxidative cleavage step are known. If the relationship between ODA and ABA biosynthesis is accepted (as the genetic evidence dictates), then the timing of the reduction of the C-6

*See 'Note added in proof' at the end of this chapter.

olefin must be considered. There are two possibilities:-

(a) Reduction at the C_{10} level - in this case it may be assumed that the initial by-product of C_{40} cleavage is a C_{10} triolefin (OTA dialdehyde). The reduction step (possibly involving the formation of ODA dialdehyde) would therefore form part of the normal pathway of metabolism for the by-product.

INDIRECT PATHWAY OF ABA BIOSYNTHESIS

VIA C40 PRECURSOR VIOLEOXANTHIN

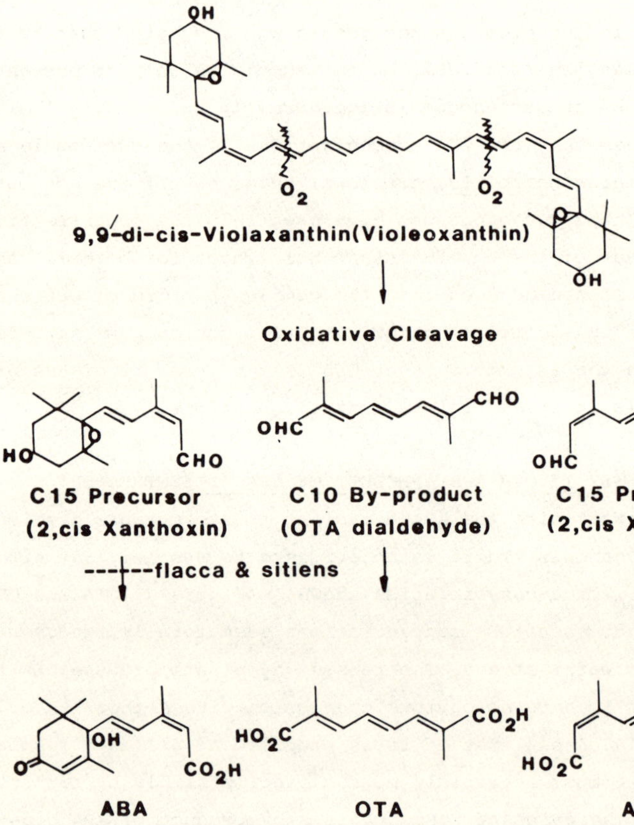

Figure 6. The ABA-deficient tomato mutant, notabilis is assigned to the C_{40} level of the biosynthetic pathway. This mutant must be acting 'upstream' of the oxidative cleavage step.

(b) Reduction at the C_{40} level - in this case it may be assumed that the initial by-product of C_{40} cleavage would be ODA dialdehyde, which is subsequently oxidised to the corresponding dicarboxylic acid. ODA may then be metabolised by some as yet unknown route. If this is correct, there must be at least one dihydrocarotenoid C_{40} precursor in the ABA biosynthetic pathway.

The latter suggestion is the more surprising of the two. Why it might be necessary to have a dihydrocarotenoid ABA precursor is something of a mystery. There seems no immediately obvious advantage in interrupting the polyene chain. It might conceivably make the molecule more flexible, thus facilitating simultaneous cleavage at two positions. Alternatively, the advantage might somehow lie in changing λ max by 'breaking' the chromophore. More detailed information on the cleavage step is required before this problem can be resolved. It would perhaps be sensible to prove whether or not a dihydrocarotenoid ABA precursor exists before worrying unduly about why it should be necessary. There is a great deal of scope for further experimental work in this area.

WHY STUDY THE BY-PRODUCTS OF ABA BIOSYNTHESIS?

It is hoped that the preceding discussion has gone some way towards answering this question. However, it is worthwhile summarising the current position by listing the following points:-

1) Evidence suggesting the existence of a C_{10} by-product automatically implies that the indirect pathway of ABA biosynthesis is correct. This helps to resolve a long running dispute.

2) The existence of a C_{10} by-product indicates that <u>both</u> C_{15} end groups of the C_{40} precursor are likely to be converted to ABA. A double oxidative cleavage event is required to produce the by-product.

3) Information can be obtained concerning the relative position of the three genetic lesions (*not*, *flc* and *sit*) with respect to the oxidative cleavage step. Specific predictions can be made concerning their ability to convert labelled Xanthoxin to ABA.

4) The possible existence of an unorthodox dihydrocarotenoid precursor of ABA has been suggested. Events at the C_{40} level may turn out to be more complex than would otherwise have been suspected.

5) Knowledge of the by-products and their metabolism is necessary to give a complete picture of the events surrounding ABA biosynthesis by the indirect pathway.

NOTE ADDED IN PROOF

S.J. Neill, A.R. Parry and R. Horgan presented a poster at this conference giving the results of a Xanthoxin feeding experiment. They synthesised ^{13}C Xanthoxin and used GC-MS to assess the ability of the wilty tomato mutants to convert Xanthoxin to ABA. They found substantial incorporation of ^{13}C label into ABA by non-mutant plants and not homozygotes (38% and 25% respectively). In contrast, both sit and flc homozygotes showed very poor incorporation of label into ABA (both around 1%). These results are exactly in line with the predictions made on the basis of the ODA accumulation pattern. They indicate that the flc and sit genetic lesions impair a step in the conversion of Xanthoxin to ABA i.e. these mutants act at the C_{15} level of the biosynthetic pathway. The results are also consistent with the prediction that notabilis acts at or before the C_{40} cleavage step and would not significantly affect conversion of Xanthoxin to ABA.

REFERENCES

Cornforth, J.W., Milborrow, B.V., Ryback, G. & Wareing, P.F. (1965) Chemistry and physiology of 'dormins' in sycamore. Identity of sycamore 'dormin' with abscisin II. Nature, 205, 1269-70.

Creelman, R.A. & Zeevaart, J.A.D. (1984) Incorporation of oxygen into abscisic acid and phaseic acid from molecular oxygen. Plant Physiol., 75, 166-169.

Firn, R.D., Burden, R.S. & Taylor, H.F. (1972) The detection and estimation of the growth inhibitor Xanthoxin in plants. Planta, 102, 115-126.

ımber, D. & Tal, M. (1970) Phenotypic reversion of flacca, a wilty mutant of tomato, by abscisic acid. Science, 169, 592-593.

Linforth, R.S.T. (1986) A genetic approach to the study of abscisic acid biosynthesis. Ph.D. Thesis. University of Nottingham.

Linforth, R.S.T., Bowman, W.R., Griffin, D.A., Hedden, P., Marples, B.A. & Taylor, I.B. (1986) 2,7-dimethyl-2,4-octadienedioic acid, a possible by-product of abscisic acid biosynthesis. Phytochem., (submitted).

Milborrow, B.V. (1971) Abscisic acid. In: Aspects of Terpenoid Chemistry and Biochemistry, ed. T.W. Goodwin, pp 137-151, London: Academic Press.

Mildner, P. & Weedon, B.C.L. (1953) Carotenoids and related compounds. Part II. C_{10} intermediates for the synthesis of carotenoids. J. Chem. Soc. (London, 1953) pp. 3294-3298.

Molnar, P., Szabolcs, J. & Radics, L. (1986) Naturally occurring di-cis-violaxanthins from Viola tricolor: isolation and identification by ^{1}H NMR spectroscopy of four di-cis-isomers. Phytochem., 25, 195-199.

Moore, R. & Smith, J.O. (1985) Graviresponsiveness and abscisic acid content of carotenoid deficient mutants of Zea mays L. Planta, 164, 126-128.

Neill, S.J. & Horgan, R. (1985) Abscisic acid production and water relations in wilty tomato mutants subjected to water deficiency. J. Exp. Bot., 36, 1222-1231.

Neill, S.J., Horgan, R., Walton, D.C. & Griffin, D. (1982b) Biosynthesis of abscisic acid. In: Plant growth substances 1982, ed. P.F. Wareing, pp 315-323. London: Academic Press.

Ohkuma, K., Addicott, F.T., Smith, O.E. & Thiessen, W.E. (1965) The structure of abscisin II. Tetrahedron Lett., 29, 2529-2535.

Oritani, T. & Yamashita, K. (1985) Biosynthesis of (+)-abscisic acid in Cercospora cruenta. Agric. Biol. Chem., 49, 245-249.

Rick, C.M. (1980) Tomato linkage survey. Rep. Tomato Genet. Coop., 30, 2-17.

Tal, M. & Nevo, Y. (1973) Abnormal stomatal behaviour and root resistance, and hormonal imbalance in three wilty mutants of tomato. Biochem. Genet., 8, 291-300.

Taylor, H.F. & Burden, R.S. (1970a) The structure and chemical transformations of Xanthoxin. Tetrahedron Lett., 47, 4071-4074.

Taylor, H.F. & Burden, R.S. (1970b) Identification of plant growth inhibitors produced by photolysis of violaxanthin. Phytochem., 9, 2217-2223.

Taylor, H.F. & Burden, R.S. (1970c) Xanthoxin a new naturally occurring plant growth inhibitor. Nature, 227, 302-304.

Taylor, H.F. & Burden, R.S. (1972) Xanthoxin a recently discovered plant growth inhibitor. Proc. Roy. Soc. B., 180, 317-346.

Taylor, H.F. & Burden, R.S. (1973) Preparation and metabolism of 2-[^{14}C]-cis, trans Xanthoxin. J. Exp. Bot., 24, 873-880.

Taylor, H.F. & Smith, T.A. (1967) Production of plant growth inhibitors from Xanthophylls: a possible source of dormin. Nature, 215, 1513-1514.

Taylor, I.B., Bowman, W.R., Griffin, D.A., Hedden, P., Kipps, M.A., Linforth, R.S.T., Marples, B.A. & Rossall, S. (1984) Abnormalities of abscisic acid accumulation in tomato mutants. In: Biochemical aspects of synthetic and naturally occurring plant growth regulators, Monograph 11, ed. R. Menhenett & D.K. Lawrence. pp. 73-90. British Plant Growth Regulator Group Wantage.

Taylor, I.B. & Rossall, S. (1982) The genetic relationship between the tomato mutants, flacca and lateral suppressor, with reference to abscisic acid accumulation. Planta, 154, 1-5.

Taylor, I.B. & Tarr, A.R. (1984) Phenotypic interactions between abscisic acid deficient tomato mutants. Theor. Appl. Genet., 68, 115-119.

Walton, D.C. (1980) Biochemistry and physiology of abscisic acid. Ann. Rev. Plant Physiol., 31, 453-489.

Wendler, N.L., Rosenblum, C. & Tishler, M. (1950) Oxidation of β-carotene. J. Am. Chem. Soc., 72, 234-239.

Zeevaart, J.A.D. (1985) Biosynthesis and catabolism of abscisic acid. Abstracts of the 12th International Conference on plant growth substances p.27. IPGSA, Heidelberg.

SEED AND FOLIAR MUTATIONS IN PISUM

Cliff L. Hedley and Trevor L. Wang,
John Innes Institute, Colney Lane, Norwich NR4 7UH, U.K.

INTRODUCTION

The pea (*Pisum sativum* L.) has been a favourite experimental tool of the plant physiologist, biochemist and geneticist for many years. Numerous scientific publications have resulted from interest in the pea plant on topics as wide apart as the basic laws of inheritance (Mendel, 1865) and the effect of nitrogen fixation on soil fertility (Askin *et al.*, 1985). One consequence of these endeavours, has been the identification and cataloguing of large numbers of genetic mutants which have led to diverse pea lines being held in gene banks throughout the world (Blixt and Williams, 1982). From the accumulated information about these mutants, Lamprecht (1948) was able to produce the first comprehensive gene map for *Pisum*.

It has been suggested (Marx, 1983) that the domestication of the pea has depended on the selection and propagation of specific mutations, especially those relating to the size and structure of the seed and to the dehiscence of the pod. Initially, this process was one of selecting and maintaining spontaneous mutations for attributes which aided the domestication process. The use of artificial hybridization between these mutants then generated a tremendous amount of phenotypic diversity from which the pea as we know it has been derived. The recent interest in redeveloping the pea plant to suit modern agricultural practices has stimulated breeders to concentrate on those mutations which affect the foliar structure of the plant and those which improve seed characteristics (Snoad, 1980). It is mutations affecting these two aspects of the plant on which we will concentrate for this review.

FOLIAR MUTATIONS

It is possible to list a large number of genes which are known to influence the foliar parts of the plant (Table 1). Genes are known which affect the size and shape of leaves, the surface structure of the leaves

Table 1. Mutations in *Pisum* affecting foliage

Adapted with additions, from Blixt, 1972, 1977 and Lamprecht, 1974. Classification and phenotypes have been simplified in many cases. Brackets around genes denotes grouped characters, around symbols denotes synonyms according to Blixt, 1977. First character in gene and symbol denotes dominance. All genes occur as spontaneous mutations except cov, crif, cris (induced only)

GENE/CHARACTER	SYMBOL	PHENOTYPE	REFERENCE
FOLIAGE STRUCTURE			
aeromaculata	fl	None or very few airspaces between epidermis and pallisade layers	Tedin & Tedin, 1926
Argenteum	Arg	Separation of epidermis and pallisade layer to give silvered leaf.	Marx, 1978, 1982
stomata number	sal to 3	Leaves have reduced number of stomata	Tavcar, 1926
(wax layer)	(bl), wa, wb, was, wel, wex, wlo, wsp	Alter amount of wax on leaves in total or individual surfaces of leaflets and stipules (bl = wsp)	Wellensiek, 1928; Lamm, 1957; Marx, 1969a; Nilsson, 1933; Lamprecht, 1939 (Blixt, 1977)
CHLOROPHYLL			
(chlorophyll types)	alb, alt, au, auv, chl to 4, chil to 23, cov, cvit, lum, o, pa, py, vac, vil, vim, xal to 5, xat, yg	Alter green colour of leaves from very light green to dark bluish green	numerous, see Blixt, 1972
FOLIAGE ONTOGENY			
afila	af	Leaflets are altered to tendrils	Goldenberg, 1965
clavicula	tl	Tendrils are altered to leaflets	White, 1917

contd/...

contd. from p.220

GENE/CHARACTER	SYMBOL	PHENOTYPE	REFERENCE
<u>cochleata</u>	<u>coch</u>	Stipules are spatula-like	Wellensiek, 1962; Gottschalk, 1971
<u>sinuate leaf</u>	<u>sil</u>	Adventitious tendrils on stipules	Marx, 1977a
<u>tendril acacia</u>	<u>tac</u>	Terminal tendril converted to leaflet	Sharma, 1972
<u>unipetiole</u>	<u>up</u>	One pair of leaflets only	von Rosen, 1944
<u>unifoliata</u>	<u>uni</u>	Leaflets converted to single leaf	Lamprecht, 1933
FOLIAGE MORPHOLOGY			
(foliage folding)	<u>cri</u>, <u>crif</u>, <u>cris</u>, <u>sb</u>	Folds/crinkling of leaves	Lamm, 1949; Gottschalk, 1964; Kellenbarger, 1952
(foliage shape)	<u>acu</u>, <u>asre</u>, <u>fo</u>, <u>fob</u>, <u>fol</u>, <u>fom</u>, <u>ins</u>, <u>sc</u>, <u>un</u>, <u>X</u>	Alter shape of leaflets, stipules or both	Lamprecht, 1958b, 1959b, 1960a, 1962c, 1964, Härstedt, 1950 Wellensiek, 1971; Matthews, 1969
(foliage size)	<u>cont</u>, <u>lat</u>, <u>st</u>, <u>stim</u>, <u>X</u>	Alter size of foliage or stipules alone	Lamprecht, 1963a, Lamm, 1957; Wellensiek, 1925; Lamprecht, 1960b, Matthews, 1969
(foliage width)	<u>ang</u>, <u>lath</u>, (<u>mare</u>), <u>nr</u>, <u>red</u>, <u>ten</u>, <u>Y</u>	Alter width of foliage (N.B. <u>mare</u> = <u>red</u>; <u>nr</u> now invalid)	Lamprecht, 1942, 1945, 1949, 1962b, 1967, Brotherton, 1923 (Marx, 1976; Blixt, 1977)

with regard to wax layers, the attachment of the epidermis, the frequency of stomata and the "colour" of the foliage, as it relates to chlorophyll content. The number of genes which affect the latter character is very large (at least 50), their effects can be on the content of chlorophyll (a or b) or carotenoids or both and may be developmentally- or spatially-related, i.e. occurring at a particular point in development or on particular areas of particular organs. Many of the effects are lethal (e.g. yellow plants, xantha), presumably since the plants cannot photosynthesise. Such plants die once the seed reserves have been exhausted, though, they will survive and produce seed when the shoots are grafted onto normal plants (personal observation). Marx (1977b) has also noted that shoots of another mutant, albino-terminalis (alt), will survive when grafted to normal plants.

The wax layer on the epidermis of leaflets, stipules, stems and pods can be reduced to varying degrees and often on individual organs. Thus, for example, wlo removes wax specifically from upper surface of leaflets whereas wa removes wax from the upper surface of leaflets and wel removes wax completely from the plant surface (see Table 1 for references). Figure 1 shows some of the variation for wax cover as seen under the scanning electron microscope.

A dominant mutation that has proved very useful to the plant physiologist and biochemist is Argenteum (Arg, Marx, 1978, 1982). Plants with the genotype Arg/Arg or Arg/arg have silvered leaves due to a limited attachment between the upper epidermis and pallisade cells. The two layers, therefore, are readily detached, a phenomenon which has proved invaluable for work on stomatal physiology where isolated epidermis is frequently required (Donkin et al., 1982; Jewer et al., 1982). Although many of the foliar mutations are similarly of academic interest, at present the genes which have been most readily exploited commercially have been associated with the shape and size of the leaves.

Changing foliage ontogeny

Normally the Pisum leaf is compound and pinnate, consisting of one (in young leaves) or more opposing pairs of leaflets with the rachis ending in tendrils. At the base of the petiole are two large, sessile stipules (Blixt, 1972; Figure 2a). There are a number of genes known to affect the development of the foliage, most interest however, has centred on three genes viz: afila (af), which converts the leaflets into tendrils (Figure 2b), stipules reduced (st), which reduces the size of the stipules (Figure 2b) and clavicula (tl), which converts the tendrils into leaflets (Figure 2c).

The afila gene was first reported by Kujala (1953) and given the symbol af by Goldenberg (1965). The gene has since been mapped to linkage group 1 (Marx, 1969c; Snoad, 1971; Kielpinski, 1982). Snoad (1974) incorporated the af gene into a breeding programme at the John Innes Institute with the ultimate objective of improving the standing ability of the dried pea crop. The increased amount of tendril on these plants caused considerable interlocking of adjacent plants in the sward, producing an open canopy which improved not only the standing ability (and thus resistance to lodging) but

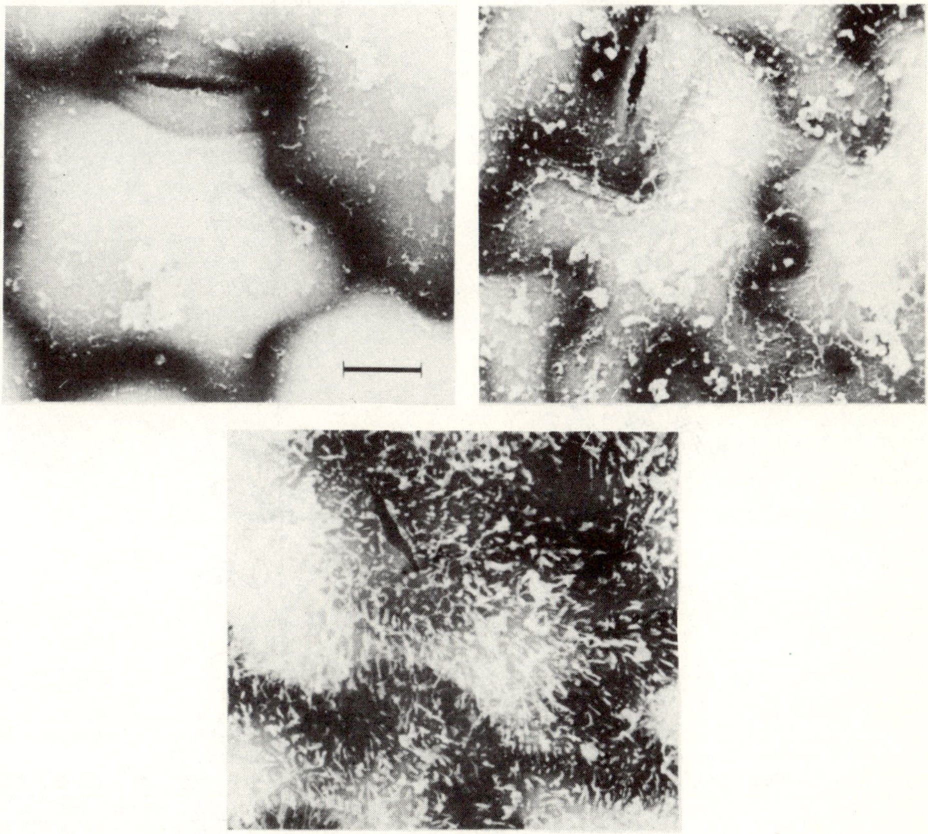

Figure 1. Variation for epidermal wax.

Scanning electron micrographs of the lower surface of leaves. Leaves of JI 35 (top left), JI 1391 (top right) and JI 38 (bottom centre) were frozen in a liquid N slush, coated with gold and examined while frozen using a Philips 501B scanning electron microscope. Bar = 10µm.

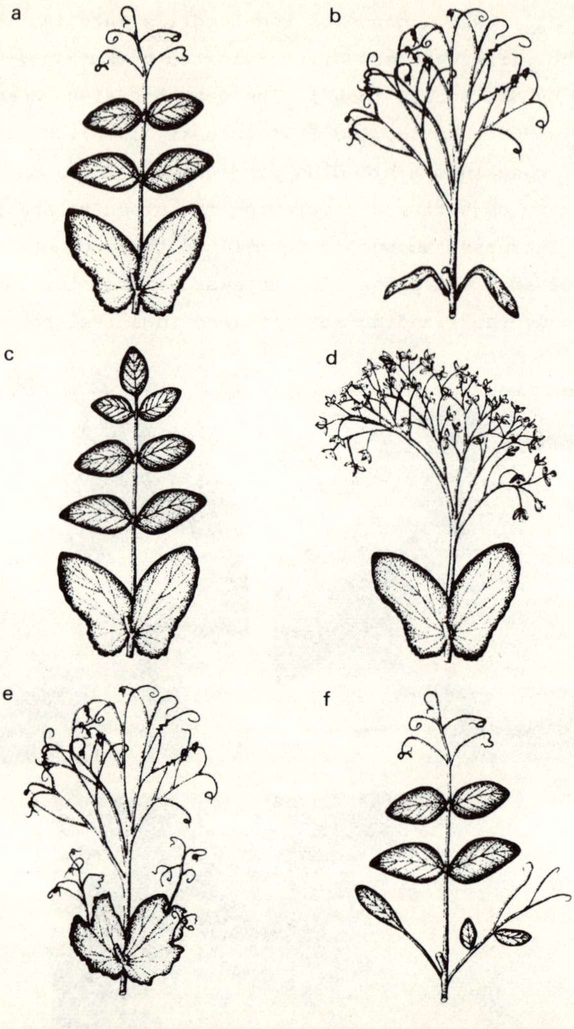

Figure 2. Pisum leaf morphology. Representations of some of the various phenotypes of pea leaves and stipules.
Genotypes: a - Af, St, Tl, Coch, Sil; b - af, st, Tl, Coch, Sil; c - Af, St, tl, Coch, Sil; d - af, St, tl, Coch, Sil; e - af, St, Tl, Coch, sil; f - Af, St, Tl, coch, Sil.

also the drying of the crop at the end of the season. This enabled quicker and more efficient harvesting.

Early plant models in the U.K. incorporated the af gene and also the st gene, which reduces the size of the stipules to vestigial organs. The st gene was first reported by Pellew and Sverdrup (1923) as a spontaneous mutant of cv. 'Duke of Albany' and was given the symbol st by Wellensiek (1925). The gene has been mapped to linkage group 3 (Lamprecht, 1948). Plants with the af/af st/st genotype formed good open canopies which stood well (Figure 3a), but the plant and crop growth rates were low compared with conventional peas (Pyke and Hedley, 1983a and b) due to poor light interception early in development (Pyke and Hedley, 1982, 1983a). It was also found that tendrils had a lower photosynthetic rate per unit dry weight than leaves (Pyke and Hedley, 1985). These differences made the plant less photosynthetically efficient but this was, to some extent, compensated by better light distribution within the canopy, which increased the net assimilation rate of the tendrils at the base of the crop (Pyke and Hedley, 1985). To overcome the problems associated with the afila/stipules reduced combination, this early model has been succeeded by others which possess larger stipules. This was achieved by incorporating afila (af/af) with normal stipules (St/St). Such af/af St/St genotypes have similar growth and photosynthetic rates to conventional peas and possess an improved standing ability (Figure 3b; Pyke and Hedley, 1983a, 1983b, 1985). These new plant models have formed the ideotype for today's dried pea crop.

Conversion of tendrils to leaflets can be achieved by another mutation clavicula (tl) which leads to replacement of the terminal and lateral tendrils with leaflets (White, 1917). Three alleles have been described at this locus, tl^{pet}, where the apical leaflets are on short petioles and tl^{w} where the leaflets are without petioles. The third allele, Tl, is completely dominant over tl^{pet}, but it is incompletely dominant over tl^{w} and the heterozygote Tl/tl^{w} is often recognizable by the presence of flattened tendrils (Blixt, 1978). These alleles have been mapped to linkage group 7 (Lamprecht, 1948). The tl^{w} gene interacts with the af gene to produce leaves which are much branched, each branch ending in a small leaflet (Figure 2d). Although plants with the $af/af\ tl^{w}/tl^{w}$ genotype, possess a very high surface area for light interception, they have a very poor standing ability since there are no tendrils to provide mutual support. There is, therefore, very little commercial interest in this gene combination.

A second gene, tendrilled acacia (tac) unlinked to tl (Sharma a

b

Figure 3. Leafless and semi-leafless peas. Experimental plots (1.6m squares) of leafless (af/af, st/st)(a) and semi-leafless (af/af, St/St)(b) varieties of Pisum sativum L.

and Aravindan, 1972; Sharma, 1972; Sharma et al., 1972) has been described. One allele of this gene, tac^B, replaces the apical tendril with a leaflet which is similar in size and shape to the lateral leaflets. The remaining lateral tendrils may be present or absent. A second mutant allele (tac^S) has been isolated (Sharma and Kumar, 1981) which has a spoon shaped terminal leaflet with a higher frequency of lateral tendrils. From the different effects induced by tl and tac, Sharma (1981) suggested that the tendril forming region of the leaf is under the control of at least two functional subunits, one responsible for the formation of lateral tendrils and another for the terminal position. He suggests that tl affects both sub-units simultaneously while tac leads to leaflet formation only at the apex while partially blocking the formation of the lateral tendrils.

There are other genes which affect the number of leaflets making up the leaf. The unipetiole gene (up, van Rosen, 1944) for example determines that only a single pair of leaflets is produced during ontogeny while the unifoliata (uni) gene (Lamprecht, 1933) produces a single simple leaf instead of the normal compound pinnate structure.

Another gene known to interact with the af gene is sinuate leaf (sil, Marx, 1977a). When combined with af, sil/sil plants have adventitious tendrils arising from clefts in the distal portion of the stipule (Figure 2e). This phenotype is only observed, however, when normal sized stipules are present (St); no adventitious tendrils are observed when af is combined with st. The development of the stipule is also affected by the gene, cochleata (coch), which converts the stipule into a spatula like object with a long petiole resembling much more a normal leaf (Figure 2f). When the cochleata gene is combined with the afila gene, plants homozygous for these genes (af/af coch/coch) have both stipules and leaves converted to tendrils. Plants which are homozygous for tl and coch (tl/tl coch/coch) have leaves without tendrils and the stipules are either strongly reduced or replaced by leaves with leaflets. When the af, tl and coch genes are present (af/af tl/tl coch/coch) the leaflets are converted to tendrils with small terminal leaflets and the stipules are replaced by branched tendrils also with small terminal leaflets (Gottschalk, 1970). It can be strongly suggested from the effects of coch and sil that the ontogeny of stipule development may be related to the remainder of the leaf.

At first sight, the genetic variation which exists for leaf ontogeny in Pisum should present great scope for the developmental botanist to investigate the control of leaf development. Unfortunately, however, there have been very few fundamental studies which have made use of these genes. Meicenheimer *et al* (1983) investigated the meristem characteristics of normal, *afila*, *clavicula* and *stipules reduced* plants and plants with combinations of these leaf phenotypes. They found that the multiple tendrils of the *afila* leaf were formed from numerous secondary branches on the leaflet primordia while the tendril-like morphology of the *afila* leaflets were derived from a radial marginal meristem characteristic of normal tendril development. Furthermore, the small leaflets formed when *af* is combined with *tl* were produced by adaxial and marginal meristems which became apparent in the distal portion of the tendril late in leaf ontogeny. The small stipules formed when *st* is present resulted from the early loss of abaxial and adaxial stipule marginal meristems. Gould *et al* (1986) found that the anatomy of the stem, petiole and rachis were not modified by the *af* or *tl* mutations and the tendrils which replaced leaves when *af* was present have normal tendril anatomy. Likewise, the terminal leaflets formed when *tl* is present had normal leaflet anatomy.

Meristems of conventional peas destined to form leaflets can be induced to form tendrils by surgical incisions (Sachs, 1969). Moreover, removal of certain primordia (P3 or P4) can induce leaflet formation at tendril sites (Gould, 1985). Application of growth regulators can also induce developmental changes albeit irregularly (Gould, 1985). The effects of triiodobenzoic acid were most significant, inducing leaflets at tendril sites on conventional peas and causing swelling of the tendrils of *afila* lines.

Models have been proposed to describe leaf development in peas. Young (1983) put forward a simple algebraic model, relying on a "size" factor, which was consistent with the findings of Sachs (1969) that conversion of leaflets to tendril may be related to the primordium size at a particular stage. It takes no account, however, of correlative influences such as those mentioned above. Gould (1985) proposed that auxin may account for such influences and the "size" factor.

Changing foliage morphology

Leaf shape is affected by a number of mutations but the character which has stimulated the most commercial interest is 'rabbit ear' (or tare) rogue. Rogue has been given the symbol *X* (Brotherton, 1923) but attempts to assign the determinant to a linkage group have so far failed because rogue does not have a general predictable pattern of inheritance (Matthews, 1969). Rogue exhibits different phenotypes in different backgrounds (Matthews, 1969).

The most obvious characteristic of 'rabbit ear' rogue plants, however, is a reduction of about one third in leaf area and about a half in stipule area, due primarily to a reduction in width (Pyke and Hedley, 1984). The rogue character was first described by Bateson and Pellew (1915), occurring spontaneously at a frequency of up to 4% in some old cultivars (Matthews, 1973). The inheritance of rogue has been described by Matthews as an example of bi-parental cytoplasmic inheritance in which rogue determinants show pronounced preferential transmission.

The reduction in foliar area seen in 'rabbit ear' rogues has been shown to be due to a reduction in the number of cells present during development and in the fully developed foliar structure and not to a reduction in cell size (Pyke and Hedley, 1984). Comparisons between leaves from the variety Feltham First and its rogue mutant, using polyacrylamide gel electrophoresis followed by staining for a number of enzymes, showed consistent qualitative and quantitative differences in peroxidase isozyme band patterns, although the significance of these differences is not known (Matthews, 1973).

The interest of plant breeders in this character has come from the increased stiffness of the plant and its tendency to remain erect for longer periods than conventional types. The rogue variety Progreta has proved to be very successful (Gent and Knight, 1978) and has replaced Maro, from which it was derived, as the most widely grown dried pea variety in the U.K.

There are several other genes known to radically alter leaf size and which may be of use to breeders. The latum gene (lat), for example, increases leaflet and stipule area by about 100%, while the reductus gene (red) reduces the width of leaflets by 65-90%. An extreme form of leaflet reduction termed maximo reductus has been described and given the gene symbol mare (Lamprecht, 1967) but this has since been shown to be synonymous with reductus (red) (Marx, 1976).

SEED MUTATIONS

Of approximately 50 genes known to influence seed characters only a few have been shown to affect the development of the embryo (Blixt, 1972). The majority of the genes affect characteristics associated with the testa, which is a tissue derived from the maternal plant. Genes are known which affect the colour, structure and texture of the testa, the size and colour of the hilum and secretions from the testa such as tragacanth (Table 2).

Table 2. Mutations in Pisum affecting seeds.

Adapted with additions from Blixt, 1972, 1977 and Lamprecht, 1974. Classification and phenotypes have been simplified in many cases. Brackets around genes denotes grouped characters, around symbols denotes synonyms according to Blixt, 1977. First character in gene and symbol denotes dominance. All genes occur as spontaneous mutations except ma (X-ray induced only).

GENE/CHARACTER	SYMBOL	PHENOTYPE REFERENCES	REFERENCES
TESTA FEATURES			
caerulicans	ar	Hilum slit-like	Tedin, 1920
Hilum black	Pl	Black pigmentation of hilum	Lock, 1907; White, 1917
Hilum major	Him	Large hilum at least half of seed diameter	Lamprecht, 1963b
foveatus	fov, sul	Deep impressions over radicula	Lamprecht, 1959c, 1960c
Gritty	Gty	Surface of testa covered by extremely small, regular and evenly distributed prominences	Marx, 1969b
testa thickness	ep1, ep2	Reduction in testa thickness (now extinct but variation for testa thickness does exist)	Kaznowski, 1926 (Alexandrov & Alexandrova, 1930, 1935)
Tragacanthum	Tra	Tragacanth present beneath testa	Lamprecht, 1956a
chenille	s	Tragacanth present outside testa causing seeds to stick together	White, 1917
SEED SHAPE			
impremere	l	Seeds have a few, large and more or less deep impressions	Tedin et al., 1925
lacunosus	di	Seeds with deep impressions	Wellensiek, 1943
minute-foveatus	mifo	Seeds have regular, close and small, shallow impressions	Lamprecht, 1962a
corrugatus	foe	Seeds have irregular, small and wrinkled impressions (may = di)	Lamprecht, 1969a (Blixt, 1977)

contd./...

contd. from p.230

GENE/CHARACTER	SYMBOL	PHENOTYPE	REFERENCES
planatus	pla, (Ro)	Seeds have flattened sides	Lamprecht, 1960c
comprimeve	com, qua	Seeds almost cubic in shape	Lamprecht, 1960c
rugosus	r, rb	Cotyledons and, therefore, seed wrinkled	White, 1917; Kooistra, 1962
SEED DIMENSIONS			
seed length	lo, lol to 3	Seed length is reduced about 0.5mm by each gene (N.B. now invalid)	Kaznowski, 1926 (Blixt, 1977)
seed size	ar, ma, sgl, sg2	Reduction in seed size and weight (N.B. ma now invalid)	Murfet, 1985; Lamprecht, 1957; Gottschalk, 1976; Blixt, 1977 (Blixt, 1977)
SEED COLOUR			
(ground colours)	a, gla, och, oh, oli, olv, sal, Umb	Alter overall colour of testa	White, 1917, Lamprecht, 1958a, 1959a, 1961a, 1961b, 1968, 1969b; Wellensiek, 1925
(colour patterns)	cor, F, Fs, gl, gri, M, mex, Obs, rag, Rf, Str, U	Alter parts of testa to give patches of pigment of varying shapes and distributions	numerous, see Blixt, 1972 (Blixt, 1977)
	Ca, cal, cat, dem, lob, mal, mp, pal, ve, z	Decolour specific parts of testa (N.B. mal now invalid)	numerous, see Blixt, 1972
COTYLEDON COLOUR			
cotyledon green	i	Cotyledons are green	White, 1917
viridis	pa	Cotyledons are dark green	Winge, 1936
orange cotyledon	Orc	Cotyledons are orange	Swiecicki & Blixt, 1984

There are, however, many problems associated with the recognition of mutations that affect seed development per se. It has been noted (Blixt, 1972), that most varieties used for mutation experiments, for commercial reasons, are recessive for the a gene (a/a) and hence lack anthocyanin pigmentation. The a allele is epistatic to many genes affecting the seed and thus many mutations may have been masked in some experiments (Blixt, 1972). Furthermore, the control of colour in the testa is complex with a large number of genes affecting ground colour, pattern colour and pattern form (Table 2). There are many interactions and the table here represents a simplified classification for the sake of space in this review. Some phenotypes can be brought about either by a single mutation or by a series of allele changes each producing a small effect; for example, the production of a yellow seed. This phenotype may be produced by recessivity at the a locus (with concomitant drastic alterations elsewhere on the plant) or by recessivity at the z, mp, pal, b, dem, cal and lob loci (Blixt, 1972).

Only a few genes affecting seed development have received commercial attention - those governing seed shape and seed colour. In this instance, however, the attention has been geared to consumer appeal rather than, as for the foliar mutations, to crop growth.

Changing seed shape

The development of the seed is due to an interaction between the maternal tissues of the testa and pod and the embryonic tissues of cotyledon and axis (Hedley and Ambrose, 1980). Many embryonic alterations may well be masked, then, by testa and other characters in mutation experiments. Likewise, some embryo characters may also indirectly affect the testa, for example cotyledon wrinkling. In this sense, the direct effects of and interactions between many shape genes are not clear. Genes such as lacunosus (di) (which may be synonymous with foe, Blixt, 1977) and minute-foveatus (mifo) (Table 2), which cause impressions in the seed (Figure 4) are classed as testa characters (Blixt, 1972) though it is far from clear if their effect is on the testa alone (Blixt, 1978). Moreover, such genes are referred to as being hypostatic (Marshall, 1963; Yarnell, 1962) to the rugosus (r) locus which causes seed wrinkling (Figure 4), since their effect cannot be observed in the presence of the r/r genotype. Similarly, the phenotypic effect of some genes e.g. those affecting seed size (see Table 2) can be brought about indirectly by other factors such as the maternal effect of the testa i.e. the size of an embryo is related to the testa in which it develops.

As mentioned above, the rugosus(r) locus affects the wrinkling of the seed. It is the best known of the genes which affect the embryo and it has proved the most important seed character as far as the pea crop is concerned. It is believed that the r mutation occurred spontaneously ca. 1600 (Lamprecht, 1956b) and Sherwood (1899) credits Knight (1799) with starting the introduction of wrinkled seeded varieties (reported in White, 1917). It was used also by Mendel (1865) in his famous experiments on peas and given the symbol r by White (1917). The most obvious effect of the r gene is that mature dry r/r seeds have a characteristic wrinkled appearance (Mendel, 1865; White, 1917). Starch grains from seeds with the round phenotype (R/R and R/r) are simple, oval shaped and tend to be larger than the complex, fissured grains of the wrinkled genotype (Gregory, 1903; Darbishire, 1908) (Figure 5). The mature r/r seeds contain approximately twice as much sugar as the round type and take up more water upon germination (Gregory, 1903, Darbishire, 1908). The commercial impact of the r locus can be readily observed, if not "tasted". All commercial vined peas, which are harvested immature, are homozygous for the r gene; the increased sweetness associated

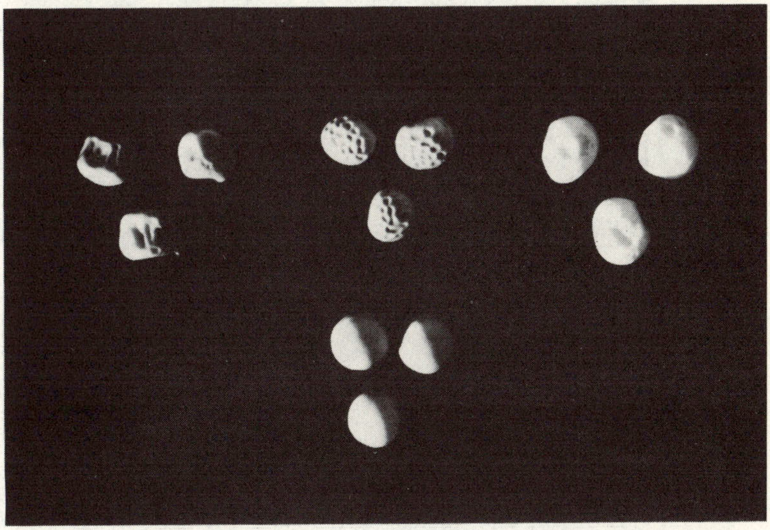

Figure 4. Seed shape in Pisum sativum. Lines of peas showing expression of the genes rugosus (r/r, top left), minute-foveatus (mifo/mifo, top centre), lacunosus (di/di, top right) and a round (R/R) line (bottom centre) near-isogenic to the r/r line.

with the presence of this gene makes the peas highly pallatable.

More recently, another type of pea seed with a wrinkled phenotype has been described (Kooistra, 1962). Test crosses have shown that a second gene is responsible for this phenotype and that it is at a different locus and in a different linkage group (3) from the r gene. This second gene was labelled r_b (Kooistra, 1962) but was changed to rb to conform to gene rules (Blixt, 1977). The r allele does not interact with rb with respect to seed shape but is epistatic with regard to starch grain shape. Combinations of the genes when homozygous at these two loci give four possible gentoypes for pea seeds: R/R Rb/Rb (round seed with simple starch grains), r/r Rb/Rb (wrinkled seed with compound starch grains), R/R rb/rb (wrinkled seed with simple starch grains) and r/r rb/rb (wrinkled seed with compound starch grains). Consequently, r/r Rb/Rb and r/r rb/rb genotypes can only be distinguished by progeny testing.

As well as affecting the shape of the starch grain, the r gene

Figure 5. Starch grains in round and wrinkled peas. Starch grains, isolated from lines near-isogenic except for the r locus, were sprinkled onto double-sided tape, coated with gold and viewed using a Philips 501B scanning electron microscope; wrinkled (r/r) above round (R/R). Bar = 10µm.

also affects the amount of starch present in the seed, the r/r genotype having 33-36% compared with 45-49% for the R/R genotype (Kappert, 1915; Kellenbarger, et al., 1951). The composition of the starch is also affected, with the R/R types having typically 20-38% amylose (Nielsen and Gleason, 1945; McCready et al., 1950; Kellenbarger et al., 1951; Schneider, 1951; Yarnell, 1962 and Colonna and Mercier, 1984) compared with 70-90% amylose in r/r types (Nielson and Gleason, 1945; Hilbert and MacMasters, 1946; McCready et al., 1950; Peat et al., 1948). The R/R types have consequently higher levels of amylopectin than the r/r types. The variation in amylose/amylopectin content of the pea starch is similar to that attributed to the amylose-extender (ae) gene in maize which has been associated with a change in starch branching enzyme (Boyer and Preiss, 1978; Boyer et al., 1980). The possibility of a comparable enzymic deficiency in pea was investigated by Matters and Boyer (1982) who found that branching enzyme activity in cotyledons of the r/r genotype was 10% the level of the R/R genotype. In addition, a single soluble starch synthase was found in the r/r extracts whereas two were found in the R/R line. Comparisons between non-isogenic r/r and R/R lines have shown that fructose-2,6-bisphosphate was more abundant in R/R than in r/r embryos during development and that this correlated with higher amounts of starch found in the R/R embryos (Edwards, 1984). Furthermore, Edwards (1984) found higher amounts of sucrose, glucose, fructose, hexose phosphates, triose phosphates and fructose-1,6-bisphosphate in r/r embryos.

The r gene does not have a consistent effect on the storage protein content of the seed (Matthews and Arthur, 1985; Shia and Slinkard, 1977) but a much greater effect on the composition of the storage protein. A greater amount of legumin, compared with the other major storage protein vicilin, is found in R/R lines compared with r/r genotypes (Davies, 1980). Wrinkled (r/r) genotypes, however, have been found to have higher levels of albumin than homozygous round (R/R) types (Schroeder, 1982). The seed of r/r genotypes have also been shown to have increased crude lipid contents of about 5% compared with about 3% for R/R lines (Coxon and Davies, 1982). There is some suggestion that the rb gene may further increase the lipid content when associated with the r gene (Coxon and Davies, 1982).

Most of the biochemical information gathered on the effects of the r gene have referred to lines which are dissimilar for other genes as well as those at the r locus. To fully understand the action of the r gene, however, it is necessary to remove effects which may be due to genes at other loci. This can be achieved by using lines which are isogenic or near-isogenic except for the r locus. Studies employing such lines have made it possible to study the specific effects of the r gene on the development of the seed. When gross effects of the r locus were analysed, in terms of changes in fresh and dry weight, there was no significant effect of the r gene prior to the disappearance of the endosperm. After the endosperm had been absorbed by the embryo the fresh weight of the r/r line was found to be consistently greater than the R/R line; dry weight tended to accumulate more rapidly in the R/R line (Hedley et al., 1986). There is a major effect, therefore, on the percentage water content of developing embryos, the r/r line at its maximum attaining about 15% more water than the R/R line. Moreover, the relationship between the fresh and dry weight of developing r/r and R/R embryos observed in vivo could be maintained in vitro when embryos were cultured on a medium containing 10% sucrose (Figure 6, Wang et al, 1986). It is the loss of this additional water during maturation which causes the characteristic wrinkled appearance in the dry seed (Kooistra, 1962). The increased water content of the r/r line is undoubtedly related to the consistently higher osmotic pressure (more negative water potential) which has been observed for r/r embryos during development. This potential may be one of the driving forces for the greater uptake of water observed in r/r seeds (Wang et al., 1986) and may itself be related to the higher sucrose content of the developing embryo induced by the effect of the r gene (Stickland and Wilson, 1983).

Changing cotyledon colour

There are several pigment genes which affect cotyledon colour, viz the chlorophyll or carotenoid content, of the cotyledons directly. These genes are also of commercial importance. Thus, embryos homozygous recessive for the i gene retain their green colour into the dry seed while the chlorophyll of I/- genotypes degrades late in development to give pale yellow seeds (White, 1917). Hence this gene is important for the dried pea industry. On the other hand, the viridis gene (pa; Winge, 1936) is relevant to the canning or freezing of peas (Marx, 1977b) since it is desirable to have lighter green peas (Pa/-) for canning than for freezing (pa/pa).

CONCLUSIONS

The empirical use by plant breeders of naturally occurring variation has served mankind well for centuries and has resulted in a steady

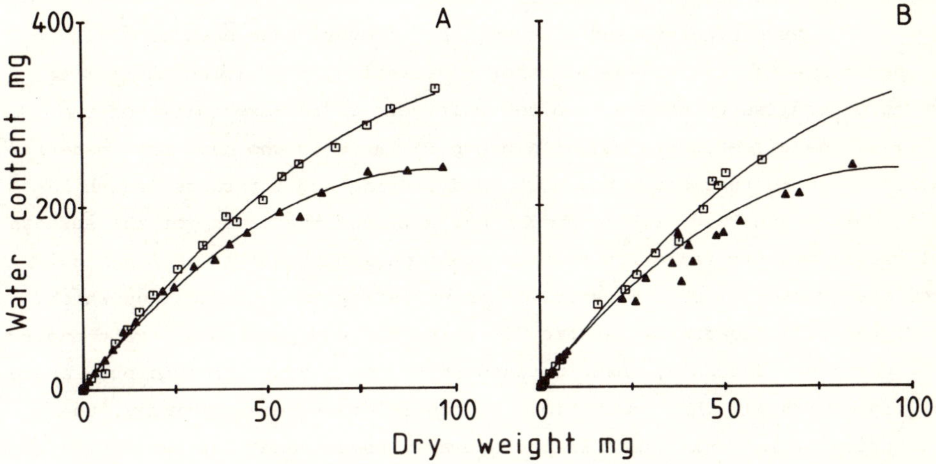

Figure 6. Relationship between embryos from round-▲(R/R) and wrinkled-□(r/r) seeded plants *in vivo* and *in vitro*.

A. Embryos were removed from seeds on plants grown in a growth cabinet (Hedley *et al.*, 1986) and their fresh and dry weights recorded.

B. Embryos were removed from plants grown as above and placed individually on sterile nutrient medium (modified from Stafford and Davies, 1979, Annals of Botany 44, 315-321) containing 10% sucrose. The fresh and dry weights were recorded after culturing the embryos for 7 days at 20°C.

improvement in most crop plants. The pea is no exception to this process and variants for vegetative and reproductive characters have been incorporated into varieties and have helped to improve the performance of the crop and to widen the uses of the seed.

Physiologists and biochemists, however, have been much slower to appreciate the usefulness of genetic variation, preferring instead to try to generalise using one or a few cultivars as representative of the species. One exception is Murfet's group in Tasmania who have used genetic variation to increase our knowledge of flowering and internode length in peas (Murfet and Reid, 1985). Murfet has outlined the basis for the successful use of genetic variation to understand physiological and biochemical problems (Murfet, 1982). Firstly, it is necessary to collect lines which cover the full breadth of genetic diversity for the particular trait under consideration. Secondly, these mutants should be used to produce pure breeding lines with all the combinations of the relevant genes. Thirdly, the production of isogenic lines is particularly useful and, lastly, it is necessary to have a standard line for reference. Furthermore, Murfet stresses the importance of a team approach with geneticists, physiologists and biochemists interacting. If Murfet's criteria are applied to the problems of understanding leaf and seed development then perhaps we can identify the reasons for our lack of progress in these areas.

With regard to studies of leaf ontogeny, it is evident that Murfet's first criterion can be satisfied since a wide breadth of genes which affect ontogeny of the leaf have been described. The second criterion has been met also, at least in part, and several workers have produced lines in which many of the genes have been combined in various ways, for example, a set of isogenic lines for the *af*, *st* and *tl* loci have existed since 1974, following a backcrossing programme by Marx (1974). It would appear, therefore, that one of the reasons for a lack of progress in this area may be the requirement of a multidisciplinary approach using the existing mutants and existing genetic information.

The problems in the use of genetic variation to understand seed development are different from those of foliage ontogeny. Firstly, very few good developmental mutants are known, or have even been sought, which affect the seed and it may be necessary, therefore, to try to widen the breadth of genetic diversity using mutagenesis. Pure lines exist for combinations of those mutations known to affect seed development, as do also isogenic lines for one mutation, the *r* locus (Hedley *et al*., 1986). Let us hope that if

Murfet's final criterion, that of a multidisciplinary approach, can be satisfied, then seed development will surrender some of its mysteries. The increasing amount of information which is being gained on the effects of the *r* gene, reported here and elsewhere (Domoney and Casey, 1985; Domoney *et al*., 1986) is perhaps the best testament to this approach.

ACKNOWLEDGEMENTS

We should like to thank our colleagues for providing information for this paper and, in particular, Mike Ambrose for the drawing of Figure 1, Dr. Peter Matthews for his excellent genetic information, Kathy Strouts, Roger Turner and S. Painter for the scanning electron micrographs. We also thank Professor Gerry Marx for his helpful comments on the manuscript.

REFERENCES

Alexandrov, W.G. & Alexandrova, O.G. (1930) On the comparative investigation of pea seeds of different origin. Proceedings of the USSR Congress of Genetics, Plant and Animal Breeding, pp. 21-46.

Alexandrov, W.G. & Alexandrova, O.G. (1935) Anatomy of flower, fruit and seed of peas. Bulletin of Applied Botany, Genetics & Plant Breeding, Series III, No. 9, pp 5-152. Leningrad: Institute of Plant Industry.

Askin, D.C., White, J.G.H. & Rhodes, P.J. (1985) Nitrogen fixation by peas and their effect on soil fertility. *In*: The Pea Crop, ed. P.D. Hebblethwaite, M.C. Heath & T.C.K. Dawkins, pp. 421-430. London: Butterworths.

Bateson, W. & Pellew, C. (1915) On the genetics of "Rogues" among culinary peas. J. Genet., *5*, 13-36.

Blixt, S. (1972) Mutation genetics in *Pisum*. Agri Hortique Genetica, *30*, 1-293.

Blixt, S. (1977) Gene symbols of *Pisum*. Pisum Newsletter, *9*, Supplement.

Blixt, S. (1978) *Pisum* gene list. Pisum Newsletter, *10*, 82-101.

Blixt, S. & Williams, J.T. (1982) Documentation of genetic resources: a model. Rome: IBPGR.

Boyer, C.D., Damewood, P.A. & Matters, G.L. (1980) Dosage effects at the *R* locus on the properties of the starch fractions. Pisum Newsletter, *12*, 7-9.

Boyer, C.D. & Preiss, J. (1978) Multiple forms of starch branching enzyme of maize; evidence for independent genetic control. Biochem. Biophys. Res. Commun., *80*, 169-175.

Brotherton, J.W. (1923) Further studies on the inheritance of "Rogue" type in garden peas (*Pisum sativum* L.). - J. Agric. Res., *24*, 815-852.

Colonna, P. & Mercier, C. (1984) Macro-molecular structure of wrinkled- and smooth-pea starch components. Carbohydrate Res., *126*, 233-248. 248.

Coxon, D.T. & Davies, D.R. (1982) The effect of the r_a and r_b loci on the lipid content of the seed of *Pisum sativum*. Theor. Appl. Genet., *64*, 47-50.

Darbishire, A.D. (1908) On the result of crossing round with wrinkled peas, with especial reference to their starch grains. Proc. Roy. Soc. B, 80, 122-135.

Davies, D.R. (1980) The r_a locus and legumin synthesis in Pisum sativum. Biochem. Genet., 18, 1207-1219.

Domoney, C. & Casey, R. (1985) Measurement of gene number for seed storage proteins in Pisum. Nucleic Acids Res., 13, 687-699.

Domoney, C., Ellis, T.H.N. & Davies, D.R. (1986) Organisation and mapping of legumin genes in Pisum. Mol. Gen. Genet., 202, 280-5.

Donkin, M.E., Travis, A.J., & Martin, E.S. (1982) Stomatal responses to light and CO_2 in the Argenteum mutant of Pisum sativum. Z. Pflanzenphysiol., 107, 201-9.

Edwards, J. (1984) Conversion of sucrose to starch in pea embryos. Ph.D Thesis. University of Cambridge.

Gent, G.P. & Knight, R.F. (1978) Development of the 'rabbit-eared' dried pea variety "Progreta". Pisum Newsletter, 10, 105-106.

Goldenberg, J.B. (1965) "Afila", a new mutation in pea (Pisum sativum L.). Boletin Genetico, 1, 27-31.

Gottschalk, W. (1964) Eine Gruppe pleitroper Gene mit weitgehend übereinstimmenden Wirkungsspektren und ein neuer Fall von Polymerie bei Pisum. Radiation Bot., 4, 267-274.

Gottschalk, W. (1970) Möglichkeiten der Blattevolution durch Mutation und Rekombination. Ein Modell für die Entwicklung und Weiterentwicklung des Leguminosenblattes). Z. Flanzenphysiol., 63, 44-54.

Gottschalk, W. (1971) Die Bedeutung der Genmutationen für die Evolution der Pflanzen. Fortsch. Evol. Forsch., Stuttgart: Fischer.

Gottschalk, W. (1976) Improvements of mutants by incorporating other mutated genes into their genomes. Pisum Newsletter, 8, 16.

Gould, K.S. (1985) Control of leaf morphogenesis in Pisum sativum L. PhD thesis. University of Manchester.

Gould, K.S., Cutter, E.G. & Young, J.P.W. (1986) Morphogenesis of the compound leaf in three genotypes of the pea, Pisum sativum L. Can. J. Bot., 64, 1268-1276.

Gregory, P.P. (1903) The seed characteristics of Pisum sativum. New Phytol., 2, 226-228.

Härstedt, E. (1950) Über die Verebung der Form von Laub- und Kelchblättern von Pisum sativum. Agri Hortique Genetica, 8, 7-32.

Hedley, C.L. & Ambrose, M.J. (1980) An analysis of seed development in Pisum sativum L. Ann. Bot., 46, 89-105.

Hedley, C.L., Smith, C.M., Ambrose, M.J., Cook, S. & Wang, T.L. (1986) An analysis of seed development in Pisum sativum. II. The effect of the r-locus on the growth and development of the seed. Ann. Bot., (in press).

Hilbert, G.C. & MacMasters, M.M. (1946) Pea starch, a starch of high amylose content. J. Biol. Chem., 162, 229-238.

Jewer, P.C., Incoll, L.D. & Shaw, J. (1982) Stomatal responses of Argenteum - a mutant of Pisum sativum L. with readily detachable leaf epidermis. Planta, 155, 146-53.

Kappert, H. (1915) Untersuchungen an Mark-, Kneifel- und Zuckerarbsen und Ihren Bastarden. Z. Induct. Abstamm. Verebungs. 13, 1-57.

Kaznowski, L. (1926) Studja nad grochem (Pisum L.) Pamietnik panstowego instytyty naukowego gospodarstwa weijskiego w pulawach, 7, 1-91.

Kellenbarger, S. (1952) Inheritance and linkage data of some characters in peas (Pisum sativum). Journal of Genetics, 51, 41-46.

Kellenbarger, S., Silveira, V., McCready, R.M., Owens, H.S. & Chapman, J.L. (1951) Inheritance of starch content and amylose content of

the starch in peas (Pisum sativum). Agron. J., 43, 337-40.
Kielpinski, M. (1982) Supplementary evidence for the linkage of afila with mutants on chromosome 1. Pisum Newsletter, 14, 30.
Knight, T.A. (1799) An account of some experiments in the fecundation of vegetables. Phil. Transact., 5, 195-204.
Kooistra, E. (1962) On the differences between smooth and three types of wrinkled peas. Euphytica, 11, 357-373.
Kujala, V. (1953) Felderbse, bie welcher die ganze Blattspreite in Ranken umgewandelt ist. Arch. Soc. Zool. Bot. Fenn., 8, 44-45.
Lamm, R. (1949) Contributions to the genetics of the Cp-chromosome of Pisum. Hereditas, 35, 203-214.
Lamm, R. (1957) Three new genes in Pisum. Hereditas, 43, 541-548.
Lamprecht, H. (1933) Ein unifoliata-Typus von Pisum mit gleichzeitiger Pistilloidie. Hereditas, 18, 56-64.
Lamprecht, H. (1939) Genstudien an Pisum sativum. IV. Über Verebung von Wachslosigkeit und ein neues Gen für lokale Ausbildung von Wachs, Wsp. Hereditas, 25, 459-471.
Lamprecht, H. (1942) Die Koppelungsgruppe Gp-Cp-Fs-Ast von Pisum. Hereditas, 28, 143-156.
Lamprecht, H. (1945) Durch Komplexmutation bedingte Sterilität und ihre Vererbung. Archiv der Julius Klaus-Stiftungfur Verebungs Forschung, Sozialanthropologie und Rassenhygiene, 20 (Supplement), pp. 126-141.
Lamprecht, H. (1948) The variation of linkage and the course of crossing over. Agri Hortique Genetica, 6, 10-48.
Lamprecht, H. (1949) Über entstehung und veresburg von schmalblattrigen typen bei Pisum. Agri Hortique Genetica, 7, 134-153.
Lamprecht, H. (1956a) Über Wirkung und Koppelung des Gens Tram von Pisum. Agri Hortique Genetica, 14, 45-53.
Lamprecht, H. (1956b) Pisum sativum L. oder P. arvense L. Eine nomenklatorische studie auf genetischer basis. Agri Hortique Genetica, 14, 1-4.
Lamprecht, H. (1957) Durch Röntgenbestrahlung von Pisum-Samen erhaltene neue und bekannte Genmutationen. Agri Hortique Genetica, 15, 142-154.
Lamprecht, H. (1958a) Eine Pisum-Mutante mit in diminutive Stammverzweigungen umgewandelte Infloreszenzen und ihre Vererbung. Agri Hortique Genetica, 16, 112-128.
Lamprecht, H. (1958b) Gekräuselte Blättchen bei Pisum und hire Vererbung. Agri Hortique Genetica, 16, 1-8.
Lamprecht, H. (1959a) Ein Gen für ockergelbe Färbung der Testa und seine Lage im Chromosom III von Pisum. Agri Hortique Genetica, 18, 9-14.
Lamprecht, H. (1959b) Das Merkmal insecatus von Pisum und seine Vererbung sowie einige Koppelungsstudien. Agri Hortique Genetica 17, 26-36.
Lamprecht, H. (1959c) Ein Gen für eingesenkte Radicula und seine Lage im Chromosom III von Pisum. Agri Hortique Genetica 17, 37-46.
Lamprecht, H. (1960a) Zur Wirkung und Koppelung des Gens Fom für Blattform von Pisum. Agri Hortique Genetica 18, 62-73.
Lamprecht, H. (1960b) Eine neue Stipelmutante bei Pisum. Agri Hortique Genetica 18, 209-213.
Lamprecht, H. (1960c) Zur Vererbung der Samenformen bei Pisum sowie über zwei neue, diese beeinflussende Gene. Agri Hortique Genetica 18, 1-22.
Lamprecht, H. (1961a) Pisum fulvum Sibth. & Sm. Geneanalytical studies with regard to the classification as species. Agri Hortique Genetica, 19, 269-297.

Lamprecht, H. (1961b) Eine neue Testafarbe von Pisum-Samen: salmoneus. Agri Hortique Genetica, 19, 213-222.

Lamprecht, H. (1962a) Über ein neues, die Form der Pisum-Samen beeinflussendes Gen sowie ein neues Gen für Teilfarbigkeit. Agri Hortique Genetica, 20, 137-155.

Lamprecht, H. (1962b) Die Entstehung der Arten und höheren systematischen Einheiten, bedingt durch eine Veränderung des Plasmas. Genetik und Paläontologie. Agri Hortique Genetica, 20, 181-213.

Lamprecht, H. (1962c) Über Symmetrie und Asymmetrie von Stipeln und Blättchen bie Pisum sowie ihre Vererbung. Agri Hortique Genetica, 20, 214-219.

Lamprecht, H. (1963a) Zur Vererbung der Hüsenlänge bei Pisum sowie über ein neues Gen für die Reduktion der Stipel. Agri Hortique Genetica, 21, 25-34.

Lamprecht, H. (1963b) Die Vererbung des ascendens-Merkmals von Pisum. Genmanifestation und Chromosomenzugehörigkeit. Agri Hortique Genetica, 21, 87-110.

Lamprecht, H. (1964) Genanalytische Studien zur Artberechtigung von Pisum transcaucasicum (Govorov) Stankov. Agri Hortique Genetica, 22, 243-255.

Lamprecht, H. (1967) Die maximo-reductus-Mutante von Pisum. Phyton, 12, 252-265.

Lamprecht, H. (1968) Neue Beiträge zur Genenkarte von Chromosom IV von Pisum. Cited in: Blixt, 1972.

Lamprecht, H. (1969a) Über ein Gen für einen auffallend grazilen Wuchs von Pisum. Phyton, 13, 161-167.

Lamprecht, H. (1969b) Die Bezeihungen zwischen Grossoverwerten und relativer Länge der chromosomen. Eine Studie im Chromosom I von Pisum. Mitteilungen Naturwissenschaftlichen Vereines fur Steiermark, 98, 195-204.

Lamprecht, H. (1974) Monographie der Gattung Pisum. Graz: Steiermärkische Landesdruckerei.

Lock, R.H. (1907) On the inheritance of certain invisible characters in peas. Proc. Roy. Soc. B, 79, 28-34.

Marshall, H.H. (1963) New genes for wrinkled or dimpled seed in peas. Can. J. Plant Sci., 43, 508-512.

Marx, G.A. (1969a) Two additional genes conditioning wax formation. Pisum Newsletter, 1, 10-11.

Marx, G.A. (1969b) A new seed gene. Pisum Newsletter, 1, 11-12.

Marx, G.A. (1969c) Linkage relations of Af. Pisum Newsletter 1, 9-10.

Marx, G.A. (1974) Stocks available. Pisum Newsletter 6, 60.

Marx, G.A. (1976) Relationships between mutants mare and red. Pisum Newsletter 8, 39.

Marx, G.A. (1977a) A genetic syndrome affecting leaf development in Pisum. Am. J. Bot., 64, 273-7.

Marx, G.A. (1977b) Classification, Genetics and Breeding. In: The Physiology of the garden pea, ed. J.F. Sutcliffe & J.S. Pate, pp.21-44, London: Academic Press.

Marx, G.A. (1978) Argenteum: a mutant under nuclear and extranuclear control. Pisum Newsletter, 10, 34-7.

Marx, G.A. (1982) Argenteum (Arg) mutant of Pisum. J. Hered., 73, 413-20.

Marx, G.A. (1983) Developmental mutants in some annual seed plants. Ann. Rev. Plant Physiol., 34, 389-417.

Matters, G.L. & Boyer, C.D. (1982) Soluble starch synthases and starch branching enzymes from cotyledons of smooth- and wrinkled-seeded lines of Pisum sativum L. Biochem. Genet., 20, 833-848.

Matthews, P. (1969) Rogue Peas. Ann. Rep. John Innes Inst., pp. 19-20.
Matthews, P. (1973) Genetic studies on spontaneous and induced rogues of Pisum sativum. PhD. Thesis. University of East Anglia.
Matthews, P. & Arthur, E. (1985) Genetic and environmental components of variation in protein content in peas. In: The Pea Crop, ed. P.D. Hebblethwaite, M.C. Heath & T.C.K. Dawkins, pp. 369-381. London: Butterworths.
McCready, R.M., Guggolz, J., Silveira, V. & Own, H.S. (1950) Method for determining starch and amylose in vegetables. Analyt. Chem., 22, 1156-1161.
Meicenheimer, R.D., Muehlbauer, F.J., Hindman, J.L. & Gritton, E.T. (1983) Meristem characteristics of genetically modified pea (Pisum sativum) leaf primordia. Can. J. Bot., 61, 3430-7.
Mendel, G. (1865) Versuche über Pflanzen-Hybriden. Verhandlungen des naturforshenden Vereins in Brünn.
Murfet, I.C. (1982) Present state of physiological and biochemical gene mapping. In: Documentation of genetic resources: a model. ed. S. Blixt & J.T. Williams, pp. 45-5. Rome: IBPGR.
Murfet, I.C. (1985) The influence of the genes ar and n on senescence in Pisum sativum L. Ann. Bot., 55, 675-83.
Murfet, I.C. & Reid, J.B. (1985) The control of flowering and internode length in Pisum. In: The Pea Crop, ed. P.D.Hebblethwaite, M.C. Heath & T.C. Dawkins. pp. 67-80. London: Butterworths.
Nielsen, J.P. & Gleason, P.C. (1945) Rapid determination of starch. Industrial and Engineering Chemistry, Analytical Edition, 17, 131-134.
Nilsson, E. (1933) Erblichkeitsversuche mit Pisum. VI-VIII. Hereditas, 17, 197-222.
Peat, S., Bourne, E.J. & Nicholls, M.J. (1948) Starches of the wrinkled and the smooth pea. Nature, 161, 206-207.
Pellew, C. & Sverdrup, A. (1923) New observations on the genetics of peas (Pisum sativum). J. Genet., 13, 125-131.
Pyke, K.A. & Hedley, C.L. (1982) Comparative studies within and between seed and seedling populations of leafed, 'leafless' and 'semi-leafless' genotypes of Pisum sativum L. Euphytica, 31, 921-31.
Pyke, K.A. & Hedley, C.L. (1983a) The effect of foliage phenotype and seed size on the crop growth of Pisum sativum L. Euphytica, 32, 193-203.
Pyke, K.A. & Hedley, C.L. (1983b) Growth and photosynthesis of mutant pea seedlings. Annals of Botany, 52, 719-724.
Pyke, K.A. & Hedley, C.L. (1984) Aspects of the rogue phenotype of peas. Plant Sci. Lett., 35, 87-90.
Pyke, K.A. & Hedley, C.L. (1985) Growth and photosynthesis of different pea phenotypes. In The Pea Crop, ed. P.D. Hebblethwaite, M.C. Heath & T.C.K. Dawkins. pp. 297-305. London: Butterworths.
Sachs, T. (1969) Regeneration experiments on the determination of the form of leaves. Israel J. Bot., 18, 21-30.
Schneider, A. (1951) Untersuchungen uber die Eignung von Erbsensorten fur Zwecke der Nasskonservierung. II. Der Zuchter, 21, 275-281.
Schroeder, H.E. (1982) Quantitative studies on the cotyledonary proteins in the genus Pisum. J. Sci Food Agric., 33, 623-633.
Sharma, B. (1972) "Tendrilled acacia", a new mutation controlling tendril formation in Pisum sativum. Pisum Newsletter, 4, 50.
Sharma, B. (1981) Genetic pathway of foliage development in Pisum sativum. Pulse Crops Newsletter, 1, 56-7.
Sharma, B. & Aravindan, K.V. (1972) Inheritance of the Tac gene in Pisum sativum. Pisum Newsletter, 4, 51.

Sharma, B., Aravindan, K.V., Tikoo, S.K. & Kant, K. (1972) Supplementary evidence about the inheritance of the Tac gene in Pisum sativum. Pisum Newsletter, 5, 52.

Sharma, B. & Kumar, S. (1981) Discovery of one more allele at the Tac locus of Pisum sativum. Pulse Crops Newsletter, 1, 21-2.

Sherwood, N.N. (1899) Garden Peas Journal of the Royal Horticultural Society, 22, 239-60.

Shia, G. & Slinkard, A.E. (1977) Relationship of seed shape and cotyledon colour to percent protein of peas. Crop Sci., 17, 183-184.

Stickland, R.G. & Wilson, K.E. (1983) Sugars and starch in developing round and wrinkled pea seeds. Ann. Bot., 52, 919-921.

Snoad, B. (1971) Linkage between Af and I on chromosome 1. Pisum Newsletter, 3, 43.

Snoad, B. (1974) A preliminary assessment of leafless peas. Euphytica, 23, 257-265.

Snoad, B. (1980) The origin, performance and breeding of leafless peas. ADAS Quarterly Review, 37, 69-86.

Sweicicki, W.K. & Blixt, S. (1984) Orange cotyledons: a new gene for cotyledon colour. Pisum Newsletter, 16, 70-1.

Tavcar, A. (1926) Die Verebung der Anzahl von Spaltoffnungen bei Pisum sativum L. Z. Pflanzenphysiol., 11, 241-259.

Tedin, H, (1920) The inheritance of flower colour in Pisum. Hereditas, 1, 68-97.

Tedin, H. & Tedin, O. (1926) Contributions to the genetics of Pisum. IV. Leaf axil colour and grey spottings on the leaves. Hereditas, 7, 102-108.

Tedin, H., Tedin, O. & Wellensiek, S.J. (1925) Note on the symbolization of flower colour factors in Pisum. Genetica, 7, 533-535.

von Rosen, G. (1944) Artkreuzung in der Gattung Pisum, insbesondere zwischen P. sativum L. und P. abyssinicum Braun. Hereditas, 30, 261-400.

Wang, T.L., Smith, C.M., Cook, S.K., Ambrose, M.J. & Hedley, C.L. (1986) An analysis of seed development in Pisum sativum. III. The relationship between the r locus, the water content and the osmotic potential of seed tissues in vivo and in vitro. Ann. Bot., (in press).

Wellensiek, S.J. (1925) Genetic monograph on Pisum. Bibliographica Genetica, 2, 343-476.

Wellensiek, S.J. (1943) Pisum-crosses VI: Seed-surface. Genetica, 23, 77-92.

Wellensiek, S.J. (1962) The linkage relations of the cochleata-mutant in Pisum. Genetica, 33, 145-153.

Wellensiek, S.J. (1971) The localization of some new mutants. Pisum Newsletter, 3, 46.

White, O.E. (1917) Studies of inheritance in Pisum. II. The present state of knowledge of heredity and variation in peas. Proc. Am. Phil. Soc., 56, 487-588.

Winge, Ö. (1936) Linkage in Pisum. Compte Rend. Trav. Lab. Carlsberg Ser. Physiol., 21, 271-393.

Yarnell, S.H. (1962) Cytogenetics of the vegetable crops. III. Legumes. A. Garden peas, Pisum sativum L. Bot. Rev. 28, 465-537.

Young, J.P.W. (1983) Pea leaf morphogenesis. Ann. Bot., 52, 311-6.

FOLIAR SENESCENCE MUTANTS AND OTHER GENETIC VARIANTS

Howard Thomas
Plants and Soils Division, Welsh Plant Breeding Station,
Plas Gogerddan, Aberystwyth, Dyfed, SY23 3EB.

INTRODUCTION

Development is the orderly expression of genetic programmes. It is the sum of ingredient processes acting in sequence or as a cascade. What does the term 'genetic programme' mean in this context? Generally it implies some kind of operon (Thomas, 1976a). The programme units of higher organisms differ from bacterial operons in that the genes from which they are built are usually scattered throughout the genome and are not linked in the physical or classical genetic sense. Thus a given gene may take part in a number of different programmes, and new programmes may arise by putting together genes in novel combinations. Moreover, a developmental programme is more than simply a coordinated transcriptional unit; expression of the component genes may be the result of inductive events acting partly or exclusively at the post-transcriptional level. To begin to understand a developmental process we need to identify the genes making up individual developmental units and the common property of these genes which makes them members of the same coterie. Then we need to know how each unit is activated in the correct place at the correct time, and works in a responsive and flexible way. A simple model developmental programme would comprise a cast of operational genes under the direction of one or more temporal, spatial and rate-modulating genes. Certain isoenzyme systems in animals and plants are regulated by such a mode of gene organization (Paigen, 1979; Scandalios and Baum, 1982).

It is possible to infer the existence of operational or regulatory genes from the purposeful nature and sheer orderliness of a given developmental event. This may be acceptable where morphogenetic processes are concerned (although it is not much help in determining how many genes there are, where they are located, what they do and how they are integrated); but in the case of senescence, which is not everyone's idea of a true developmental process, the characteristics of predictability, seasonal punctuality and systematic progress are inadequate evidence for genes at work. It may be argued, for example, that if a human being ingests cyanide, death will be the inevitable outcome. Cyanide is fatal because the proteins of the respiratory chain have a particular reactivity towards it as a consequence of their genetically-determined structures. But no-one would call this genetically-programmed death. Much of the evidence for and against a genetic basis for ageing and death is open to this kind of criticism. It is important to distinguish between developmental processes that proceed under the active direction of specific genes, and changes that occur because the materials from which living tissue is made are intrinsically evanescent or happen to have particular chemical reactivities towards xenobiotics and other non-physiological factors. It is generally true that gerontological processes appear to have a genetic basis, but in many (perhaps most) cases this is simply because macromolecules have a finite 'shelf-life' which is genetically determined only in the sense that durability may not be a high priority in their design specifications. There are good physiological reasons for believing that leaf senescence is not a gerontological process of the built-in obsolescence kind and these are considered in the next sections. As to the question of the identity of senescence-specific genes, there is no substitute for formal genetic analysis, an approach made possible by the existence of the variants and mutants that are the subjects of the remainder of this discussion.

SENESCENCE AS A PHASE OF LEAF DEVELOPMENT

Is senescence truly a developmental process? One way of understanding the developmental significance of senescence is by looking at its effect on leaf function. If we regard carbon fixation as being the primary role of the leaf then the onset and rate of senescence become important determinants of the contribution made by the leaf to the C economy of the plant. By way of illustration, Figure 1 presents a C budget for a bean leaf from birth to death (Čatsky et al., 1976). Gross C income is given by the photosynthesis curve. Carbon expenditure is the sum of dark and photorespiration plus the CO_2 equivalent of dry matter accumulated during leaf growth. The net C contribution to the plant as a whole is given by the 'carbon credit contour' - the area enclosed by income and expenditure curves (shaded area - Figure 1). The right-hand boundary of the contour is defined by the onset and progress of senescence. By changing the time of initiation (equivalent to the inflexion point) and rate of the senescence process (downward slope of the boundary line) sensitive control may be exercised over the net C benefit the plant receives from that particular leaf.

A similar argument may be employed concerning the left-hand boundary of the C contour. In this case, it is the expansion of the photosynthetic surface which establishes the limit. On a crop scale leaf expansion and senescence are quantified as leaf area index (LAI) and leaf area duration (LAD). It is significant that physiologists have long recognized these indices to be the prime yield determinants in many crops (Hunt, 1982). Indeed, variations in these parameters, rather than in maximal photosynthetic or respiratory capacity (which set the vertical dimensions of the C credit contour) can account for virtually all variation in the yields of these crops. As an example, Figure 2 shows how LAD and yield are very closely correlated in several wheat varieties grown under a range of environmental conditions (Evans et al., 1975).

Photosynthesis is not the only function of leaf tissue. In making a leaf, the plant invests a great deal of raw material, some of which may be in limited supply, and will at a later stage repossess this material

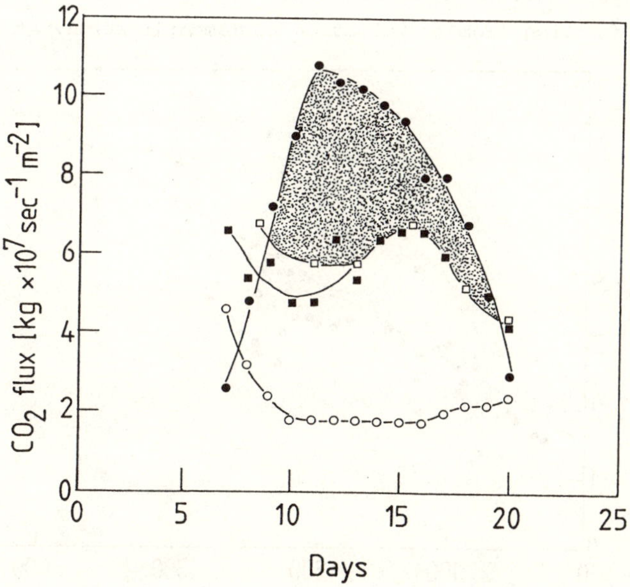

Figure 1. The carbon economy of a *Phaseolus* leaf during ontogeny. ● photosynthesis; o dark respiration; ■ dark + photorespiration; □ respiration + dry matter accumulated. The shaded area delimits the 'carbon credit contour'. Data from Čatsky et al. (1976).

and move it elsewhere to make new tissue. Modular construction and an orderly discarding and renewing of organs is a characteristic feature of plant development (Thomas, 1984). The constant internal recycling of nutrients is part of this process. In this sense, leaves are also storage organs and senescence can be considered to be the phase of nutrient re-mobilization. The major element subject to this cycle of deposit and with-drawal is nitrogen. Figure 3 demonstrates the flux of N through total pro-tein and the major photosynthetic protein ribulose-1,5-bisphosphate carboxylase (RuBPC) in rice leaves (Makino et al., 1984). During senescence, the leaf becomes an active net exporter of N. It has been pointed out (Huffaker and Peterson, 1974) that RuBPC has many of the characteristics of a storage protein as well as an enzyme. For a full discussion of the significance and regulation of N redistribution in senescence, see Thomas (1983a). It is no coincidence that senescence shares many of the features of other, more typical, mobilization phases in the plant life-cycle such as seed germination (Thomas, 1984). It follows that senescence is as much a developmental process as germination is.

We may conclude that initiation of senescence signifies a tran-sition in leaf function from assimilation to remobilization. At the cellular

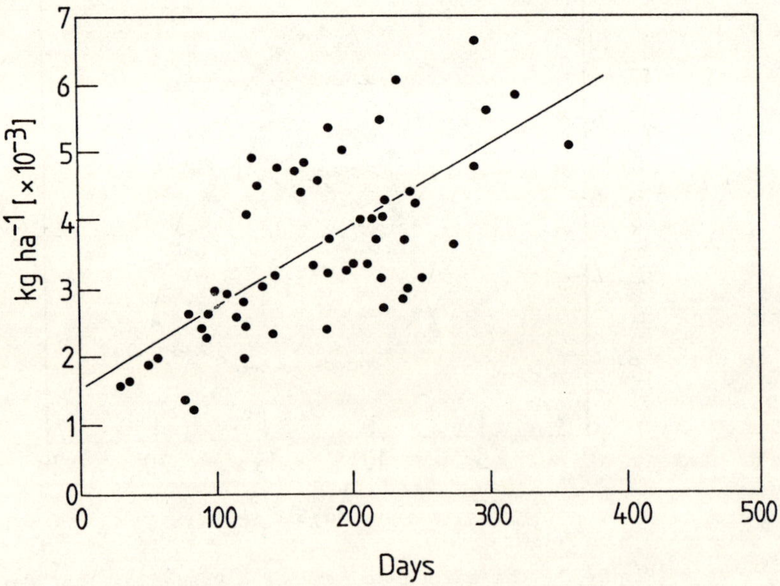

Figure 2. Grain yield as related to duration of leaf area after ear emergence for Triticum crops in a number of different environments (after Evans et al., 1975).

level this means termination of the chloroplast's photosynthetic career and realization of its role as a storage organelle. And at the molecular level it represents a change in the status of RuBPC and other components of the photosynthetic apparatus from catalytic to reserve proteins. The utility of leaf senescence in strategies of plant morphogenesis and adaptation (Stoddart and Thomas, 1982; Thomas, 1984) can be directly ascribed to the controlled operation of this functional switch.

GENETIC REGULATION OF FOLIAR SENESCENCE

Leaf senescence may have the appearance of a developmental process; but can it be shown to be a PROGRAMMED process also? The best evidence for genetic programming of senescence is the existence of defined mutations. These will be discussed later. Here we consider some of the less direct signs of a positive control over senescence exerted by specific genes.

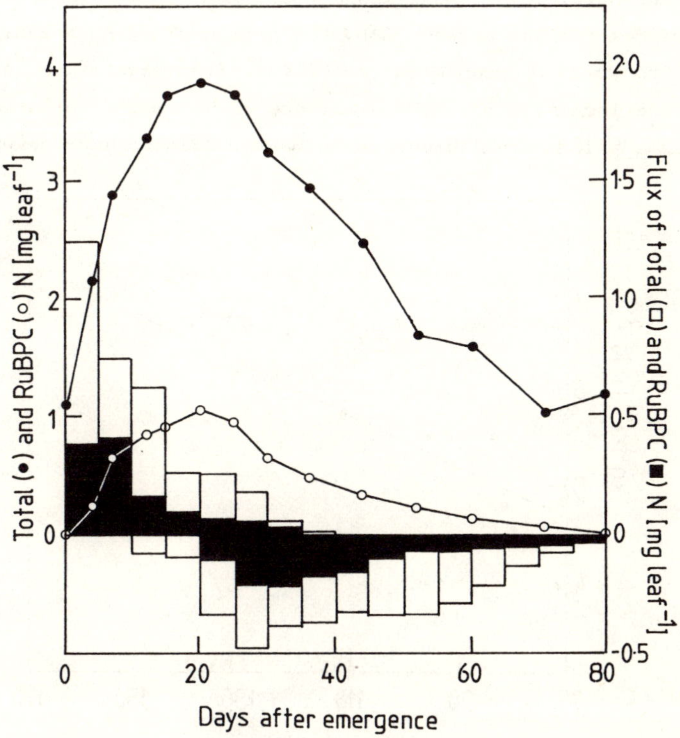

Figure 3. Levels of nitrogen in, and fluxes through, the total N and ribulose-1,5-bisphosphate carboxylase fractions of *Oryza* leaves during ontogeny. Data from Makino *et al.* (1984).

We have already seen that LAD is strongly correlated with yield in crops such as wheat (Figure 2). That LAD may be, at least in part, genetically determined is suggested by observations of consistent varietal differences in certain cereals, including Sorghum (Duncan et al., 1981) and Zea. Figure 4 presents the LAD/yield relationship for ten Zea hybrids over two growing seasons (Tollenaar and Daynard, 1978). The relationship between cereal grain formation and vegetative senescence is more complex than the striking correlations of Figure 2 and Figure 4 might suggest, however. Unlike other seeds, such as pulses or oilseeds, cereal grains are relatively rich in C compared with N. Prolonging the assimilatory phase of cereal leaf development might be expected to lead directly to an increase in yield. It might also be predicted that removing the developing grain would not necessarily delay leaf senescence and might even precipitate it. The latter response has, indeed, been observed in Zea (Allison and Weinmann, 1970) and Hordeum (Mandahar and Garg, 1975). Maximizing yield and quality in cereals requires a series of compromises between LAD, the period of C and N assimilation and the extent to which reduced N is remobilized from vegetative tissues (Stoddart and Thomas, 1982). The existence of heritable variation for these factors has enabled cereal breeders to make continued improvements in the

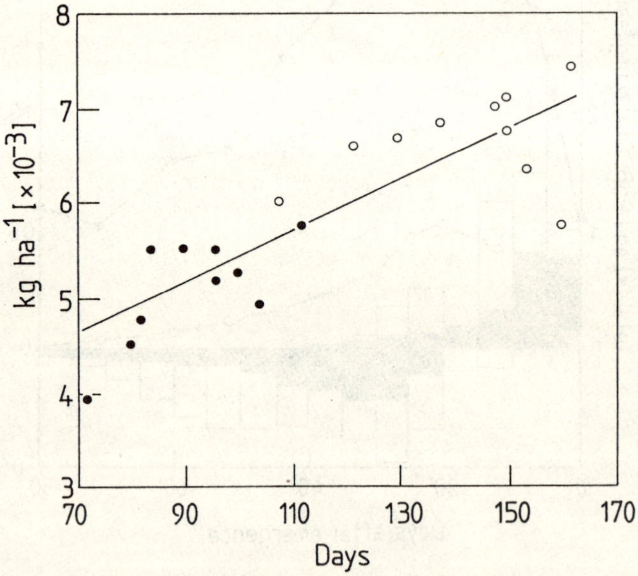

Figure 4. The relationship between leaf area duration and grain yield in ten Zea hybrid genotypes over two growing seasons. ● 1969; o 1970. (Tollenaar & Daynard, 1978).

crop, and provides evidence for genetic control of senescence in these species, as the following examples will show.

Christensen et al. (1981) observed that removing developing ears of Zea resulted in a pronounced reduction in the rate of dry weight and reduced N accumulation in vegetative parts. They concluded that the onset of senescence is related to the magnitude of nitrate uptake by, or nitrate flux through, the leaves. In subsequent publications (Crafts-Brandner et al., 1984a,b) these studies were applied to a range of hybrids, including two with extended LAD (Swank et al., 1982). One of these genotypes, FS854 (which is credited with the world record yield for non-irrigated maize), was especially notable for the degree to which greenness was retained even when ears were removed (Crafts-Brandner et al., 1984a). 'Stay green' or SGIA (stays green in August) is a heritable trait which has attracted a lot of interest from maize breeders and is a feature of some commercial hybrid varieties (Duvick, 1984). The character is physiologically and genetically complex, however (Crafts-Brandner et al., 1984b), and there are strong indications that more than one kind of genetic background can produce the extended greenness phenotype.

A great deal of genetic variation exists for grain protein content in Triticum. For example a number of cultivars with up to 25% more protein than normal soft winter wheats have been derived from Atlas 66, a high protein line described by Middleton et al. (1954). Rao and Croy (1972) observed that three varieties of Triticum with Atlas 66 in their parentage exhibited higher post-anthesis levels of leaf proteolytic activity than did Triumph 64, a low N variety. In a later study, Dalling et al. (1976) reported that acid proteinase was highly correlated with the rate of N redistribution from senescing leaves in two Triticum cultivars. An interesting approach to the genetics of leaf senescence in Triticum is seen in some work by Boyd and Walker (1972). They measured the amount of chlorophyll in segments of wheat flag leaves at the time of excision and after incubation on moist filter paper in darkness. Chlorophyll (chl) stability was expressed as the percentage of pigment remaining after 4 days. Stabilities ranging from less than 10% to more than 50% were observed in 19 varieties surveyed. The authors stated that a high level of chl in fresh leaf segments and low chl stability were characteristic of cultivars with high grain protein N. One low chl stability and three high stability varieties were selected for genetic analysis in a partial diallel cross (Table 1). 'Stable' x 'stable' crosses gave 'stable' F_1 progeny; crosses with the 'unstable' line produced progeny with reduced chl stability.

Thus there is certainly a strong genetic element in the relationship between grain fill and leaf senescence in cereals; but most of this variability is evidently continuous, qualitative and polygenic in nature, precluding the identification of specific loci.

At another level of biological organization, there is biochemical evidence that changes in gene expression underlie senescence. Some genes are turned off, notably those located in the chloroplast such as rbcL which codes for the large subunit of RuBPC, and coordinated nuclear genes such as that for the RuBPC small subunit (Speirs and Brady, 1981). Other (mostly nuclear) genes become expressed although it is by no means certain that activation of these genes occurs at the transcription level. For a detailed discussion of modulations in gene expression during leaf ageing see Thomas and Stoddart (1980) and Woolhouse (1984). One of the most dramatic and elegant illustrations of nuclear control exercised over chloroplast senescence is provided by the work of Yoshida (1961). When cells of Elodea densa fronds were plasmolysed in $CaCl_2$ solution, some of the protoplasts became separated into a nucleated and an enucleated half. Subsequently, chloroplasts were observed under the microscope over a period of 8 days in culture. In the absence of a nucleus chloroplasts enlarged, remained green and accumulated increasing amounts of starch. In nucleated protoplasts, on the other hand, the plastids underwent a senescence-like shrinkage and yellowing. Sometimes plasmolytic separation of two half-protoplasts was incomplete and they remained connected by a narrow 'plasma bridge'. Now the chloroplasts both sides of the bridge became senescent simultaneously. Yoshida also showed that treatment with ribonuclease inhibited the necrotic influence of

Table 1. Inheritance of chlorophyll a stability in a partial diallel cross between four Triticum varieties (data from Boyd and Walker, 1972)

	Chlorophyll stability (%)				
Variety	Ramona	WU29	WU30	Timgallen	Parental means
Ramona	42	46	37	16	35
WU29		46	46	13	35
WU30			45	11	37
Timgallen				20	12

the nucleus.

To date, cytological and biochemical studies of nucleus-chloroplast interactions have provided little better than oblique glimpses of regulatory processes operating in leaf senescence. These days there are much more direct ways to tackle the question of alterations in gene expression associated with the changing function of the ageing chloroplast. In particular, there is growing interest in the role of specific degradative processes in controlling chloroplast organization (Mishkind *et al.*, 1985). Similar mechanisms may well be engaged in dismantling the plastids of senescing leaf tissue.

SOYBEAN

Green and yellow cotyledons were among the characters Mendel studied in his experiments on inheritance in *Pisum*. According to Darbishire (1911), Bunyard was the first to show that immature seeds of Mendel's green and yellow lines are green and that the lines differ in the presence or absence of a factor which causes the pigments to fade as the seeds mature. If we accept that cotyledons are essentially modified leaves, then pigmentation changes during seed maturation may be likened to the yellowing process in foliar senescence. Green and yellow cotyledon types are found in *Glycine* also (Woodworth, 1921) and in some cases these characters are associated with variations in the senescence pattern of leaves (Table 2). A maternally-inherited factor, originally described by Terao (1918) and Veatch and Woodworth (1930), and now designated *cyt-G*, prevents yellowing not only of embryos but also of pods and leaves. Two recessive nuclear loci d_1 and d_2 exert a similar influence over greenness in fruits and foliage (Bernard and Weiss, 1973). There are a number of nuclear and cytoplasmic genes giving rise to yellow leaves in soybeans (e.g. Shoemaker *et al.*, 1985). One of these, y_3, is expressed as early senescence in relation to pod development and is counteracted by *G*, a nuclear dominant which, in the absence of y_3, prevents yellowing of the seed coat but not of any other tissue (Bernard and Weiss, 1973).

Apart from one short report (Kahanak *et al.*, 1978) there have been no studies of the physiological or metabolic consequences of possessing these genes. On the other hand, Phillips and co-workers have described in some detail the genetics, agronomy and physiology of soybean lines exhibiting the delayed leaf senescence (DLS) trait (Abu-Shakra *et al.*, 1978; Pierce *et al.*, 1984; Phillips *et al.*, 1984). Interest in DLS arises because of the complex interdependence of fruiting and foliar senescence and its relationship to yield in soybeans. Before genetic variation in the process is discussed, the major general features of leaf senescence in soybean are considered.

Figure 5 is a schematic representation of important physiological events in the life of the soybean crop. In the early part of crop growth photosynthetic activity and N reduction are maximal. As N reduction declines, the emphasis changes to a high rate of N fixation. Levels of N in vegetative tissues achieve a maximum at the onset of fruiting and fall rapidly thereafter, along with photosynthesis, as leaves senesce and nitrogen becomes concentrated in the growing seeds. The association of senescence with seed formation in soybeans differs in a fundamental way from the relationship in cereals because yellowing is greatly delayed in depodded or male sterile bean plants (Leopold *et al.*, 1959; Burton *et al.*, 1979). Hypotheses concerning the nature of the trigger for monocarpic senescence have been critically considered by Wang and Woolhouse (1982), Thomas (1983a) and Noodén (1984) and need not be presented again here. But just how inflexible is the correlation between fruiting and vegetative senescence in soybeans? In 1975, Sinclair and de Wit published an analysis of yield as a function of N requirement for

Table 2. Some genetic loci determining senescence behaviour in soybeans

Locus	Mode of inheritance	Phenotypic expression	Effective genotype	Ref.
Cyt-G	cytoplasmic	Embryos, pod walls foliage stay green		a,b
$\underline{d}_1\underline{d}_2$	nuclear, recessive	Embryos, pod walls foliage stay green	$\underline{d}_1\underline{d}_1\underline{d}_2\underline{d}_2$	c
G	nuclear, dominant	Seed coat remains green (but not foliage or embryo)		a,b
$\underline{Y}_3$	nuclear, recessive	Premature foliar yellowing	Not expressed in combination with *G*	b,d
$\underline{dt}_1\underline{dt}_2\underline{e}_1\underline{e}_2$	nuclear, recessive	Delayed leaf senescence	$\underline{dt}_1\underline{dt}_1\underline{dt}_2\underline{dt}_2\underline{e}_1\underline{e}_1\underline{e}_2\underline{e}_2$	e

References: a - Terao, 1918; b - Bernard and Weiss, 1973; c - Veatch and Woodworth, 1930; d - Terao and Nakatomi, 1929; e - Pierce *et al.*, 1984.

the major seed crop species. Soybean emerged as unique in demanding a high input of N to support seed production - so high that the authors postulated that soybean plants must mobilize vegetative N during pod fill to such an extent that they 'self-destruct'. This analysis was based on a number of assumptions about the sizes of N and C fluxes in the soybean crop. A subsequent re-examination of the problem by Sheehy (1983), using more recent data concerning the C and N economy of nodulated legumes during fruiting, suggests that there is no theoretical reason why soybeans should need to self-destruct. The existence of DLS soybean lines strongly supports this conclusion.

In 1976, a survey of experimental soybean lines growing at the University of California, Davis was undertaken with the object of identifying individuals showing altered senescence patterns at fruiting time (Abu-Shakra et al., 1978). Five plants with green leaves subtending brown pods were identified out of several thousand examined and the one with the highest N-fixation activity was used as the subject of continued study. This

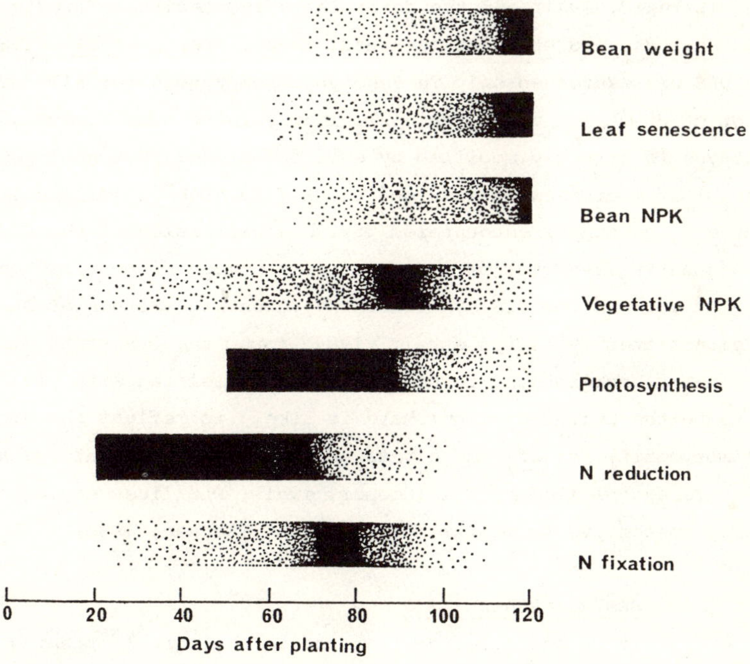

Figure 5. Functional and compositional changes during ontogeny of the Glycine crop.

individual was an F_2 derivative of a cross between Lee 68 and L63-1097. Of 14 F_3 progeny produced subsequently, 9 senesced normally and the remainder exhibited DLS. The latter were shown to retain chlorophyll, leaf protein, RuBPC activity and nodule N fixation at high levels compared with normal siblings. Leaves of progeny from the elite F_3 plant contained 60-210% more protein and chlorophyll than those of normal F_4 plants and their root nodules fixed N over an extended period. Grafting experiments showed that the DLS trait is determined by the genotype of the shoot and not the root (Phillips _et al._, 1984).

The analysis was continued by Pierce _et al._ (1984). Two individuals from progeny populations showing uniform DLS phenotype were selected from the F_6 generation and designated 6N3 and 8N2. F_7 plants derived from these lines were employed in crosses with a number of normal varieties of well-defined genotype. Four loci were found to be decisive in the expression of DLS. All DLS plants are of the early flowering, determinate type. Flowering time in soybean is determined by two major genes, E_1, e_1 and E_2, e_2 (Bernard, 1971) and the determinate/indeterminate growth habit by the genes Dt_1, dt_1 and Dt_2, dt_2 (Bernard, 1972). Pierce _et al._ (1984) showed that DLS was expressed only in genotypes homozygous for all four loci, that is, $dt_1dt_1dt_2dt_2e_1e_1e_2e_2$ (Table 2). They pointed out, however, that the DLS phenotype is strongly modified by environment and that analyses of F_2 and F_3 material from crosses between 6N3 and Elf (a high-yielding cultivar with $dt_1dt_1e_1e_1$ genotype) encountered variable expression of the DLS character and a quantitative mode of segregation. A number of modifier genes appear to be at work in the DLS syndrome. The agronomic virtues of DLS remain to be demonstrated. Phillips _et al._ (1984) found the DLS trait, in two different genetic backgrounds, to be negatively correlated with yield. Any practical value the character might have is likely to reflect the yield advantages of homozygosity for dt_1 and e_1, as seen in the short-stature, high-performance cultivars exemplified by Elf (Cooper, 1981); DLS lines may also be particularly responsive to foliar fertilization (Kannan, 1980).

FESTUCA PRATENSIS

It is characteristic of the morphological organization of grass plants that they respond to a transient stress by discarding existing leaves and rapidly deploying new leaves 'stored' in the apical meristem region. A population of _Festuca pratensis_ growing at the Welsh Plant Breeding Station in 1969 was inadvertently subjected to a short period of water stress.

During recovery it was observed that a single individual in the selection did not exhibit the normal foliar yellowing reaction shown by the rest of the population. A phenotypically non-yellowing F_2 line derived from the original plant was designated Bf 986 in 1972. Bf 993, the subject in most of the published genetic and biochemical work, was the result of a cross between Bf 986 and the cultivar Rossa, the usual wild-type comparison in these studies.

It was quickly established that the non-yellowing character is not just confined to the moisture stress response but is evident whenever chlorophyll degradation occurs in the normal life-cycle of Festuca plants. The lower leaves of vegetative plants do not pass through a yellowing stage during sequential senescence; foliage, haulms and grains of ripening mutant plants remain green; pathogen-infected plants do not exhibit necrotic yellowing. Especially convenient for biochemical investigations is the fact that the contrast between mutant and normal tissue is also expressed in detached leaves. It also became clear early in the crossing programme that the green mutant character is not expressed in hybrids with normal yellowing varieties.

Formal genetic analysis of the mutation was undertaken using isoenzymes of cytosolic phosphogluco-isomerase (PGI-2) to monitor the crossing process. At least four allelic variants (designated a,b,c,d - Jones et al., 1980) of the Pgi-2 locus in species of the Lolium-Festuca complex have been identified and hybrids between individuals possessing different isoforms of PGI-2 are readily identified by starch gel electrophoresis. Populations of mutant line Bf 993 and wild-type Rossa were screened for PGI phenotype. As is common in the predominantly outbreeding forage grass species, considerable heterogeneity at the Pgi-2 locus was observed, even in the apparently uniform commercial variety Rossa. By surveying only a relatively small number of individuals, normal and mutant lines with distinct PGI phenotypes were identified. PGI is a dimeric enzyme. Subunits coded by different alleles of Pgi-2 associate in pairs to produce isoforms of the holoenzyme: a single isoenzyme band indicates homozygosity at the corresponding Pgi locus. Bf 993 and Rossa individuals with genotype Pgi-$2^{b/b}$ and Pgi-$2^{c/c}$ were selected from populations growing in the summer of 1982.

In the 1983 season, after vernalization, Bf 993, Pgi-$2^{b/b}$ and Bf 993 Pgi-$2^{c/c}$ were crossed with Rossa Pgi-$2^{c/c}$ and Rossa Pgi-$2^{b/b}$ respectively. There was no evidence of a maternal influence on the inheritance of either PGI or the green mutation. Over 96% of the progeny from these Pgi-$2^{b/b}$

x Pgi-$2^{c/c}$ crosses were heterozygous (Pgi-$2^{b/c}$). Four percent self-fertility is not atypical for this species. Each of the homozygous individuals (product of selfing) was identical in yellowing phenotype with the corresponding parent; all F_1 hybrids were yellowing. In 1984 hybrids were backcrossed with parents. Others were intercrossed to produce the F_2 generation. The progenies were analyzed for PGI-2 and senescence phenotype as before. The results are summarized in Figure 6. Pgi-$2^{b/b}$, Pgi-$2^{b/c}$ and Pgi - $2^{c/c}$ segregated in the F_2 generation in a ratio very close to 1:2:1 (Figure 6A). The ratio of 2.14:1 for yellow and green phenotypes is not significantly different from 3:1. The backcross between F_1 and the recessive parent Bf 993 gave a 1:1 ratio (Figure 6B). Interestingly, the crosses

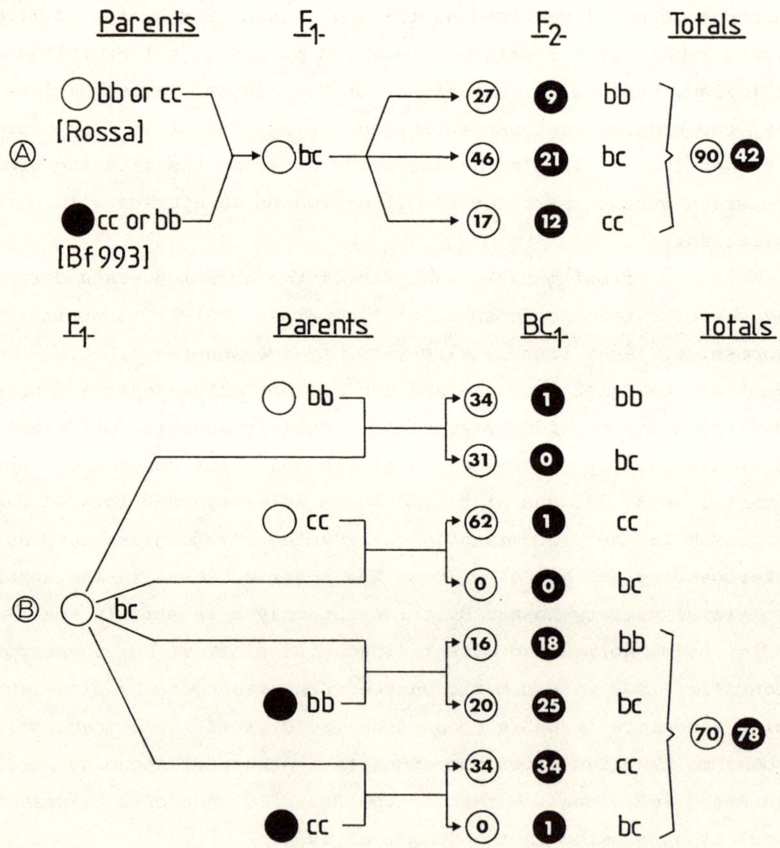

Figure 6. Inheritance of the b and c alleles of Pgi-2 and of the yellowing (o) and mutant non-yellowing (●) senescence phenotypes in *Festuca pratensis*.

Pgi-$2^{c/c}$ (Bf 993 or Rossa) x Pgi-$2^{b/c}$ produced virtually no Pgi-$2^{b/c}$ progeny. There exists an incompatibility (S) locus which, in the related genus Lolium, is known to be linked to Pgi-2 (Cornish et al., 1980) but this appears not to account for the anomalous inheritance figures in this experiment. The possibility that a Pgi-linked lethal locus exists in Festuca is under further study. We believe that the deviation from 3:1 observed in the F_2 generation, although not statistically significant, may be accounted for by the influence of this locus. The general picture to emerge from these investigations is clear. The green locus in Festuca pratensis is inherited as a classical Mendelian recessive and is not linked to Pgi-2.

The biochemical consequences of the mutation have been established in a number of previous studies. Much of the senescence syndrome proceeds quite normally in non-yellowing individuals. Table 3A is a catalogue of the changes occurring during senescence of detached Festuca leaves upon which the mutation has no apparent influence. Underlying visible expression of the genetic lesion, a range of pleiotropic effects is apparent (Table 3B), a common feature of which is an extreme persistence of components forming the hydrophobic core of thylakoid membranes. Pigments, glycolipids and intrinsic membrane polypeptides appear to be metabolized in a coordinated fashion, so that a change in the catabolism of one of the partners in the association will be answered by changes in the behaviour of the other two. What do we know about the routes of pigment, protein and lipid breakdown when the fabric of the leaf is being dismantled and remobilized? The answer is - not much. It is highly likely that the missing link in the chain of cause and effect from mutant gene to altered phenotype will be discovered in one of the catabolic pathways that are only now becoming understood. Thomas (1986) has reviewed the evidence supporting a central role for fatty acids released from chloroplast lipids in the integration of lipid, pigment and protein degradation during senescence. The biochemical basis of the non-yellowing trait is proposed to be an impairment in glycolipid turnover (Thomas et al., 1985). It has not been possible to demonstrate a deficiency in the acyl hydrolase activity of mutant leaf tissue; but there is no reason to believe that this low-specificity, non-chloroplastic enzyme is any more likely to have a normal physiological function in chloroplast disassembly than are the acid endopeptidases of the vacuole (Thomas, 1984). The Festuca mutant is proving to be a unique asset for the study of degradative processes contributing to turnover and senescence in leaf development (Thomas, 1983a, 1984; Hilditch et al., 1986b).

Table 3. Summary of the senescence characteristics of non-yellowing mutant and wild type Festuca pratensis Huds.

A. Similarities in senescence behaviour between normal and mutant Festuca genotypes

Decline in total soluble protein[a]
Acceleration and inhibition of protein loss by abscisic acid and kinetin respectively[a]
Appearance of novel esterase isoenzymes[a]
Cycloheximide-sensitive activation of peroxidase[b]
Inactivation of glutamate-pyruvate aminotransferase[b]
Decline in Fraction 1 protein[c]
Decline in PSI and PSII electron transport capacity,[c,d]
Decline in LSU and SSU of RuBPC[e]
Decline in extrinsic thylakoid polypeptides[e]
Increase in acid proteinase activity[f]
Fatty acid complement of thylakoid acyl lipids retained[g]
Reduction in rate of senescence-associated changes by light[h]
Unchanging number of chloroplasts per cell[i]
Decline in thylakoid-associated chlorophyll bleaching activity[j]
Decline in tissue photosynthesis measured as O_2 evolution[d]
Slow downward drift in rate of dark respiration measured as O_2 uptake[d]

B. Differences in senescence behaviour between normal and mutant Festuca genotypes

Mutant retains chlorophyll, even when treated with abscisic acid[a]
Even at extreme senescence, mutant chloroplasts retain recognizable thylakoids, though in an unstacked and loose condition[c]
Much reduced accumulation of plastoglobuli in senescing mutant plastids[c]
Retention of light-harvesting chlorophyll protein in mutant tissue[c,e,h,k]
Retention of P_{700} protein complex of PSI in senescing mutant thylakoids[e,h]
Reduced rate of acyl lipid degradation during senescence of mutant chloroplasts[g]
Mutant chloroplasts retain a greater number of functional PSII reaction centres than normal chloroplasts during senescence[d]
Light-independent (but not the rapid light-dependent) degradation of the Q_B protein of PSII is disabled in the mutant[l]

References:
a: Thomas and Stoddart, 1975; b: Thomas, 1976b; c: Thomas, 1977; d: Hilditch et al., 1986a; e: Thomas, 1982a; f: Thomas, 1982b; g: Harwood et al., 1982; h: Thomas, 1983a; i: Thomas, 1983b; j: Thomas et al., 1985; k: Hilditch, 1986; l: Hilditch et al., 1986b.

For practical purposes, Festuca is self-sterile. Varieties of most perennial temperate grass species are usually genetically heterogeneous populations. Conventional procedures for producing isogenic lines (by repeated backcrossing to the wild-type parent, for example) are at best inefficient and usually run into insuperable problems of self-incompatibility and inbreeding depression when applied to the obligately outcrossing grasses. The non-yellowing mutation is at present maintained in the non-uniform genetic background of a typical F.pratensis population, essentially similar to that of the wild-type comparison. Two alternative approaches to the production of isogenic lines have been adopted. One is to transfer the non-yellowing trait into F.pratensis var. apennina. This is a high-altitude fescue ecotype from Switzerland which is unusual in being better than 70% self-fertile (Bean, 1974). Apennina is tetraploid (Clarke et al., 1976), but the triploid progeny of a cross with diploid F.pratensis are viable and fertile and, in progeny of backcrosses, euploidy is strongly reasserted. Secondly, intergeneric crosses have been made between non-yellowing F. pratensis and Lolium temulentum. The latter is highly self-fertile and does not have a complex vernalization requirement for flowering, a characteristic of F.pratensis which, in practice, limits crossing programmes to one generation per year. We expect to see the establishment of normal and mutant lines with substantially isogenic backgrounds within two to five years.

CONCLUSIONS

Plant development is a sequence of responses to a reiterated question: does the survival value of maintaining a given organ exceed the cost of discarding that organ and making a new (and possibly different) structure? This question is addressed directly to the genotype, which answers by expressing the appropriate combination of component genes. When the answer is 'discard' then the senescence programme is run. We have seen that this programme, as it concerns leaves, comprises genes of different kinds. There are discrete loci which appear to control defined steps in senescence-specific metabolic pathways (exemplified by yellowing/non-yellowing in Festuca). There are genes and gene combinations altering the temporal and allometric relationships of the senescence syndrome (such as those determining the DLS trait in soybean and stay-green in maize). And there are genes influencing the rates at which processes proceed during senescence (N mobilization in cereals, for instance). Mutations provide the clearest possible evidence for the developmental regulation of higher plant senescence by operational, spatio-temporal and rate-controlling genes.

REFERENCES

Abu-Shakra, S.S., Phillips, D.A. & Huffaker, R.C. (1978) Nitrogen fixation and delayed leaf senescence in soybeans. Science 199, 973-975.

Allison, J.C.S. & Weinmann, H. (1970) Effect of absence of developing grain on carbohydrate content and senescence of maize leaves. Plant Physiol. 46, 433-436.

Bean, E.W. (1974) Seed production characteristics of tetraploid meadow fescue. Rep. Welsh Pl. Breed. Stn for 1973, p. 51.

Bernard, R.L. (1971) Two major genes for time of flowering and maturity in soybeans. Crop Sci. 11, 242-244.

Bernard, R.L. (1972) Two genes affecting stem termination in soybeans. Crop Sci. 12, 235-239.

Bernard, R.L. & Weiss, M.G. (1973) Qualitative genetics. In: Soybeans: Improvement, Production and Uses. Ed. B.E. Caldwell, pp. 117-154. Madison: American Society of Agronomy.

Boyd, W.J.R. & Walker, M.G. (1972) Variation in chlorophyll a content and stability in wheat flag leaves. Ann. Bot. 36, 87-92.

Burton, J.W., Wilson, R.F. & Brim, C.A. (1979) Dry matter and nitrogen accumulation in male-sterile and male-fertile soybeans. Agron. J. 71, 548-552.

Čatský, J., Tichá, I. & Solárová, J. (1976) Ontogenetic changes in the internal limitations to bean-leaf photosynthesis. I. Carbon dioxide exchange and conductances of carbon dioxide transfer. Photosynthetica 10, 394-402.

Christensen, L.E., Below, F.E. & Hageman, R.H. (1981) The effects of ear removal on senescence and metabolism of maize. Plant Physiol. 68, 1180-1185.

Clarke, J., Chandrasekharan, P. & Thomas, H. (1976) Studies in *Festuca* 9. Cytological studies of *Festuca pratensis* var. *apennina* (De Not.) Hack. (2n=28) Z. Pflanzenzücht. 77, 205-214.

Cooper, R.L. (1981) Development of short-statured soybean cultivars. Crop Sci. 21, 127-131.

Cornish, M.A., Hayward, M.D. & Lawrence, M.J. (1980) Self incompatibility in ryegrass. III. The joint segregation of S and PGI-2 in *Lolium perenne*. Heredity 44, 55-62.

Crafts-Brandner, S.J., Below, F.E., Harper, J.E. & Hageman, R.H. (1984a) Differential senescence of maize hybrids following ear removal. I. Whole plant. Plant Physiol. 74, 360-367.

Crafts-Brandner, S.J., Below, F.E., Wittenbach, V.A., Harper, J.E. & Hageman, R.H. (1984b) Differential senescence of maize hybrids following ear removal. II. Selected leaf. Plant Physiol. 74, 368-373.

Dalling, M.J., Boland, G. & Wilson, J.H. (1976) Relation between acid proteinase activity and redistribution of nitrogen during grain development in wheat. Aust. J. Plant Physiol. 3, 721-730.

Darbishire, A.D. (1911) Breeding and the Mendelian Discovery. New York: Cassell.

Duncan, R.R., Bockholt, A.J. & Miller, F.R. (1981) Descriptive comparison of senescent and nonsenescent sorghum genotypes. Agron. J. 73, 849-853.

Duvick, D.N. (1984) Genetic contribution to yield gains of U.S. hybrid maize 1930-1980. In: Genetic Contributions to Yield Gains of Five Major Crop Plants (CSSA Special Publication 7). pp. 15-47. Madison: Crop Science Society of America.

Evans, L.T., Wardlaw, I.F. & Fischer, R.A. (1975) Wheat. In: Crop Physiology. Ed. L.T. Evans, pp. 101-149. Cambridge: University Press.

Harwood, J.L., Jones, A.V.H.M. & Thomas, H. (1982) Leaf senescence in a non-yellowing mutant of Festuca pratensis. III. Total acyl lipids of leaf tissue during senescence. Planta 156, 152-157.

Hilditch, P. (1986) Immunological quantification of the chlorophyll a/b binding protein in senescing leaves of Festuca pratensis Huds. Plant Sci. 45, 95-99.

Hilditch, P., Thomas, H. & Rogers, L. (1986a) Leaf senescence in a non-yellowing mutant of Festuca pratensis: photosynthesis and photosynthetic electron transport. Planta 167, 146-151.

Hilditch, P., Thomas, H. & Rogers, L. (1986b) Two processes for the breakdown of the Q_B protein of chloroplasts. FEBS Lett. (in press).

Huffaker, R.C. & Peterson, L.W. (1974) Protein turnover in plants and possible means of its regulation. Ann. Rev. Plant Physiol. 25, 363-392.

Hunt, R. (1982) Plant Growth Curves. London: Arnold.

Jones, T.W.A., Hodge, T.P. & Bean, E.W. (1980) Environmental and developmental effects on isoenzymes in a tetraploid hybrid ryegrass. Z. Pflanzenzücht. 85, 53-58.

Kahanak, G.M., Okatan, Y., Rupp, D.C. & Noodén, L.D. (1978) Hormonal and genetic alteration of monocarpic senescence in soybeans. Plant Physiol. 61, s138.

Kannan, S. (1980) Mechanisms of foliar uptake of plant nutrients: accomplishments and prospects. J. Plant Nutr. 2, 717-735.

Leopold, A.C., Niedergang-Kamien, E. & Janick, J. (1959) Experimental modification of plant senescence. Plant Physiol. 59, 1136-1140.

Makino, A., Mae, T. & Ohiro, K. (1984) Relation between nitrogen and ribulose-1,5-bisphosphate carboxylase in rice leaves from emergence through senescence. Plant Cell Physiol. 25, 429-437.

Mandahar, C.L. & Garg, I.D. (1975) Effect of ear removal on sugars and chlorophylls of barley leaves. Photosynthetica 9, 407-409.

Middleton, G.K., Bode, C.E. & Bayles, B.B. (1954) A comparison of the quantity and quality of protein in certain varieties of soft wheat. Agron. J. 46, 500-502.

Mishkind, M.L., Jensen, K.H., Branagan, A.J., Plumley, F.G. & Schmidt, G.W. (1985) Roles of proteases in chloroplast biogenesis. In: Current Topics in Plant Biochemistry and Physiology Volume 4. Ed. D.D. Randall, D.G. Blevins, R.L. Larson & T. Kagawa. pp. 34-50. Columbia: University of Missouri.

Noodén, L.D. (1984) Integration of soybean pod development and monocarpic senescence. Physiol. Plant 62, 273-284.

Paigen, K. (1979) Acid hydrolases as models of genetic control. Ann. Rev. Genet. 13, 417-466.

Phillips, D.A., Pierce, R.O., Edie, S.A., Foster, K.W. & Knowles, P.F. (1984) Delayed leaf senescence in soybeans. Crop Sci. 24, 518-522.

Pierce, R.O., Knowles, P.F. & Phillips, D.A. (1984) Inheritance of delayed senescence in soybean. Crop Sci. 24, 515-517.

Rao, R.O. & Croy, L.I. (1972) Protease and nitrate reductase seasonal patterns and their relation to grain protein production of 'high' vs 'low' protein wheat varieties. J. Agr. Food Chem. 20, 1138-1141.

Scandalios, J.G. & Baum, J.A. (1982) Regulatory gene variation in higher plants. Adv. Genet. 21, 347-370.

Sheehy, J.E. (1983) Relationships between senescence, photosynthesis, nitrogen fixation and seed filling in soybean Glycine max (L.). Merr. Ann. Bot. 51, 679-682.

Shoemaker, R.C., Cody, A.M. & Palmer, R.G. (1985) Characterization of a cytoplasmically-inherited yellow foliar mutant (cyt-Y_3) in soybean. Theor. Appl. Genet. 69, 279-284.

Sinclair, T.R. & de Wit, C.R. (1975) Photosynthate and nitrogen requirements for seed production by various crops. Science 189, 565-567.

Speirs, J. & Brady, C.J. (1981) A coordinated decline in the synthesis of subunits of ribulose bisphosphate carboxylase in ageing wheat leaves. II. Abundance of messenger RNA. Aust. J. Plant Physiol. 8, 603-618.

Stoddart, J.L. & Thomas, H. (1982) Leaf senescence. In: Encyclopedia of Plant Physiology. New Series Volume 14A. Nucleic Acids and Proteins in Plants I. Ed. D. Boulter & B. Parthier, pp. 592-636. Berlin: Springer Verlag.

Swank, J.C., Below, F.E., Lambert, R.J. & Hageman, R.H. (1982). Interaction of carbon and nitrogen metabolism in the productivity of maize. Plant Physiol. 70, 1185-1190.

Terao, H. (1918) Maternal inheritance in the soybean. Am. Nat. 52, 51-56.

Terao, H. & Nakatomi, S. (1929) On the inheritance of chlorophyll colorations of cotyledons and seed-coats in the soybean. Jap. J. Genet. 4, 64-80.

Thomas, H. (1976a) The structure of nucleic acids and proteins and their function in subcellular assembly. In: Plant Structure, Function and Adaptation. Ed. M.A. Hall, pp. 1-48. London: MacMillan.

Thomas, H. (1976b) Leaf growth and senescence in grasses. Rep. Welsh Pl. Breed. Stn for 1975, pp. 133-138.

Thomas, H. (1977) Ultrastructure, polypeptide composition and photochemical activity of chloroplasts during foliar senescence of a non-yellowing mutant genotype of Festuca pratensis. Planta 137, 53-60.

Thomas, H. (1982a) Leaf senescence in a non-yellowing mutant of Festuca pratensis. I. Chloroplast membrane polypeptides. Planta 154, 212-218.

Thomas, H. (1982b) Leaf senescence in a non-yellowing mutant of Festuca pratensis. II. Proteolytic degradation of thylakoid and stroma polypeptides. Planta 154, 219-223.

Thomas, H. (1983a) Proteolysis in senescing leaves. Brit. Plant Growth Reg. Group Monogr. 9, 45-59.

Thomas, H. (1983b) Leaf senescence in a non-yellowing mutant of Festuca pratensis. IV. Senescence in the light. Photosynthetica 17, 506-514.

Thomas, H. (1984) Cell senescence and protein metabolism in the photosynthetic tissue of leaves. In: Cell Ageing and Cell Death (SEB Seminar Series 25). Ed. I. Davies & D.C. Sigee, pp. 171-188. Cambridge: University Press.

Thomas, H. (1986) The role of polyunsaturated fatty acids in senescence. J. Plant Physiol. 123, 97-105.

Thomas, H., Lüthy, B. & Matile, Ph. (1985) Leaf senescence in a non-yellowing mutant of Festuca pratensis Huds. Oxidative chlorophyll bleaching by thylakoid membranes during senescence. Planta 164, 400-405.

Thomas, H. & Stoddart, J.L. (1975) Separation of chlorophyll degradation from other senescence processes in leaves of a mutant genotype of meadow fescue (Festuca pratensis). Plant Physiol. 56, 438-441.

Thomas, H. & Stoddart, J.L. (1980) Leaf senescence. Ann. Rev. Plant Physiol. 31, 83-111.

Tollenaar, M. & Daynard, T.B. (1978) Leaf senescence in short-season maize hybrids. Can. J. Plant Sci. 58, 869-874.

Veatch, C. & Woodworth, C.M. (1930) Genetic relations of cotyledon color types of soybeans. J. Am. Soc. Agron. 22, 700-702.

Wang, T.L. & Woolhouse, H.W. (1982) Hormonal aspects of senescence in plant development. Brit. Plant Growth Reg. Group Monogr. 8, 5-25.

Woodworth, C.M. (1921) Inheritance of cotyledon, seed-coat, hilum and pubescence colors in soy-beans. Genetics 6, 487-553.

Woolhouse, H.W. (1984) The biochemistry and regulation of senescence in chloroplasts. Can. J. Bot. 62, 2934-2942.

Yoshida, Y. (1961) Nuclear control of chloroplast activity in Elodea leaf cells. Protoplasma 54, 476-492.

PLANT GROWTH SUBSTANCES AUTONOMY IN TISSUE CULTURES

Roger Horgan,
Department of Botany and Microbiology, The University College of Wales, Aberystwyth SY23 3DA, U.K.

INTRODUCTION

Plant growth substance autonomy in tissue cultures refers to the ability of certain tissue cultures to grow on defined media in the absence of externally applied cytokinins and/or auxins. Autonomous tissue cultures may be broadly divided into three types:

(1) Crown gall tissues.

(2) Tissues derived from 'genetic tumors'.

(3) Habituated tissues, in which the cytokinin and/or auxin requirement has been lost. Such tissues may have no growth substance requirement when initially isolated or they may gradually lose the requirement during culture.

Knowledge of the molecular basis of growth substance autonomy, at both the biochemical and genetic level, varies considerably for these three types of tissue culture. The molecular genetic and biochemical basis of cytokinin and auxin production in crown gall tissues is now well understood. In contrast little is known of the mechanisms underlying cytokinin and auxin autonomy in genetic tumors and habituated tissues. This article will examine the current state of knowledge of the molecular basis of growth substance autonomy in the three types of habituated tissues with regard to their growth substance status and where possible the underlying genetics of growth substance production.

CROWN GALL TISSUES

An important feature of all crown gall tissues is their ability to grow on defined media in the absence of externally supplied cytokinins and auxins. The demonstration that this is due to the presence on the T-DNA of genes directing the synthesis of (Δ^2-isopentenyl)adenosine-5'-monophosphate ([9R-5'P]iP) and IAA has been an outstanding achievement for plant molecular biology. Considerable research effort by several groups utilising transposon mutagenesis revealed the presence of two loci on the T-DNA which markedly affected tumor morphology both on plant-borne tumors and to a lesser extent in tissue cultures. These loci were designated _tms_ and _tmr_ (Hooykaas _et al._, 1982). Insertional inactivation of the _tms_ locus gave rise to tumors exhibiting a preponderance of leafy shoots whereas inactivation of the _tmr_ locus produced tumors consisting mainly of roots. The phenotypic similarity between these tissue types and the effects of varying the amounts and the ratio of auxin and cytokinin in normal, growth substance-requiring, tobacco tissue cultures led to the suggestion that the _tms_ and _tmr_ loci controlled the ratio of endogenous auxin to cytokinin in crown gall tissue. Inactivation of the _tmr_ locus led to reduction in the cytokinin to auxin ratio and hence led to the formation of 'rooty' tumors. Inactivation of the _tms_ locus increased the endogenous cytokinin to auxin ratio and thus the tumors exhibited a 'shooty' phenotype. These suggestions were supported by correction of the mutant phenotypes by external applications of auxin or cytokinin. Initial investigations suggested that both genes may be involved in cytokinin metabolism. Measurements, by radioimmunoassay (RIA), of gross 'zeatin-like' cytokinin levels in _tms_ and _tmr_ mutants of tobacco crown gall tissue clearly showed a very large reduction in the level of 'zeatin-like' cytokinins in the _tmr_ tissues when compared with a wild-type tumor. _Tms_ mutant tissue on the other hand had considerably elevated levels of 'zeatin-like cytokinins. Simultaneous measurements of IAA in these tissues revealed a four-fold decrease in level in _tms_ tissues (Morris _et al._, 1982; Akiyoshi _et al._, 1983).

The results of these studies clearly pointed towards the involvement of the _tmr_ gene in cytokinin biosynthesis although the exact function of the _tms_ gene was not clearly revealed. In addition it was tenatively demonstrated that cell-free extracts of _tmr_ mutant tissue were unable to catalyse the reaction between 5'-AMP and Δ^2-isopentenylpyrophosphate to form the cytokinin [9R-5'P]iP (Morris _et al._, 1982). This reaction which is the key step in the biosynthesis of cytokinins is shown in Figure 1.

The _tmr_ region of the nopaline Ti plasmid pTi T37 was cloned and the nucleotide sequence determined (Goldberg _et al._, 1984). An investigation of this sequence suggested that it coded for a protein with a molecular weight of 27 kD. A protein of this size had also been suggested as the product of the same region of an octopine Ti plasmid. A 27 kD protein had been shown to be expressed from the octopine _tmr_ region in cell-free extracts of _E.coli_ and _A.tumefaciens_ and in _E.coli_ minicells. Barry _et al._ (1984) constructed a plasmid, designated pMON230, which contained the _tmr_ sequence downstream from a _trp_ promoter. This enabled them to obtain sufficiently high expression of the _tmr_ gene to observe both the synthesis of [^{35}S]-labelled protein in _E.coli_ maxicells and the production of active enzyme by _E.coli_ cultures. Cell free extracts of _E.coli_ carrying the pMON230 plasmid were shown to catalyse the reaction between 5'-AMP and Δ^2-isopentenylpyrophosphate to give a mixture of [9R-5'P]iP and (Δ^2-isopentenyl)adenosine ([9R]iP). The latter compound presumably arose from the presence of phosphatases in the extract. [9R-5'P]iP was identified by coelution on reverse phase HPLC with a synthetic standard and by its conversion to [9R]iP by alkaline phosphatase followed by identification of this product by HPLC. Cell free extracts of _E.coli_ not harbouring the pMON230 plasmid showed no ability to synthesise [9R-5'P]iP. These results, which have been independently confirmed by two other groups (Akiyoshi _et al_., 1984; Buchmann _et al_., 1985), unequivocally establish the function of the _tmr_ gene. Although of bacterial origin, this is the first biosynthetic gene for a plant growth substance to have been cloned and expressed _in vitro_.

The precise function of the _tms_ locus in controlling the auxin/cytokinin ratio in crown gall tissues has proved more difficult to establish. Initially it was demonstrated by transposon mutagenesis that the _tms_ locus contained at least two genes. RNA blotting of polyadenylated transcripts showed that the locus encoded two RNAs. Sequence analysis of the genes involved indicated a protein product from gene 1 of 83.7 kD and from gene 2 of 49.6 kD (Klee _et al_., 1984). It was suggested that these genes may function

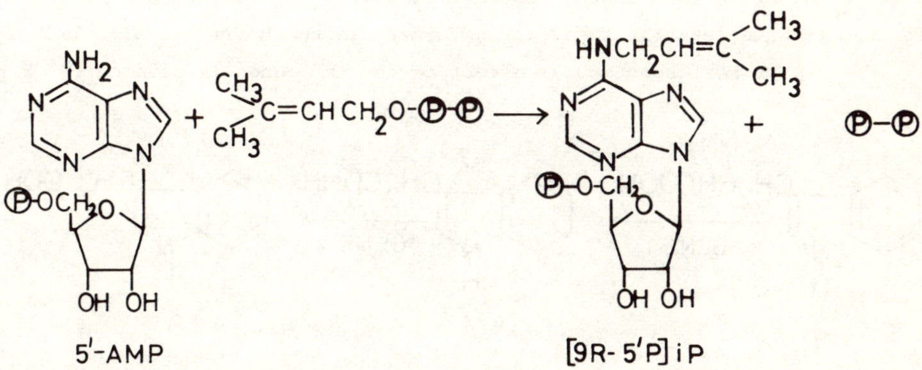

Figure 1. Pathway of cytokinin biosynthesis in crown gall tissue as controlled by the T-DNA _tmr_ gene.

in an analagous manner to the auxin biosynthesis genes of Pseudomonas savastanoi. In this organism tryptophan is converted to indole-3-acetamide (IAM) by tryptophan-2-monooxygenase and this is converted to IAA by an amido-hydrolase as shown in Figure 2. A vector plasmid/gene 2 construct was inserted into E.coli. The resulting cells expressed a 49 kD protein which converted radiolabelled IAM into IAA (Schroder et al., 1984). Similar enzymatic activity was detected in A.tumefaciens containing intact gene 2 and in tobacco crown gall tissue. A tumefaciens containing Ti plasmids mutagenised in the region of gene 2 and a hormone-independent (non-crown gall) line of tobacco callus tissue did not convert IAM to IAA. Tobacco plants which were transformed by Agrobacterium carrying gene 1 as the only functional growth substance gene on the T-DNA accumulated 500-1000 times more IAM than transformed plants lacking this gene or untransformed plants (Van Onckelen et al., 1985). In addition it has recently been demonstrated that cell-free extracts of tissues carrying a functional gene 1 are able to convert tryptophan to IAM via tryptophan-2-monooxygenase. Extracts from untransformed tissues or transformed tissues lacking this gene showed greatly reduced activity (Van Onckelen et al., 1986).

The above evidence, although somewhat less direct than that for cytokinin biosynthesis, strongly supports the view that transformation of plant tissues by the T-DNA results in introduction of a novel biosynthetic pathway to IAA. The role of this pathway in controlling the development of crown gall tumors is, however, far from clear. Auxin levels in tumor tissues appear to be, at the most, doubled by the presence of both genes 1 and 2 relative to untransformed pith tissue. However, tms mutant tissue is reported to have less IAA than untransformed tissue (Morris et al., 1982).

The most spectacular effect of the tms genes is seen on cytokinin

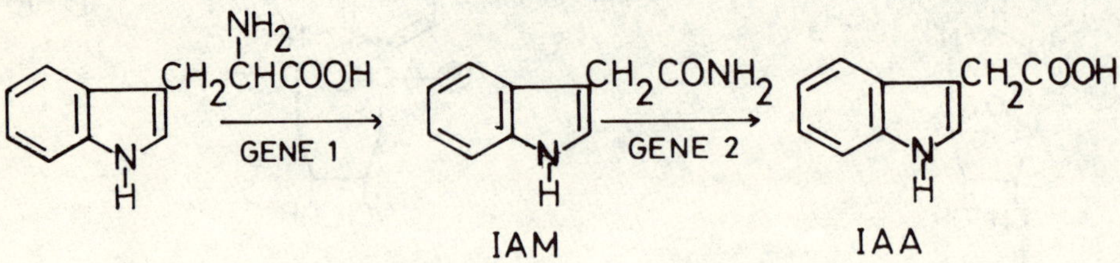

Figure 2. Pathway of IAA biosynthesis in crown gall tissue as controlled by the T-DNA tms genes 1 and 2.

levels which are greatly elevated in tissues in which they have been inactivated (Table 1). This strongly suggests that the developmental changes associated with insertional mutagenesis of the T-DNA predominantly controlled by cytokinin levels. The role of gene 4 in this process is clearly established. The influence of genes 1 and 2, their products, or the IAA level on cytokinin levels in tms mutants has yet to be explained.

GENETIC TUMOURS

It has been known for over 100 years that tumorous outgrowths of various types can occur in certain interspecific hybrids of several plant genera e.g. *Brassica*, *Bryophyllum*, *Gossypium*, *Lycospersicon*, *Datura* and *Lilium*. The most detailed studies of the genetics of tumor formation have been carried out with the genus *Nicotiana*. In *Nicotiana*, which comprises 64 species, about 300 interspecific hybrids are known, of which 30 are tumorogenic. Since the initial work of Kostoff (1930) on the combination *N.langsdorffii* x *N.glauca*, this hybrid has been the most studied. The normal interspecific hybrid carries one complete set of chromosomes from each parent in the genomic proportions of 1:1. Various genomic combinations of this hybrid have been produced varying from two *N.langsdorffii* genomes with one *N.glauca* genome to one *N.langsdorffii* genome with three *N.glauca* genomes. All these hybrids produced tumors although the genomic ratios influenced the growth of excised tumors in tissue culture (Kehr and Smith, 1954).

Naf (1958) demonstrated that tumor-prone hybrids arise by the

Table 1. Cytokinin levels in tobacco crown gall tissues carrying mutations in the tms (auxin) and/or tmr (cytokinin) tissues (McGaw, Horgan, Wullems and Schilperoot, unpublished data). Measurements were made by isotope dilution mass spectrometry.

	Cytokinin levels ($ng.g^{-1}$ fresh wt.)			
Tissue	[9R-5'P]Z	[9R]Z	Z	[7G]Z
LBA 4001 (wt)	25	18	22	11
LBA 4210 (tmr)	10	19	34	2
LBA 4060 (tms)	184	63	75	779
LBA 1501-1 (tms)	527	78	34	316
TSO 119 (tmr, tms)	11	27	35	8
SRI 4013-3	92	51	32	70
SRI (control)	11	16	7	4

combination of genetically-distant genomes and classified the species involved in these combinations into two groups called "plus" and "minus". It has been suggested that members of the "plus" group such as N.langsdorffii and N.longiflora carry nuclear genes for a factor (factor 1) which regulates tumor induction, whereas members of the "minus" group carry genes for a factor (factor ee) which controls the degree of tumor expression (Ahuja, 1968). Hybrids of members in the same group are not tumorous whereas inter-group hybrids produce tunors. Naf's hypothesis was found to fit 45 out of 49 combinations tested. Repeated backcrossing of N.longiflora and N.debneyitabacum resulted in a tumor prone hybrid which contained only a single N.longiflora chromosome (Ahuja, 1962). This result indicated that the factor 1 genes are located on a particular chromosome. The molecular natures of factors 1 and ee are unknown although it has been suggested that in combination they produce some form of biochemical and/or physiological imbalance leading to spontaneous tumorigenesis. It is clear that one component of this so-called 'imbalance' must be an increase in auxin and cytokinin levels in the tumor tissues, since in vitro culture of genetic tumors does not require the addition of these growth substances to the culture medium. Indeed, the first demonstration of the production of interspecific plant hybrids via protoplast fusion was achieved by the fusion of N.glauca and N.langsdorffii protoplasts (Carlson et al., 1972). It proved possible to select for the heterokaryons by the ability of these cells to proliferate in the absence of growth substances. Mature plants regenerated from callus tissue exhibited an identical tumor-prone phenotype to those produced sexually. In spite of considerable circumstantial evidence for the involvement of endogenous auxin and cytokinins, in at least tumor proliferation, knowledge of the biochemistry of these growth substances in genetic tumors is very limited. There is some evidence that auxin biosynthesis is increased in pith explants from N.glauca x N.langsdorffii relative to pith explants from non-tumorous parent plants (Cheng, 1972). However the effect of this on IAA levels in genetic tumors is unknown.

HABITUATED TISSUES

The term 'habituation' refers the loss of the exogenous requirement for auxin and/or cytokinins by plant tissues and cells in culture. In general, plant tissue cultures require an external supply of both auxin and cytokinin for growth. Some cultures require only one of these growth regulators and very few will grow in the absence of both. Often the substitution of the generally used synthetic auxin naphthaleneacetic acid by 2,4-D leads to a loss of the cytokinin requirement. In the context of this article 'habituation' will refer to the gradual loss of the requirement for a growth substance by cultures which initially require it. This phenomenon has been extensively studied by Meins and his co-workers in tobacco tissues (Meins, 1982). They have chosen to use the somewhat outmoded and confusing term 'epigenetic' to describe the process of habituation. This is based mainly on the heritable yet reversible nature of hormone autonomy in certain tobacco callus tissues. Meins and Foster (1985) have described in considerable detail the cytokinin requirement of cell lines isolated from leaves, pith and stem cortex of the tobacco variety "Havana 425". Cortical cells from this variety are cytokinin-autonomous when initially isolated and are said to exhibit the 'constitutive' phenotype. Pith tissues consist of 'inducible cells', which habituate in response to kinetin or incubation at 35°C, and 'noninducible cells', which do not respond to these treatments. Cultures derived from explants of leaf laminae consist only of 'noninducible' cells. Pith tissue from plants regenerated from cloned lines of 'noninducible' cells contain both 'inducible' and 'noninducible' cells. 'Inducible' cells arise at rates greater than 5×10^{-3} per cell generation which is 100-1000 times faster than the reported rates of somatic mutations in tobacco species. This together with the fact that during the regeneration process 'inducible' cells can also give rise to 'noninducible' cells, i.e. the process is reversible, is taken as evidence for the 'epigenetic' nature of the process of habituation.

A clearer picture of the genetics underlying one aspect of cytokinin autonomy has emerged from studies on cultured leaf tissues derived from "Havana 425" and on callus cultures of _Phaseolus vulgaris_. In the course of the type of study described above, Meins and his coworkers (Meins _et al._, 1983) found that plants regenerated from cloned leaf and cortex tissue of one particular plant of "Havana 425" differed in cytokinin habituation. Leaf tissues derived from leaf cells exhibited the normal nonhabituated phenotype, whereas leaf tissues derived from cortex cells were habituated. F_1 hybrids between cortex derived plants and normal plants exhibited an intermediate degree of habituation. The F_2 generation contained both parental and intermediate phenotypes in a ratio consistent with cytokinin autonomy in this system being controlled by a single Mendelian gene, exhibiting incomplete dominance. Further backcrossing confirmed this finding.

Mok _et al._ (1980) have studied the cytokinin requirement of hypocotyl tissue cultures from plants derived from crosses between a genotype of P.vulgaris (P.I.200960), which exhibits cytokinin dependence in tissue culture, and two cytokinin-independent genotypes (G 50 and P.I. 286303). Hypocotyl tissue cultures from F_1 plants exhibited an intermediate degree of cytokinin autonomy. The F_2 generation contained both parental and intermediate phenotypes. The frequency distribution of cytokinin requiring progeny in F_2 and backcross populations suggested, as for the tobacco system described above, the presence of a single set of alleles controlling cytokinin autonomy.

POSSIBLE MECHANISMS OF GROWTH SUBSTANCE AUTONOMY

It can be seen from the foregoing brief description of auxin and cytokinin autonomy in tissue cultures that, with the exception of crown gall tissues, there is little understanding of the biochemical mechanisms responsible for this phenomenon. Whilst the crown gall system has been completely described with regard to the genetics of auxin and cytokinin biosynthesis, some features of its auxin and cytokinin biochemistry remain unexplained; for example, the influence of the tms (auxin) genes on cytokinin levels. However, plant growth substance autonomy in crown gall is easily explained by the presence and expression of oncogenes coding for auxin and cytokinin biosynthetic enzymes. In the case of the other examples of plant growth substance autonomy described above, understanding is severely limited by lack of information at both the biochemical and genetic level. It is particularly unfortunate that little detailed information exists on the auxin and cytominin contents of genetic tumors and habituated tissues.

Ames (1974) has measured the IAA content of plants of the tumor prone amphiploid N.suaveolens x N.langsdorffii using the indole-α-pyrone method. The highest level of ca. 100 ng/g fresh weight was observed 11 days after sowing and the level declined to ca. 20 ng/g fresh weight after 27 days. It was noted that the maximum incidence of tumor production coincided with the period when the IAA level was at its lowest and it was suggested that a reduction in IAA level acts as a trigger for tumorogenesis. Unfortunately there are no reliable data on IAA and cytokinin levels in genetic tumor tissues grown in vitro. Thus any biochemical interpretation of the previously described genetics of these systems is purely speculative.

There are basically two likely explanations for the acquisition of auxin/cytokinin autonomy by plant tissues. Firstly an increase in pool size of these growth substances relative to that in hormone-dependent tissues.

This in turn may be brought about by the 'switch on', or an increased rate, of biosynthesis and/or a reduced rate of metabolism. Secondly the tissue may acquire an increased 'sensitivity' to the growth substance(s) in question. The term 'sensitivity' is not really definable within our current knowledge of plant growth substance physiology. In the absence of any knowledge of the precise mode of action of cytokinins and auxin it must be taken simply to mean the ability of a tissue to respond to these hormones. Since hormone autonomy represents a loss of requirement for externally applied hormones, changes in tissue sensitivity towards auxin and/or cytokinins are unlikely to be the cause of this phenomenon. It is most likely that growth substance autonomy is caused by biochemical changes involving the biosynthesis and/or metabolism of the compounds concerned.

In the case of the cytokinins the crown gall system provides an interesting qualitative and quantitative model. If it is assumed that the pathway of biosynthesis of cytokinins in normal plant tissues and crown gall are the same (this is by no means proven beyond doubt) then from the qualitative point of view the T-DNA may simply be introducing an additional copy of an already-existing plant gene into the cell. However from the quantitative point of view the production of cytokinins via the T-DNA gene is probably much greater than via the plant gene(s) under any circumstances. Using bioassays Nakajima et al. (1979) estimated the [9R]Z content of tobacco crown gall cells to be 40-80 ng kinetin equivalents per g.f.wt. and that of habituated cells to be less than 0.1 ng kinetin equivalents per g.f.wt. Using mass spectrometry and stable isotope dilution Scott and Horgan (1984) found levels of several zeatin-related cytokinins in tobacco crown gall callus tissue to be in the 10 to 300 ng per g.f.wt. range, whereas in an autonomous, non-crown gall tissue they were at least 100 fold lower and barely detectable by the method used.

If cytokinins do in fact function as one level of control of cell division in plants it is reasonable to assume that their biosynthesis will be carefully regulated. The T-DNA gene does not appear to have any in-built regulatory mechanism and so its introduction into a plant cell probably results in a gross over production of cytokinins, far in excess of the amounts required to maintain cell division. The other forms of cytokinin autonomy described in this article may result from relatively subtle changes in the plant's own mechanism for regulating cytokinin levels. The cytokinin autonomy of genetic tumors and habituated tissues may be explicable in terms of changes in, as yet unidentified, genes which exert a regulatory function on the gene or genes coding for the biosynthetic enzymes. Loss of function of these 'regulator' genes may 'switch on' cytokinin production. This would seem a likely explanation for the cytokinin autonomy of the Phaseolus hypocotyl and tobacco leaf tissue cultures described above. In these cases the loss of function may be simply due to a 'classical' mutation. In the case of genetic tumors it may arise via recombination.

It is of interest to note that it has recently been demonstrated that expression of the vir region of the Ti plasmid, which also contains a cytokinin biosynthetic gene, is activated by acetosyringone (1) and α-hydroxyacetosyringone (2) (Figure 3; Stachel et al., 1985). These molecules are released from plant cells on wounding. It may be hypothesised that the formation of wound callus results from the activation of the plant's own cytokinin biosynthesis genes by wound-derived compounds. This model may also explain the cytokinin activity of dihydroconiferyl alcohol (3) (Lee et al., 1981) which is structurally similar to the syringones but has no structural similarity to the cytokinins. As has been pointed out by Meins the reversible nature of some forms of habituation precludes the involvement of classical mutations and he has used the term 'epigenetic' to describe such phenomena. This term has little usage in the context of modern molecular genetics where more constructive hypotheses may be put forward to explain habituation. For example, the properties of habituated tissues are consistent with the involvement of a transposable element carrying a gene involved in the regulation of cytokinin biosynthesis.

Whilst the above speculations are based on the premise that cytokinin levels in plant tissues are controlled at the biosynthetic level the possibility of their regulation via metabolism cannot be ruled out. The key enzyme in cytokinin metabolism appears to be cytokinin oxidase which irreversibly inactivates cytokinins by side chain cleavage (McGaw and Horgan,

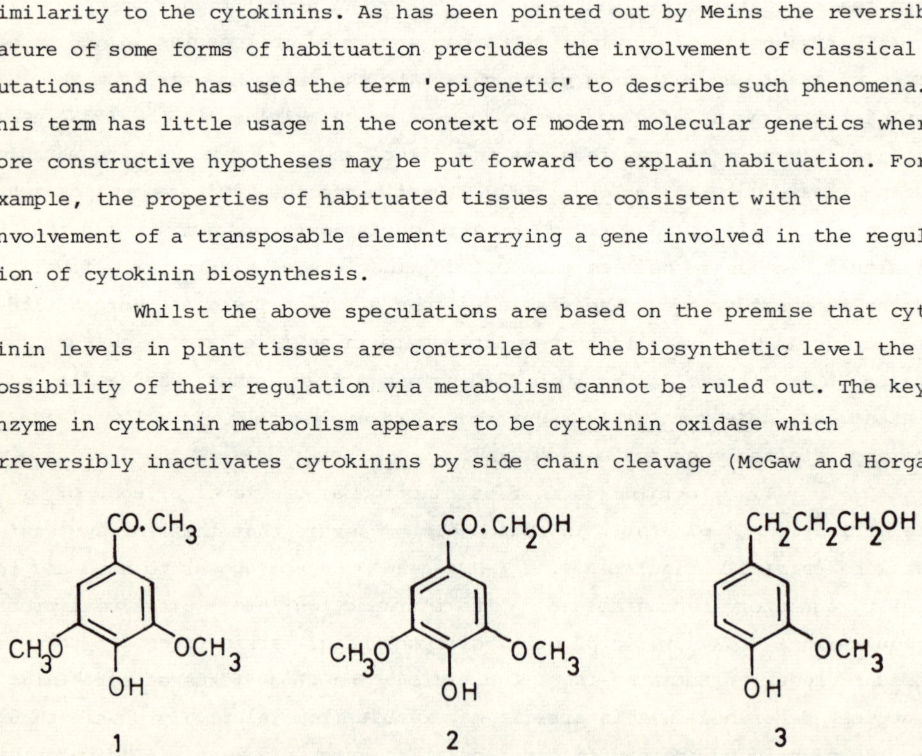

Figure 3. Structure of (1) acetosyringone,
(2) α-hydroxyacetosyringone and (3) dihydroconiferyl alcohol.

1983). Unfortunately there is no direct evidence that endogenous cytokinin levels can be modulated by cytokinin oxidase. However Laloue (personal communication) has shown that diphenylurea derivatives which are active as cytokinins in tissue culture systems are potent inhibitors of cytokinin oxidase from what germ. Mok et al. (1982) have shown that the response of callus tissue from two different genotypes of Phaseolus to externally-applied cytokinin is related to the rate of side chain cleavage. Thus genetic differences in cytokinin inactivation do exist but the role of this process in cytokinin autonomy of tissue cultures in unknown.

Lack of knowledge of the biochemical processes regulating IAA levels makes it difficult to appraise the limited information on IAA levels in habituated tissues. Pengelly and Meins (1983) have shown, by RIA, that crown gall tissues and habituated tissues of 'Havana 425' tobacco contain comparable levels of IAA. However, IAA levels in crown gall tissues continued to rise during several weeks in culture whereas IAA production in habituated tissues was limited to the first few days of culture. As with the cytokinins one can invoke several plausible hypotheses to account for IAA autonomy. However, since much less is known about the biochemistry of IAA turnover in tissue cultures than about cytokinin turnover there would seem little value in this exercise at present.

It can be seen from the above account that plant growth substance autonomy in tissue cultures is a veritable 'encyclopedia of ignorance'. Unlike the other systems discussed in this volume it is not even possible to ascribe physiological differences to defined mutations or to precise biochemical processes. Nevertheless autonomous tissue cultures do represent auxin and cytokinin mutants, in the most catholic sense. Their further study, utilising modern methods of molecular biology and plant growth substance biochemistry, should reveal a great deal about the mechanisms of control of cytokinin and auxin production in plants.

ACKNOWLEDGEMENT

I am grateful to Dr Mike Young for helpful advice and discussion.

REFERENCES

Ahuja, M.R. (1962) A cytogenetic study of heritable tumors in Nicotiana species hybrids. Genetics, 47, 865-880.

Ahuja, M.R. (1968) An hypothesis and evidence concerning the genetic components controlling tumor formation in Nicotiana. Mol. Gen. Genet. 103, 176-184.

Akiyoshi, D.E., Morris, R.O., Hinz, R., Mischke, B.S., Kosuge, T., Garfinkle, D.J., Gordon, M.P. & Nester, E.W. (1983) Cytokinin auxin balance in crown gall tumors is regulated by specific loci in the T-DNA. Proc. Nat. Acad. Sci. USA, 80, 407-411.

Akiyoshi, D.E., Klee, H., Amasino, R.M., Nester, E.W. & Gordon, M.P. (1984) T-DNA of Agrobacterium tumefaciens encodes an enzyme of cytokinin biosynthesis. Proc. Nat. Acad. Sci. USA, 81, 5994-5998.

Ames, I.H. (1974) Endogenous levels of auxin and tumorigenesis in a Nicotiana amphidiploid. Plant Physiol., 54, 953-955.

Barry, G.F., Rogers, S.G., Fraley, R.T. & Brand, L. (1984) Identification of a cloned cytokinin biosynthetic gene. Proc. Nat. Acad. Sci. USA, 81, 4776-4780.

Buchmann, I., Marner, F.J., Schroder, G., Waffenschmidt, S. & Schroder, J. (1985) Tumor genes in plants: T-DNA encoded cytokinin biosynthesis. EMBO J., 4, 853-859.

Carlson, P.S., Smith, H.H. & Dearing, R.D. (1972) Parasexual interspecific plant hybridization. Proc. Nat. Acad. Sci. USA, 69, 2292-2294.

Cheng, T.-Y. (1972) Induction of indoleacetic acid synthesis in tobacco pith explants. Plant Physiol., 50, 723-727.

Goldberg, S.B., Flick, J.S. & Rogers, S.G. (1984) Nucleotide sequence of the tmr locus of Agrobacterium tumefaciens pTi T37 T-DNA. Nucleic Acids Res., 12, 4665-4690.

Hooykaas, P.J.J., OOms, G. & Schilperoort, R.A. (1982) Tumors induced by different strains of Agrobacterium tumefaciens. In: Molecular Biology of Plant Tumors, ed. G.Kahl & J.S. Schell, pp. 373-390. New York: Academic Press.

Kehr, A.E. & Smith, H.H. (1954) Genetic tumors in Nicotiana hybrids. Brookhaven Symp. Biol., 6, 55-76.

Klee, H., Montoya, A., Horodyski, F., Lichtenstein, C., Garfinkle, D., Fuller, S., Flores, C., Peschon, J., Nester, E. & Gordon, M. (1984) Nucleotide sequence of the tms genes of the pTiA6NC octopine Ti plasmid: Two gene products involved in plant tumorigenesis. Proc. Nat. Acad. Sci. USA, 81, 1728-1732.

Kostoff, D. (1930) Tumors and other malformations on certain Nicotiana hybrids. Zentralbla. Bakteriol., Parasitenkd., Infektionskr. Hyg., Abt. 1.: Orig., 81, 244-280.

Lee, T.S. Purse, J.G., Pryce, R.J., Horgan, R. & Wareing, P.F. (1981) Dihydroconiferyl alcohol - a cell division factor from Acer species. Planta, 152, 571-577.

McGaw, B.A. & Horgan, R. (1983) Cytokinin oxidase from Zea mays kernels and Vinca rosea crown gall tissue. Planta, 159, 30-37.

Meins, F., Jr. (1982) Habituation of cultured plant cells. In: Molecular Biology of Plant Tumors, ed. G. Kahl & J.S. Schell, pp. 3-31. New York: Academic Press.

Meins, F., Jr., Foster, R. (1985) Reversible, cell-heritable changes during the development of tobacco pith tissues. Devel. Biol., 108, 1-5.

Meins, F., Jr., Foster, R. & Lutz, J.D. (1983) Evidence for a Mendelian factor controlling the cytokinin requirement of cultured tobacco cells. Devel. Genet., 4, 129-141.

Mok, M.C., Mok, D.W.S., Armstrong, D.J., Rabakoarihanata, A. & Kim, S-G. (1980) Cytokinin autonomy in tissue cultures of Phaseolus: A genotype-specific and heritable trait. Genetics, 94, 675-686.

Mok, M.C., Mok, D.W.S., Dixon, S.C., Armstrong, D.J. & Shaw, G. (1982) Cytokinin structure activity relationships and the metabolism of N^6 - (Δ^2-isopentenyl)adenosine-8-^{14}C in Phaseolus callus tissues. Plant Physiol. 70, 173-178.

Morris, R.O., Akiyoshi, D.E., MacDonald, E.M.S., Morris, J.W., Regier, D.A. & Zaerr, J.B. (1982) Cytokinin metabolism in relation to tumor induction by Agrobacterium tumefaciens. In: Plant Growth Substances 1982, ed. P.F. Wareing, pp. 175-183. London: Academic Press.

Naf, U. (1958) Studies in tumor formation in Nicotiana hybrids. I. The classification of parents into two etiologically significant groups. Growth, 22, 167-180.

Nakajima, H., Yokota, T., Matsumoto, T., Noguchi, M. & Takahashi, N. (1979) Relationships between hormone content and autonomy in various autonomous tobacco cells cultured in suspension. Plant Cell Physiol., 29, 1489-1499.

Pengelly, W.L. & Meins, F., Jr. (1983) Growth, auxin requirement, and indole-3-acetic acid content of cultured crown-gall and habituated tissues of tobacco. Differentiation, 25, 101-105.

Schroder, G., Waffenschmidt, S., Weiler, E.W. & Schroder, J. (1984) The T-region of Ti plasmids codes for an enzyme synthesising indole-3-acetic acid. Eur. J. Biochem., 138, 387-391.

Scott, I.M. & Horgan, R. (1984) Mass spectrometric quantification of cytokinin nucleotides and glycosides in tobacco crown gall tissues. Planta, 161, 345-354.

Stachel, S.E., Messens, E., Van Montagu, M. & Zambryski, P. (1985) Identification of the signal molecules produced by wounded plant cells that activate T-DNA transfer in Agrobacterium tumefaciens. Nature, 318, 624-629.

Van Onckelen, H., Rudelsheim, P., Inze, D., Follin, A., Messens, E., Horemans, S., Schell, J., Van Montagu, M. & De Greef, J. (1985) Tobacco plants transformed with the Agrobacterium T-DNA gene 1 contain high amounts of indole-3-acetamide. FEBS Lett. 181, 373-376.

Van Onckelen, H., Prisen, E., Inze, D., Rudelsheim, P., Van Lijsebettens, M., Follins, A., Schell, J., Van Montagu, M. & De Greef, J. (1986) Agrobacterium T-DNA gene 1 codes for tryptophan-2-monooxygenase activity in tobacco crown gall cells. FEBS Lett., 198, 357-360.

INDEX

SOCIETY FOR EXPERIMENTAL BIOLOGY SEMINAR SERIES

A series of multi-author volumes developed from seminars held by the Society for Experimental Biology. Each volume serves not only as an introductory review of a specific topic, but also introduces the reader to experimental evidence to support the theories and principles discussed, and points the way to new research.

1. Effects of air pollution on plants Edited by T.A. MANSFIELD
2. Effects of pollutants on aquatic organisms Edited by A.P.M. LOCKWOOD
3. Analytical and quantitative methods Edited by J.A. MEEK and H.Y. ELDER
4. Isolation of plant growth substances Edited by J.R. HILLMAN
5. Aspects of animal movement Edited by H.Y. ELDER and E.R. TRUEMAN
6. Neurones without impulses: their significance for vertebrate and invertebrate systems Edited by A. ROBERTS and B.M.H. BUSH
7. Development and specialisation of skeletal muscle Edited by D.F. GOLDSPINK
8. Stomatal physiology Edited by P.G. JARVIS and T.A. MANSFIELD
9. Brain mechanisms of behaviour in lower vertebrates Edited by P.R. LAMING
10. The cell cycle Edited by P.C.L. JOHN
11. Effects of disease on the physiology of the growing plant Edited by P.G. AYRES
12. Biology of the chemotactic response Edited by J.M. LACKIE and P.C. WILLIAMSON
13. Animal migration Edited by D.J. AIDLEY
14. Biological timekeeping Edited by J. BRADY
15. The nucleolus Edited by E.G. JORDAN and C.A. CULLIS
16. Gills Edited by D.F. HOULIHAN, J.C. RANKIN and T.J. SHUTTLEWORTH
17. Cellular acclimatisation to environmental change Edited by A.R. COSSINS and P. SHETERLINE
18. Plant biotechnology Edited by S.H. MANTELL and H. SMITH
19. Storage carbohydrates in vascular plants Edited by D.H. LEWIS
20. The physiology and biochemistry of plant respiration Edited by J.M. PALMER
21. Chloroplast biogenesis Edited by R.J. ELLIS
22. Instrumentation for environmental physiology Edited by B. MARSHALL and F.I. WOODWARD
23. The biosynthesis and metabolism of plant hormones Edited by A. CROZIER and J.R. HILLMAN
24. Coordination of motor behaviour Edited by B.M.H. BUSH and F. CLARAC
25. Cell ageing and cell death Edited by I. DAVIES and D.C. SIGEE
26. The cell division cycle in plants Edited by J.A. BRYANT and D. FRANCIS
27. Control of leaf growth Edited by N.R. BAKER, W.J. DAVIES and C. ONG
28. Biochemistry of plant cell walls Edited by C.T. BRETT and J.R. HILLMAN
29. Immunology in plant science Edited by T.L. WANG
30. Root development and function Edited by P.J. GREGORY, J.V. LAKE & D.A. ROSE
31. Plant canopies: their growth, form and function Edited by G. Russell, B. MARSHALL & P.G. JARVIS (forthcoming)
32. Developmental mutants in higher plants Edited by H. THOMAS & D. GRIERSON